Eighth Edition
Fire Hose Practices

Barbara Adams
Project Manager/Editor

Thomas P. Ruane
Senior Technical Editor

Validated by the International Fire Service Training Association

Published by
Fire Protection Publications
Oklahoma State University

Cover photo courtesy of Rick Montemorra, Mesa (AZ) Fire Department.

The International Fire Service Training Association

The International Fire Service Training Association (IFSTA) was established in 1934 as a "nonprofit educational association of fire fighting personnel who are dedicated to upgrading fire fighting techniques and safety through training." To carry out the mission of IFSTA, Fire Protection Publications was established as an entity of Oklahoma State University. Fire Protection Publications' primary function is to publish and disseminate training texts as proposed and validate by IFSTA. As a secondary function, Fire Protection Publications researches, acquires, produces, and markets high-quality learning and teaching aids as consistent with IFSTA's mission.

The IFSTA Validation Conference is held the second full week in July. Committees of technical experts meet and work at the conference addressing the current standards of the National Fire Protection Association and other standard-making groups as applicable. The Validation Conference brings together individuals from several related and allied fields, such as:

- Key fire department executives and training officers
- Educators from colleges and universities
- Representatives from governmental agencies
- Delegates of firefighter associations and industrial organizations

Committee members are not paid nor are they reimbursed for their expenses by IFSTA or Fire Protection Publications. They participate because of commitment to the fire service and its future through training. Being on a committee is prestigious in the fire service community, and committee members are acknowledged leaders in their fields. This unique feature provides a close relationship between the International Fire Service Training Association and fire protection agencies which helps to correlate the efforts of all concerned.

IFSTA manuals are now the official teaching texts of most of the states and provinces of North America. Additionally, numerous U.S. and Canadian government agencies as well as other English-speaking countries have officially accepted the IFSTA manuals.

ISBN 0-87939-238-X *Library of Congress Control Number: 2004106555*

Eighth Edition, First Printing, August 2004 *Printed in the United States of America*

10 9 8 7 6 5 4 3 2

If you need additional information concerning the International Fire Service Training Association (IFSTA) or Fire Protection Publications, contact:

Customer Service, Fire Protection Publications, Oklahoma State University
930 North Willis, Stillwater, OK 74078-8045
800-654-4055 Fax: 405-744-8204

For assistance with training materials, to recommend material for inclusion in an IFSTA manual, or to ask questions or comment on manual content, contact:

Editorial Department, Fire Protection Publications, Oklahoma State University
930 North Willis, Stillwater, OK 74078-8045
405-744-4111 Fax: 405-744-4112 E-mail: editors@osufpp.org

Table of Contents

Preface

This manual is the eighth publication of an IFSTA manual dedicated specifically to the construction, care, and use of fire hose. This manual (originally titled **Unit III, Hose Practices**) can trace its roots back to the early Oklahoma A&M College (now Oklahoma State University) **Unit Manuals for Fire Service Training.** The material contained in the manuals then (as it remains today) is not the opinion of one person, but it represents the consensus of a group of subject-matter experts who validate each new edition. The fire hose practices contained within the manual should not be viewed in the way one would view a *standard,* except, of course, when a specific standard or code is identified. Rather, view them as *recommended practices* that will provide safe and efficient means for fire and emergency service organizations to use their fire hose. With this concept in mind, **Fire Hose Practices,** 8th Edition, was written to provide the latest information about how fire hose, appliances, and fittings are constructed and maintained. Additionally, the manual describes numerous methods for loading and deploying fire hose. No one method is recommended over another. Each method is presented so that an organization can select the method or methods that most suit the needs of its community.

As a foundation for the development of portions of this manual, the following National Fire Protection Association (NFPA) standards were referenced and used:

- NFPA 1001, *Standard for Fire Fighter Professional Qualifications* (2002 edition)
- NFPA 1901, *Standard for Automotive Fire Apparatus* (2003 edition)
- NFPA 1961, *Standard on Fire Hose* (2002 edition)
- NFPA 1962, *Standard for the Inspection, Care, and Use of Fire Hose, Couplings, and Nozzles and the Service Testing of Fire Hose* (2003 edition)
- NFPA 1963, *Standard for Fire Hose Connections* (2003 edition)
- NFPA 1964, *Standard for Spray Nozzles* (2003 edition)

As with any training manual, a tremendous amount of work is required to prepare the technical information through research and acquire photographs and drawings as well as perform the other numerous important detailed tasks. This finished product is possible only because of the efforts of many very dedicated individuals — from recruit firefighters who loaded and deployed hose for photos to editorial, production, and library professionals who put the many pieces together. The writer gratefully acknowledges this assistance and thanks them for their efforts. Fire Protection Publications (FPP) also gratefully acknowledges and thanks the dedicated and hard-working IFSTA committee members for their continuing input and diligent work.

IFSTA Fire Hose Practices Validation Committee

Chair
Paul Valentine
Mount Prospect Fire Department
Mount Prospect, IL

Vice Chair
Stanley Gibson
Coos Bay Fire and Rescue
Coos Bay, OR

Secretary
Terrence Leen
Alameda County Fire Department
Sonoma, CA

Committee Members

Ronald Ballentine
St. Johns Fire Department
Johns Island, SC

David R. Dean
Marshall Fire Department
Marshall, TX

Sam Goldwater
Northtree Fire International
Conifer, CO

Charles Hall
Rhode Island Fire Academy
North Kingstown, RI

Robert Leigh
Aurora Fire Department
Denver, CO

Loren Lippincott
Blackhawk Technical College
Janesville, WI

William Streuber
North Charleston Fire Department
North Charleston, SC

Many thanks go to committee members Sam Goldwater and Paul Valentine for providing photographs and ideas for illustrations for this manual. FPP is also grateful for permissions for text and illustrations from the following individuals/organizations:

The Library of Congress, Washington, D.C.

National Fire Protection Association, Quincy, MA

Ansul, Inc., Marinette, WI

Akron Brass Company, Wooster, OH
Bill Mack

Emergency One, Inc., Ocala, FL

Cicul-Air Corporation, Warren, MI

Rice Hydro, Inc., Carson City, NV

Elkhart Brass Manufacturing Company, Inc., Elkhart, IN
Paul Albinger, Jr.
Phil Turner

Conoco, Inc., Ponca City, OK

Jaffrey Fire Protection Company, Inc., Jaffrey, NH

Hydra-Shield Manufacturing, Inc., Irving TX

Task Force Tips, Inc., Valparaiso, IN

Snap-tite, Inc., Erie, PA

Kochek Company, Inc., Putnum, CT

Hall of Flame Museum, Phoenix, AZ

Niedner Limited, Coaticook, Quebec, Canada
Tom Troutner

Keith Flood

Frank T. Garza

Chicago (IL) Fire Department

Gilbert (AZ) Fire Department

Goodyear (AZ) Fire Department

Augustus Fire Tool, Island Heights, NJ

Jeff Fine, Goodyear, AZ

Lakeside (AZ) Fire Department

Mesa (AZ) Fire Department
 Mary Cameli
 Rick Montemorra (Photographer)

Morgantown (WV) Fire Department
 David Fetty

Peoria (AZ) Fire Department

Phoenix (AZ) Fire Department
 Kevin Roche

George Anagnost, Peoria, AZ

Rocky Grove Fire Department, Franklin, PA

Savoy (IL) Fire Department
 Michael Forrest

Saint Vincent College Fire Department, Latrobe, PA
 Patrick Lacy, Chief (Ret.)

South Union Fire Department, Uniontown, PA
 Adam Buchheit, Chief

Illinois Fire Service Institute, University of Illinois, Champaign, IL

Margaretville (NY) Fire Department

Pennsylvania State Fire Academy, Lewistown, PA
 Fred C. McCutcheon

West Virginia Fire Service Extension, West Virginia University, Morgantown, WV

National Fire Academy, Emmitsburg, MD

Oklahoma State Fire Service Training, OSU, Stillwater, OK
 Dan Knott
 Jason Louthan
 Philip Pope
 Richard Teeter
 Bryan West

Stillwater (OK) Fire Department
 Robert Black
 Bill Bunch
 Mike Cotton
 Don Dominick
 Dave Greer
 Tom Hart
 Rick Hauf
 James Holley
 Todd Jones
 Zach Logan
 Rex Mott
 Dale Parrish
 Jake Rhoades
 Gary Stanton
 Dusty Stokes
 Thomas Tharp
 Virginia Wallace

Tulsa (OK) Fire Department
 Steve Abernathy
 Trent Brennen
 Chris Creekmore
 R. B. Ellis
 Ray Evins
 Kevin Horner
 Kevin McLarty
 Mike Rusher
 Abren Williams
 R. D. Works

Midwest City (OK) Fire Department
 Jack Fry
 Matthew Morgan
 Doug Howard
 Greg Vernon

Washington (MO) Fire Department
 W. H. (Bill) Halmich
 Tim Frankenberg
 Jeff Aholt
 Patrick Eckelkamp
 Mile Holtmeier
 Shane Stephens
 Tim Borgmann
 Judy Halmich

Last, but certainly not least, gratitude is extended to the members of the Fire Protection Publications Hose Practices Project Team whose contributions made the final publication of this manual possible.

Fire Hose Practices Project Team

Project Manager/Editor
Barbara Adams, Senior Editor

Technical Contributors/Reviewers
Fred Stowell, Senior Technical Editor
Jeff Fortney, Senior Technical Editor

Photography
Fred Stowell, Senior Technical Editor
Jeff Fortney, Senior Technical Editor
Tara Gladden, Editorial Assistant
Foster Cryer, Research Technician
Nathan York, Research Technician

Proofreader/Editorial Contributor
Melissa Noakes, Curriculum Editor
Cynthia Brakhage, Senior Editor

Liaison Assistant
Susan S. Walker, Instructional Development Coordinator

Editorial Assistant
Tara Gladden

Research Technicians
Nathan York
Foster Cryer

Production Manager
Don Davis

Illustrators and Layout Designers
Ann Moffat, Production Coordinator
Ben Brock, Senior Graphic Designer
Clint Parker, Senior Graphics Designer
Lee Shortridge, Senior Graphic Designer
Missy Reese, Senior Graphic Designer
Jeff Mitchell, Graphics Technician

Library Researchers
Susan F. Walker, Librarian
Shelly Magee, Senior Data Control Technician

Introduction

The use of fire hose during fire suppression is taken for granted by firefighters/emergency responders and the public alike. Yet, it is as essential to any fire-suppression operation as any pumping apparatus (also known as *pumper* or *engine*) or fire hydrant to which it is connected. In its simplicity, fire hose is fundamental to the accomplishment of the primary mission of the fire and emergency services: the preservation of lives and property. Fire hose provides a quick, flexible means of bringing water from a source (fire hydrant, lake, reservoir, stream, or water tenders/mobile water supply vehicles) and delivering it to a fire or combustible/flammable liquid spill incident.

For something that is in essence a series of hollow tubes connected together and filled with water or other extinguishing agent, fire hose provides a remarkably versatile tool for fire-fighting and combustible/flammable liquid spill control operations. The very nature of the job it is called upon to perform requires that it be resilient, easy to deploy, and, above all, reliable. The varied fire-fighting/emergency-response tasks are often accomplished under challenging and adverse conditions. Although fire hose is often abused during fire-suppression or emergency-response efforts, it is expected to continue near flawless service even when it has sustained damage.

Fire hose must be strong enough to harness the high pressures generated by modern fire pumps, yet light enough for the emergency-response team to maneuver and operate. Fire hose needs to be strong enough to resist cuts, damage from abrasion and vehicle tires, chemical attack, mold and mildew attacks, and general abuse at the incident site while ceaselessly providing an uninterrupted flow of water or other extinguishing agent.

Historically, fire hose evolved as an alternative and improvement to the "bucket-brigade" system (passing buckets back and forth between a water-supply source and a fire). Greater water flows were needed to supply the ever-improving hand-powered fire pumps and steam engines of the time (toward the end of the 19th century). A first generation of fire hose was leather-constructed, assembled with riveted copper seams (**Figure I.1**). Hose constructed in this manner was very heavy and difficult to maneuver. Even with these deficiencies, the advent of this hose revolutionized the fire service.

Figure I.1 Example of early leather-jacketed riveted hose.

Couplings (permanently attached fittings used to connect other hoselines or devices) had yet to be invented, so the hose ends were inserted into each other to form a continuous waterway. Although it leaked badly, was unwieldy, and was very heavy, this first hose was a vast improvement over the bucket-brigade system used earlier.

As technology improved, so did fire hose construction. Further development was done by James Boyd in 1821 with his invention of rubber-lined, cotton-webbed fire hose accompanied by Charles Goodyear's vulcanization process in 1839, which allowed for the construction of rubber hose reinforced with a cotton ply. These developments continued to improve the versatility of fire hose. The invention of the circular loom by Reddaways of Manchester in the mid 1800s allowed for the manufacture of canvas tubes. These tubes proved to be much lighter than leather and far more maneuverable. Although the newer, cheaper, and lighter canvas or cotton hose was available in 1850, riveted leather hose remained popular and in use in large city fire departments until the late 1890s.

As fire hose technology improved, the development of the hose coupling also progressed simultaneously. Iron-threaded couplings were followed by copper or brass metal couplings, which quickly became the attachment of choice. The use of brass for the coupling along with an expansion ring to keep the coupling attached to the fire hose provided a superior method for connecting and extending these hoselines. Light aluminum couplings were not common until the years before the Second World War because of the high electrical needs of aluminum production. The advent of anodized aluminum and high-strength plastics continue to improve fire service couplings today.

Purpose

The purpose of this manual is to provide the fire and emergency services responder with the information needed to fulfill the requirements of National Fire Protection Association (NFPA) 1001, *Standard for Fire Fighter Professional Qualifications* (2002 edition), that relate to fire hose and also to serve as a comprehensive source of information about fire hose and its use. Other NFPA standards addressed in this manual are as follows:

- NFPA 1961, *Standard on Fire Hose* (2002 edition)
- NFPA 1962, *Standard for the Inspection, Care, and Use of Fire Hose, Couplings, and Nozzles and the Service Testing of Fire Hose* (2003 edition)
- NFPA 1963, *Standard for Fire Hose Connections* (2003 edition)
- NFPA 1964, *Standard for Spray Nozzles* (2003 edition)

Several basic types of fire hose are defined by these standards and discussed in this text: attack hose, supply hose, occupant-use hose, forestry hose, and suction hose. Additionally, specialty fire hoses used in other fire-suppression systems are also presented. Each of these fire-hose types has a specific design quality that makes it unique for the service it provides.

Scope

The scope of this manual includes fire hose types and construction methods, proof (acceptance) and service testing, fire hose maintenance, fire hose nozzles and related appliances, general fire hose loads and deployment procedures, and fire hose-handling techniques. Because so many "right ways" of accomplishing fire-hose evolutions are available, no single technique is presented as the best method. This manual is a reference to the fire and emergency services to assist each fire or emergency services organization in selecting those methods and procedures that are appropriate for fire hose use.

Book Organization

The text is divided into seven chapters. Each chapter describes a general area of fire-hose knowledge and is presented in a step-by-step methodology to allow the emergency responder/student an easy transition to new material in each subsequent section or chapter. A comprehensive glossary, appendices, and index follow the body of the text where the reader can find easy references to the material presented.

Chapter 1, Fire Hose and Couplings, discusses fire hose and fire hose connections. A description of the requirements of national and international standards for each is presented. The various types of hose and the differences found in fire hose couplings are defined. An explanation of each category and type of fire hose, accompanied with its design capabilities and limitations, are given. Lastly, the methods of attaching fire hose couplings are presented for each of the common coupling types described in national and international fire hose standards.

Chapter 2, Fire Hose Care, Maintenance, and Service Testing, presents the causes of fire hose and coupling damage and outlines damage-prevention procedures. Proper methods of cleaning, maintaining, and storing fire hose and couplings are discussed. Fire hose service testing and record keeping methods are also described for each type of fire hose available. Daily and after-use inspections of fire hose and fire couplings are also described. Easy-to-follow skill sheets at the end of the chapter give steps for routine care tasks, service testing, and recoupling procedures.

Chapter 3, Fire Hose Nozzles, gives a brief discussion of the types and uses of fire and emergency services nozzles. The various types of solid stream, fog, and foam nozzles are presented. Several special-purpose nozzles for special situations are also described.

Chapter 4, Fire Hose Appliances and Hose Tools, provides an overview of fire hose appliances and the tools associated with them. Various types of valves and valve devices are described along with instructions on how they are deployed and operated. Other fire hose fittings, foam appliances, and fire hose tools are shown along with operating procedures for their use.

Chapter 5, Basic Methods of Handling Fire Hose, presents methods for making fire hose connections, rolling hose, carrying hose, and deploying hoselines. Procedures fundamental for the initial deployment and advancement of attack fire hose are presented. Hose rolls, carries, and operating hoseline techniques are described. Skill sheets are provided at the end of the chapter as a quick reference as well as an easy-to-follow guide for connecting/disconnecting hose/couplings, connecting to fixed fittings, attaching nozzles, and loosening tight connections plus steps for rolling, deploying, and advancing fire hose.

Chapter 6, Supply Hose Loads and Deployment Procedures, presents descriptions of supply fire hose and the methods for loading and deploying it during emergency operations. Selecting and making basic fire hose loads are described along with supply hose deployment procedures. These procedures are accompanied by skill sheets at the end of the chapter for convenience and consistency during training. Also contained in this chapter is a description of the process of connecting supply hose to a fire hydrant.

The final chapter prepares the firefighter/emergency responder to safely and effectively deploy attack hose at the emergency scene. Chapter 7, Attack Hose Loads, Finishes, Hose Packs, and Deployment Procedures, describes the various methods of placing attack hose after unloading it from the pumping apparatus. Skill sheets accompany each of these evolutions at the end of the chapter. From making basic "attack-line" loads and finishes to using high-rise, wildland, or standpipe fire hose packs, each procedure is presented as a quick reference to the firefighter/emergency responder or training officer. Additionally, the chapter outlines "special" fire-hose use during ladder operations or while hoisting attack and supply hose during fire-fighting operations in large structures.

Skill sheets are provided following Chapters 2, 5, 6, and 7. These sheets have been developed as a guide to completing hose maintenance and repair procedures, proper cleaning and storage, hose loading techniques, deploying supply lines, and deploying attack lines. Many acceptable methods are available to accomplish these tasks. Refer to local rules and procedures manuals to select the approved methods of your department/organization.

Key Information

Various types of information in this book are given in shaded boxes marked by symbols or icons. See the following examples:

Fire Hose Sidebar

Sidebars give additional relevant information that is more detailed, descriptive, or explanatory than that given in the text.

Fire Hose Information

Information boxes give facts that are complete in themselves but belong with the text discussion. It is information that may need more emphasis or separation. They can be summaries of points, examples, calculations, scenarios, or lists of advantages/disadvantages.

As mentioned earlier, this edition also uses skill work sheets to describe the step-by-step procedures for the skills covered in the text. These sheets are referenced in the appropriate chapters and appear at the end of those chapters.

Three key signal words are found in the text and within the skill work sheets: **WARNING, CAUTION,** and **NOTE**. Definitions and examples of each are as follows:

- **WARNING** indicates information that could result in death or serious injury to fire and emergency services personnel. See the following example:

WARNING

Conduct all hoseline fire-suppression activities with a minimum of two firefighters. For interior structural fire-suppression activities, an emergency backup team of two firefighters must be available for immediate service.

- **CAUTION** indicates important information or data that fire and emergency service responders need to be aware of in order to perform their duties safely. See the following example:

CAUTION

Glass shards and other sharp debris may become embedded in fire hose during fire-suppression operations. Ensure that all emergency responders wear gloves during hose and cleanup evolutions to avoid injury.

- **NOTE** indicates important operational information that helps explain why a particular recommendation is given or describes optional methods for certain procedures. See the following example:

NOTE: Some fire and emergency services organizations, depending upon the hose loads that they use, may store hose with the male coupling on the outside of the hose roll.

Chapter 1
Fire Hose and Couplings

Chapter 1
Fire Hose and Couplings

Fire hose moves water from one location to another. Because fire hose is used for a number of functions during fire-fighting or other emergency-response operations, many types are manufactured, depending on their intended purposes. Fire hose is available in different lengths and diameters, in natural or synthetic materials, and with different types and sizes of couplings (fittings that connect hose length ends together). Fire hose is made of very durable materials because it is used frequently and must withstand the wear that occurs with almost daily use. Municipal fire departments usually limit the diameters of their fire hose to 6 inches (152 mm) or less, which provides them with an adequate waterway for fire-suppression activities while allowing enough hose to be loaded on the pumping apparatus to meet local deployment requirements. Industrial fire brigades, on the other hand, may use hose that is 12 inches (305 mm) in diameter or larger, depending upon the fire flow required to protect their manufacturing, chemical, or hazardous processes (see sidebar).

Any discussion about fire hose must also include the fire hose coupling. The coupling is integral to the hose system, and it must complement the fire hose in strength and resiliency, yet be light enough to maneuver easily. Its design must allow for easy coupling and uncoupling under the worst of conditions; and at the same time, the connection must be tight enough to withstand the pump pressures generated by the pumping apparatus while meeting the needs for fire-suppression or other emergency operations (including hazardous materials incidents). Couplings come in a full range of sizes corresponding to the standard available fire hose sizes.

A thorough knowledge of fire hose is necessary in order to use it more effectively and care for it properly. This knowledge includes understanding the methods of fire hose and coupling construction and the types of fire hose along with the respective uses for and limitations of each. This chapter begins with a brief history of the standardization of fire hose and couplings. It goes on to describe basic fire hose classifications, hose performance requirements, and hose construction. The last part of the chapter discusses couplings — types, construction, and methods by which they are attached to hose.

Fire Flow

Fire flow is defined as the quantity of water available for fire-fighting operations in a given area. It is calculated in addition to the normal water consumption in the area. Water flow testing procedures are used to determine the rate of water flow available for fire-fighting purposes at various points within the water distribution system.

The National Fire Protection Association (NFPA) defines the term as the flow rate of a water supply that is available for fire-fighting operations, measured at 20 psi (138 kPa) {1.38 bar} residual pressure (the pressure acting on a point in the system with a flow being delivered). See NFPA 1141, *Standard for Fire Protection in Planned Building Groups* (2003), NFPA 14, *Standard for the Installation of Standpipes, Private Hydrants, and Hose Systems* (2000), and NFPA 1410, *Standard on Training for Initial Emergency Scene Operations* (2000), for more information).

History of the Standardization of Fire Hose and Couplings

One of the great dilemmas facing the fire and emergency services in North America at the end of the nineteenth century was that fire and emergency service organizations were using fire hose and fittings/couplings of different sizes and thread types. At one time, there were over 2,000 different fire hose coupling threads in the United States and Canada. The problem was most apparent when fire and emergency response organizations were required to join forces during major fires and conflagrations. More often than not, it was impossible for pumping apparatus from one organization to connect to hydrants and pumping apparatus of neighboring organizations. This situation, of course, resulted in a greatly impaired fire-fighting effort. As a result, needless lives and property were lost because water could not be efficiently moved to the fire scene. For example, during the Boston fire of 1872 and the Baltimore, Maryland, "Big Fire" of February, 1904, assisting fire departments could not make connections to city fire hydrants. In Baltimore, many firefighters resorted to opening fire hydrants, allowing water to flow into ditches and holes, and then drafting (drawing) from them. Eighty-six blocks of downtown Baltimore were lost, including the so-called "fireproof" City Hall **(Figure 1.1)**.

Recognizing the increasing problems caused by incompatible fire hose couplings, the newly organized International Association of Fire Engineers (IAFE) (now called the International Association of Fire Chiefs or IAFC) passed its first resolution in 1873, which was to set standards for fire hose screw threads. The standard that was developed represented an initial step in a process that continues today. This ongoing process ensures that regardless of origin, a pumping apparatus will be able to make a connection to a water source for fire suppression or other emergency situation response. Although there is no law that requires a fire or emergency services organization to use a certain fire hose screw thread, the threads on all pumping apparatus inlets and discharges meet this standard when they are built. Local thread gauges may be fitted on a pumping apparatus by connecting an adapter (fitting with dissimilar threads but the same inside diameter) over the national standard threads that are installed on all 2½-inch (65 mm) or larger discharges when a fire pump is manufactured. See National Fire Protection Association (NFPA) 1901, *Standard for Automotive Fire Apparatus* (2003), for specific requirements.

Numerous organizations, crossing many professional disciplines, realized the benefit of encouraging a national standard for fire hose couplings. The insurance industry suffered enormous losses because fire pumps could not access available municipal water supplies. Engineers struggled with local thread standards. Fire and emergency service organizations often had to respond into areas that had different fire hose threads. All of these groups wanted a standardized thread. Concerted efforts began late in the 19th century and continue today. Although not fully recognized by all jurisdictions, tremendous progress has been made. The continued support of many national organizations, including the International Fire Service Training Association (IFSTA), NFPA, and American Standards

Figure 1.1 Eighty-six blocks of downtown Baltimore, Maryland, were lost in the "Big Fire" of 1904. *Courtesy of The Library of Congress, Washington, D.C.*

Association (ASA), strive for the eventual adoption of a standard for all fire hose couplings.

NFPA Coupling Screw Thread Specifications

As early as 1898, NFPA began developing thread specifications for fire hose couplings. Prompted by the 1904 Baltimore fire, this work led to the appointment of an NFPA committee in 1905 to standardize not only couplings, but also fire hose, nozzles, and accessories. This NFPA committee developed general screw thread specifications covering the 2½-, 3-, 3½-, and 4½-inch (65 mm, 77 mm, 90 mm, and 115 mm) hose and appliance sizes. An earlier report from IAFE produced in conjunction with the active cooperation of the American Water Works Association (AWWA) provided the basis for this standard.

For 2½-inch (65 mm) couplings, the principle dimensions of 7½ threads per inch (TPI) and 3$\frac{1}{16}$ inches (78 mm) outside diameter of the external thread were selected to facilitate conversion of existing couplings (see sidebar). Many municipal fire and emergency service organizations had adopted either 7 or 8 threads per inch and a 3- or 3½-inch (77 mm or 90 mm) outside diameter external thread standard. It was also not uncommon for municipalities to adopt a thread standard based upon local industrial fire hose specifications. Industrial sites, railroads, chemical facilities, and other manufacturing facilities had a great influence on local fire and emergency service organizations and the coupling standard that was used. For example, 6 threads per inch is commonly found in many eastern communities influenced by the coal and steel industries, which was also a standard thread size used by the Baltimore and Ohio (B&O) Railroad. In essence, the hose thread size was found in rail yards and other associated sites. Many communities simply adopted the thread standard of these large local industries.

By mutual agreement, the fieldwork of the NFPA committee concerned with encouraging adoption and application of the screw thread standard was taken over by a Committee on Fire Prevention and Engineering Standards of the National Board of Fire Underwriters (NBFU) in 1917. At the same time, NFPA organized a Committee on Small Hose Couplings to develop standards on fire hose screw threads in sizes from ½- to 2-inch (13 mm to 50 mm) nominal diameters. NFPA 194, *Standard for Small Hose Coupling Screw Threads* (now NFPA 1963, *Standard for Fire Hose Connections,* 2003), which covered these coupling sizes, was developed and adopted in 1922. These small-sized couplings had the same general thread-design characteristics as the standard couplings for 2½-inch (65 mm) and larger hose.

In October, 1923, NBFU, NFPA, and American Society of Mechanical Engineers (ASME) requested that ASA approve and designate this standard as an *American Standard* for fire hose threads. Shortly after, ASA assigned joint sponsorship for this project to NBFU, AWWA, and ASME. Through the cooperation of a group of gauging experts, including members of the National Screw Thread Commission (NSTC), the limiting dimensions (maximum and minimum coupling thread tolerances during machining) were also added to the original thread specifications. The standard for fire hose coupling screw threads for sizes 2½ inches (65 mm) and larger was approved by the ASA in May, 1925.

Standardized thread is called the *American National Fire Hose Connection Screw Thread* (abbreviated *NH*). Because it is also commonly

Threads per Inch (TPI)

Canada and Great Britain continue to use the threads per inch method for fire hose couplings even though they converted from the English or Customary System of measurement to the International System of Units or SI (metric system). Notably, metric conversions to threads per millimeter (TPMM) from the TPI method have been made in several Canadian provinces, but they are metric approximations of the original thread standard. There is no Canadian national hose thread standard, which results in numerous thread standards being adopted by the individual provinces across the country, paralleling the experience of the United States. Canada has, however, adopted nationwide a threadless/sexless coupling for forestry applications. Of particular interest is the similar occurrence in Great Britain, who also continues to use the TPI standard for threaded fire hose couplings instead of adopting a metric thread standard.

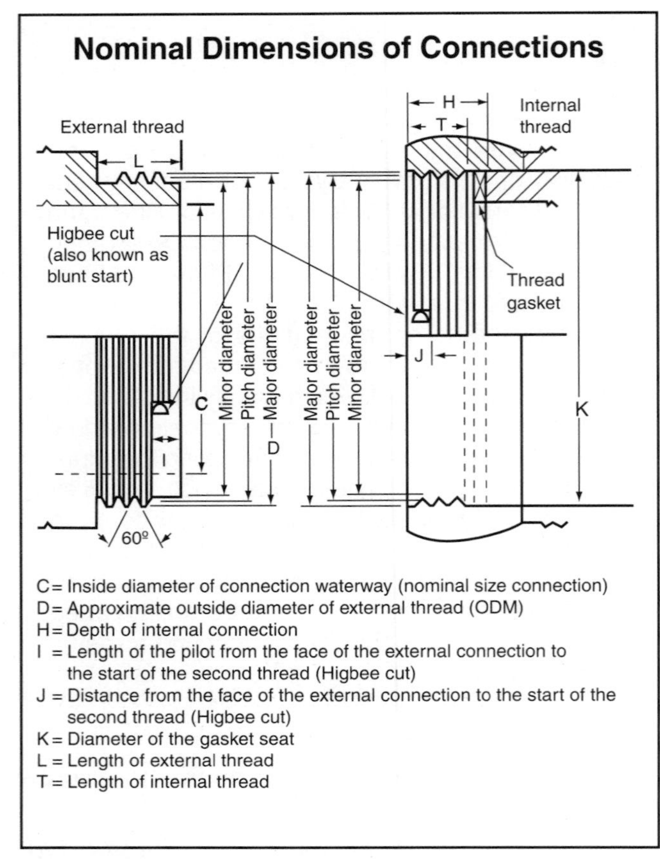

Figure 1.2 Specifications for a 2½-inch (65 mm) coupling with National Standard Threads. *Reprinted with permission from NFPA 1963, Standard for Fire Hose Connections, Copyright © 2003, National Fire Protection Association, Quincy, MA 00269. This printed material is not the complete and official position of the National Fire Protection Association on the referenced subject, which is represented only by the standard in its entirety.*

referred to as the *National Standard Thread,* it is often abbreviated as *NST* or *NS.* A detailed drawing of the National Standard Thread is shown in **Figure 1.2.**

ASA Sectional Committee

In January, 1927, ASME requested that ASA authorize the organization of a sectional committee to complete the standardization of fire hose couplings and attempt to unify and complete the specifications for dimensions of small hose couplings. A sectional committee was organized in October, 1928, under the sponsorship of ASME to prepare specifications for screw threads for small hose couplings ranging from ½ to 2 inches (13 mm to 50 mm) nominal size. Data on these small threads were then published by ASA. These specifications varied somewhat from those recommended by NFPA, but ultimately ASA adopted the NFPA recommendations for a heavier thread for fire hose, which allowed for easier coupling connections during emergency operations.

In 1961, the duties of the ASA Sectional Committee were transferred to a newly established subcommittee of the ASA Sectional Committee on the Standardization of Pipe Threads for which ASME and American Gas Association (AGA) were joint sponsors. The subcommittee was organized to deal with threads for fire hose couplings and fittings.

Coupling Thread Standardization Program

In 1955, NFPA adopted standards for threads on 4-, 5-, and 6-inch (100 mm, 125 mm, and 150 mm) supply hose because almost every fire pump manufacturer was using a different thread on certain sizes of pumping apparatus. Unfortunately, the supply hose from one pumping apparatus could not be used on another pumping apparatus at the same time (see Supply Hose under Fire Hose Classifications section).

NFPA adopted dimensions for gaskets from ¾ to 6 inches (20 mm to 150 mm) for standard fire hose couplings as well as provided data on the required gasket seat dimensions in 1956. Gaskets are essential features of a fire hose coupling standard because hose connections have swivel or *female* fittings with rubber gaskets that provide watertight seals when the couplings are connected to the opposing male threaded coupling (see Threaded Couplings under Coupling Classifications section). NFPA also prepared a text showing the suggested application of the standard to various items of fire-fighting equipment because experience had shown that the wrong size of standard thread was sometimes used, causing leaks and limiting the effectiveness of the water supply operation.

A survey conducted by NFPA in 1965 showed that 65 percent of the fire and emergency service organizations serving U.S. communities of over 20,000 population used standard fire hose coupling screw threads on all sizes of fire hose. The approximate percentages of communities using standardized threads on the various coupling sizes were as follows:

- ¾-inch and 1-inch (20 mm and 25 mm) threads — 95 percent
- 1½-inch (38 mm) threads — 84 percent
- 2½-inch (65 mm) threads — 73 percent

The degree of standardization is believed to be considerably higher in small communities, many of which organized their fire and emergency service organizations after the adoption of the standard. Most large cities (those with long established fire and emergency service organizations) implemented fire hose thread standards before the creation of the National Standard Thread. Often other local suburban fire and emergency service organizations that were located close to these metropolitan areas adopted the same thread standard. In essence, this "local standard" method functions well as long as other, outside fire and emergency service organizations are not expected to seamlessly integrate with the metropolitan fire defenses. Approximately half of the states in the U.S. have regulations, codes, or laws supporting fire hose thread standardization. As mentioned previously, Canada has no law or standard that requires Canadian fire and emergency service organizations to adopt a "national thread," and none of the provincial governments have adopted a standard based upon NFPA 1963. Although adopting a standard thread is recommended (much the same as it is for many communities in the United States), Canadian provinces and local municipal fire and emergency service organizations are free to determine and adopt their own thread specifications. However, Canada has adopted nationwide a threadless/nonthreaded coupling for forestry applications.

At its 69th annual meeting in 1965, NFPA passed a resolution to intensify its efforts to accomplish complete standardization of fire hose screw threads throughout the country by asking for assistance from fire chiefs, fire and emergency service organizations, manufacturers, and governmental agencies. NFPA, IAFC, International Association of Firefighters (IAFF), American National Standards Institute (ANSI), AWWA, and many others have assisted with the standardization program.

In 1967, NFPA revised NFPA 194, *Standard for Screw Threads and Gaskets for Fire Hose Couplings,* to include the new material available from the ASA subcommittee. Several editorial and style changes were adopted in the standard in 1968 and 1974. In 1979 the numeric designation of the document was changed from NFPA 194 to NFPA 1963. The 1993 edition of the standard reorganized and expanded the scope of the standard to include the general requirements for couplings and adapters as well as the inclusion of a chapter to cover nonthreaded connections in the 4- and 5-inch (100 mm and 125 mm) sizes (see Nonthreaded and Sexless Couplings under Coupling Classifications section) **(Table 1.1).**

Table 1.1
American National Fire Hose Connection Screw Treads per Inch for Various Sizes of Threaded Hose Couplings

Size of Hose Coupling	Threads per Inch (TPI)
¾ inch	8
1 inch	8
1½ inches	9
2 inches	None listed in NFPA 1963
2½ inches	7.5
3 inches	6
3½ inches	6
4 inches	4
4½ inches	4
5 inches	4
6 inches	4

Based on information found in Section 5.2.2 in NFPA 1963, *Standard for Fire Hose Connections*, 2003 Edition.

National Fire Hose and Coupling Standards

National standards defining the performance of materials and construction for fire hose and couplings are now written by several organizations, including the following (See **Appendix A,** Fire Hose Standards by Various Organizations):

- National Fire Protection Association (NFPA)
- Rubber Manufacturers Association (RMA)
- American Society for Testing and Materials (ASTM)
- United States Department of Agriculture (USDA)
- United States Forest Service (USFS)

American National Standards Institute (ANSI) assists in the process of writing standards by coordinating the efforts of other related organizations to produce standards.

Agencies that test fire hose components are as follows:

- Underwriters Laboratories Inc. (UL)
- Underwriters' Laboratories of Canada (UL Canada)
- Factory Mutual (FM)

Fire Hose Classifications

The fire and emergency services describe or classify fire hose in a number of ways. This manual places the many types of hose into six broad categories that are based on the way the hose is constructed and used. Descriptions of these types are given in the sections that follow. Fire extinguisher hose and special cold-resistant hose are also described. See the information boxes for definitions of frequently used fire hose terms. The six categories are as follows:

- Attack hose
- Supply hose
- Occupant-use fire hose
- Forestry fire hose
- Suction hose
- Booster fire hose

General Fire Hose Definitions

Attack hose — Hose that is used by trained firefighters or fire brigade members to combat fires beyond the incipient fire stage

Supply hose — (1) Hose that is designed for the purpose of moving water between a pressurized water source and a pump that is supplying attack hoselines or (2) hose that is used to maintain a water system (either as a continuous conduit or by connecting water supply sources)

Suction hose — Intake hose that connects pumping apparatus or portable pump to a water source; two types are as follows:

- ***Hard suction (or hard sleeve)*** — Rigid, noncollapsible hose that operates under vacuum conditions without collapsing, allowing a pumping apparatus or portable pump to "draft" or draw water from static or nonpressurized sources (lakes, rivers, wells, etc.) that are below the level of the fire pump; usually available in 10-foot (3 m) sections; when carried on a pumping apparatus, 2 sections are required
- ***Soft sleeve (often referred to as soft suction)*** — Collapsible hose that connects a pumping apparatus to a pressurized water supply source (generally a short length of supply hose with female couplings on both ends); not designed for drafting operations; NFPA 1901 requires that a minimum of 15 feet (4.6 m) be carried on pumping apparatus

Forestry fire hose — Small-sized, lightweight hose that is designed to meet the specialized requirements for fighting wildland fires

Occupant-use fire hose — Hose used by a building's (or ship's) occupants to fight incipient-stage fires before the arrival of trained firefighters or fire brigade members

Booster hose — Noncollapsible, preconnected hose having an elastomeric or thermoplastic tube, a braided or spiraled reinforcement, and an outer protective cover; used for small, incipient-stage exterior fires; often carried on pumping apparatus on reels

Small diameter hose — Fire hose that is between ½ and 3 inches (13 mm and 77 mm) in diameter and most often used as attack hose (but also used for booster hose)

Large diameter hose (LDH) — Fire hose that is 3½ inches (90 mm) or larger in diameter (used for supply and suction hose)

Fire Hose Design and Test Terms

Service Test — Test conducted by the manufacturer following production and then annually by the user to ensure fire hose still meets minimum operational standards (See Chapter 2, Fire Hose Care, Maintenance, and Service Testing.)

Design Service Test Pressure — Pressure equal to approximately 110 percent of the operating pressure; test conducted by the manufacturer

Annual Service Test Pressure — Test pressure used when the fire and emergency service organization conducts the annual service test; test pressure is marked on the hose

Burst Test — Destructive test on a 3-foot (0.9 m) length of hose at three times its labeled service test pressure to determine its maximum strength; test conducted by the manufacturer

Burst Test Pressure — Pressure equal to at least three times the service test pressure

Proof Test (formerly called acceptance test) — Test conducted by the manufacturer to ensure fire hose meets standards; measures movement or leaking of couplings in a supply connection when pressurized to a pressure two times the service test pressure

Kink Test — Test conducted by folding the hose over on itself, securing it to maintain the kink, and pressurizing the hose to ensure performance; test conducted by the manufacturer

Elongation — Amount of additional length created by pressurizing a hose; specifically the percent a length of hose increases in length from an initial measurement with the hose pressurized at 10 psi (69 kPa) {0.7 bar} to a final measurement with the hose pressurized to an established test pressure or its proof test pressure (See Chapter 2, Fire Hose Care, Maintenance, and Service Testing.)

Reinforcement — Structural support in the form of woven, braided, or spiraled yarn within a rubber-covered hose

Rise — Distance that a hose elevates above the test table while under pressure; specifically the maximum distance (measured from the test table to the underside of the hose) that a length of hose lifts off the table when it is pressurized to an established test pressure or its proof test pressure

Trade Size — Generally recognized nominal size of hose, the minimum internal diameter of the waterway as described by the manufacturer, or more accurately, the internal diameter of the hose when pressurized at operating pressures

Twist — Rotation of hose under pressure; specifically the number of revolutions a fixed length (50 feet or 15 m) of hose rotates when it is pressurized from an initial pressure of 10 psi (69 kPa) {0.7 bar} to an established test pressure (or its proof test pressure) with one end held stationary and the other end allowed to move freely

Warp — (1) Yarns (called ends) running lengthwise down a hose; (2) maximum distance any portion of a fixed length (50 feet or 15 m) of hose deviates from a straight line drawn from the center of the fitting at one end to the center of the fitting at the other end with the hose pressurized to an established test pressure or its proof test pressure

Weft — Yarns (filler yarn) running circumferentially around a hose; measured in picks per inch

Attack Hose

Attack hose is any hose supplying water between the attack pumping apparatus and the nozzle that is used by trained firefighters and fire brigade members to directly control and extinguish fires that have grown beyond the incipient stages of fire development **(Figure 1.3)**. This group of fire hose is by far the broadest because there are many types of attack fire hose on the market.

Attack hose may have a fabric jacket or be a rubber-covered type. Woven fabric-jacket hose may be single-jacketed or multiple-jacketed. Most fabric-jacket hose is double-jacketed and lined for maximum durability. Newer thermoplastic, ethylene-propylene-diene-terpolymer (EPDM) lined or extruded nitrile rubber hose is rapidly replacing the heavier double-jacketed styrene-butadiene rubber (SBR) latex compound attack hose prevalent in the past. Rubber-covered hose may be jacketed and lined or constructed so that rubber and fabric are extruded into a single, inseparable unit. Each of these attack hose types is manufactured in a variety of sizes. Although many of the characteristics of attack hose are similar to that of forestry hose (described later), forestry hose usually has a heavier and more durable construction.

Attack hose is differentiated from other categories of hose by its minimum service test pressure

Figure 1.3 Preconnected attack hoselines are commonly stored in arrangements above or below the pump panel on pumping apparatus. *Courtesy of Chief W. H. (Bill) Halmich, Washington (MO) Fire Department, Judy Halmich photographer.*

of 300 psi (2 068 kPa) {20.7 bar}. The manufacturer labels hose test pressure on the outside jacket or cover of each hose section. See Chapter 2, Fire Hose Care, Maintenance, and Service Testing.

All new fire hose must successfully pass several performance tests to ensure that it is manufactured in compliance with the hose standard, NFPA 1961, *Standard on Fire Hose* (2002). Additionally, successful completion of these tests and annual service testing are indicators that the fire hose will continue to perform at the highest level demanded of it for its service life. However, even though a length of hose passes the tests, it could still fail during use because of mechanical damage, chemical exposure, depreciation due to age, or degradation due to sunlight or high heat. Attack fire hose performance requirements during the tests conducted by the manufacturer as described in NFPA 1961 are as follows:

- Design service test performed with a minimum test pressure of 300 psi (2 068 kPa) {20.7 bar}
  - — The manufacturer performs the design service test.
  - — The purchasing agency conducts a service test before placing the hose into service and then conducts annual service tests.
- Proof test at a pressure of two times the minimum service test pressure, 600 psi (4 137 kPa) {41 bar}, without movement of couplings, leakage, or breaking of any threads in the reinforcement (structural support)
- Burst test using a short piece of test hose, 3 feet (0.9 m) in length with a pressure of three times that of the labeled service test pressure
- Kink test using a full section of hose and a test pressure of 1.5 times the minimum service test pressure; hose is sharply kinked 18 inches

(457 mm) from the free end by tying the hose back against itself as close to the fittings as practicable

- Total elongation of hose cannot exceed 8 percent for 2½-inch (65 mm) or smaller hose, 10 percent for 3-inch (77 mm) hose, and 13 percent for 3½-inch (90 mm) or larger hose
- Hose may not twist more than 4¼ turns in a 50-foot (15 m) section (coupled piece or length) for 2-inch (50 mm) or smaller hose
  - — Final twist is in the direction that tightens the coupling connection.
  - — Warp of the hose is no more than 20 inches (508 mm).
  - — No rise is permitted when tested.

Occasionally, fire and emergency service organizations use attack hose of small diameters (2½ or 3 inches [65 mm or 77 mm]) in dual capacities of either attack or supply hose during fire-suppression operations (see Supply Hose section). These small diameter hoses can attain adequate flows for short distances and are easy to maneuver. These supply hose evolutions often require dual hoseline deployments to provide the quantity of water needed while reducing the overall friction loss in the supply hose system.

A serious problem when using 1½- to 2-inch (38 mm to 50 mm) diameter attack fire hose over long distances, however, is that the pressure loss caused by water friction on the inner lining of the hose seriously reduces the water flow, often below safe and acceptable levels. Flow reduction caused by friction loss is variable from hose manufacturer to hose manufacturer and between hose styles. Each type and size of fire hose exhibits different friction-loss characteristics. When large volumes of water must be delivered with 2½- or 3-inch (65 mm or 77 mm) diameter hose, several parallel hoselines may be required or several pumping apparatus must be positioned at intervals to "relay" the water along the length of hose. This relay method effectively "boosts" the water pressure to overcome the pressure losses experienced with small supply hose when it is deployed long distances (over 500 feet [152 m]).

Modern fire hose with the benefits of improved construction and manufacturing methods and materials has made many of the "old" Freeman-based friction-loss formulas and tables developed in 1888 obsolete. New hose materials and construction methodology have improved the performance of modern fire hose; however, the physical laws of hydraulics continue to affect friction in hoses and pipes in the following ways:

- Friction loss in hose varies directly with the length of the hoseline deployed. Friction loss for each section of hose while maintaining a constant flow is cumulative. Example:
  - — If a given hose diameter at a particular rate of flow has a friction loss of 10 psi (69 kPa) {0.7 bar}, the friction loss is doubled if the length of hose is doubled.
- When hoses are the same size, friction loss increases dramatically as the pressure (flow velocity) is increased. Friction loss develops much faster than the change in pressure (flow velocity). Example:
  - — A length of 3-inch (77 mm) hose flowing 200 gpm (757 L/min) has a friction loss of 3.2 psi (22 kPa) {0.22 bar}. As the flow doubles from 200 to 400 gpm (757 L/min to 1 514 L/min), friction loss increases four times to 12.8 psi (88 kPa) {0.88 bar}.
- As the diameter of the hose is increased while the discharge remains constant, friction loss is reduced in proportion to the diameter of the hose. This principle readily proves the advantage of large diameter hose.
- For a given flow velocity, the friction loss experienced is approximately the same regardless of the water pressure in the hoseline. Friction loss is the same when hoses at different pressures flow the same amount of water. Example:
  - — If 100 gpm (379 L/min) passes through a 3-inch (77 mm) hose within a certain time, the water must travel at a specified velocity (feet per second [meters per second]). For the same rate of flow to pass through a 1½-inch (38 mm) hose, the velocity must be greatly increased. Four 1½-inch (38 mm) hoses are needed to flow 100 gpm (379 L/min) at the same velocity required for a single 3-inch (77 mm) hose.

See IFSTA's **Pumping Apparatus Driver/ Operator Handbook,** Chapter 6, for a comprehensive discussion of friction loss and how to calculate its effect. Modern fire hose friction-loss calculations are also included.

Freeman's Formulas

Engineer John R. Freeman developed and conducted many of the early investigations of hydraulics for the fire service in the late 1880s and early 1890s. The resulting formulas were used extensively until recent advancements in fire hose construction made them obsolete. He measured friction losses in fire hose and recommended the use of 250-gpm (946 L/min) fire streams. These tests led to the employment of 3-inch (77 mm) hose by most fire and emergency service organizations.

Supply Hose

Supply hose is hose that moves large volumes of water between a pressurized water source and a pump that is supplying attack hoselines. It also is used to maintain a water system as a continuous conduit or by connecting supply sources. Usually the pressures of supply hoselines are lower than those used for attack fire hose. Supply hose is also generally larger than attack hose, having a minimum diameter of 3½ inches (90 mm) and a minimum service test pressure of 200 psi (1 379 kPa) {13.8 bar} **(Figure 1.4).**

Intake hose is supply hose that connects a pumping apparatus or a portable pump to either a pressurized or nonpressurized water source. See Suction Hose section for information on the hard suction type. *Soft sleeve supply hose* (often referred to as a *soft suction*) is a nonrigid, collapsible intake hose that transfers water from a pressurized water source such as a fire hydrant to the fire pump intake on the pumping apparatus; it is not used for drafting from a static water source. In most instances, this hose is a short length of large diameter hose (5- or 6-inch diameter [125 mm or 150 mm]) at least 15 feet [4.6 m] in length with female couplings on both ends. It usually has multiple jackets with a lined construction or is manufactured from extruded nitrile rubber. Soft sleeve supply hoses are available in sizes ranging from 3½ to 10 inches (65 mm to 254 mm) in diameter and are typically constructed of sections of regular supply hose.

Figure 1.4 Supply hose is available in sizes from 3½ to 12 inches (90 mm to 305 mm) in diameter. Sizes 3½, 4, and 5 inches (90 mm, 100 mm, and 125 mm) are shown. *Courtesy of Sam Goldwater.*

The emergence of synthetics and advanced rubber technology allows for the production of the large-sized hoses. With ever-increasing demand for greater fire flow over long distances, large diameter hose (LDH) accomplishes this task efficiently and effectively. LDH has gained popularity in the modern fire and emergency services, having evolved during World War II as the result of the huge firebombing raids of European cities. In essence, LDH became an aboveground water main, supplying tremendous quantities of water over long distances, which enabled fire-suppression forces to mount massive fire-fighting efforts.

Large diameter hose was also developed to overcome the pressure-loss problems of small diameter supply hose. Available in both woven-jacket and rubber-covered versions, these large supply hoses are usually 3½, 4, 4½, 5, 6, 8, 10, and 12 inches (90 mm, 100 mm, 115 mm, 125 mm, 150 mm, 225 mm, 250 mm, and 300 mm) in diameter. Most commonly; however, supply hose for fire and emergency use seldom exceeds 6 inches (150 mm) in diameter and 100 feet (30 m) in length. The larger supply hoses (often referred to as *aboveground water mains*) are typically found in manufacturing, industrial, or chemical production facilities where the required fire flows far exceed those of most municipalities. They are also found in drinking water supply and military applications. These hoses range from 6 to 12 inches (150 mm to 300 mm) in diameter and are commonly found in lengths up to 660 feet (201 m) to minimize the cost of numerous couplings.

Although not identified as a supply hose diameter, 2½-inch (65 mm) hose was once considered the minimum inside diameter size of supply hose. However, this size lost popularity because many fire and emergency service organizations adopted larger hose (3½-inch [90 mm] or larger diameters).

Supply hose must have a minimum service test pressure of 200 psi (1 379 kPa) {13.8 bar}. Supply hose must meet the following minimum design characteristics when tested by the manufacturer:

- Minimum trade size of 3½ inches (90 mm)
- Minimum design service test pressure of 200 psi (1 379 kPa) {13.8 bar}
- Withstand a test pressure of two times its design service test pressure without movement of the couplings, leakage, or suffering any thread failure
- Pass a burst test of three times the design service test pressure without failure
- Withstand a kink test pressure of 1½ times the design service test pressure with no failure of its reinforcing elements when using a full section of hose
- Maximum elongation no more than 10 percent when measured from its initial length when pressurized at 10 psi (69 kPa) {0.7 bar} and final measurement taken when the proof test pressure is reached
- Maximum twist of 1¾ turns in a 50-foot (15 m) hose length when proof tested
  - Final twist is in the direction that tightens the coupling connection.
  - Warp of the hose is no more than 20 inches (508 mm).
  - No rise is permitted when tested.

Occupant-Use Fire Hose

Occupant-use fire hose (also known as *standpipe hose, house line fire hose, or shipboard/marine fire station hose*) is hose used by a building's (or ship's) occupants to initiate fire-suppression activities at incipient-stage fires before the arrival of trained firefighters or fire brigade members. The minimum trade size for occupant-use hose is 1½ inches (38 mm). Occupant-use hose is designed and tested by the manufacturer to have the following minimum design qualities:

- Minimum service test pressure of 150 psi (1 034 kPa) {10.3 bar}
- Withstand a proof test pressure of two times its design service test pressure without movement of couplings, leakage, or breaking of any thread in the reinforcement
- Withstand minimum burst test pressure of three times the design service test pressure without failure of test sample
- Withstand minimum kink test pressure of 1½ times the design service test pressure without breaking any thread in the reinforcement of full length of hose
- Maximum elongation not exceeding 10 percent when measured from its initial length when pressurized at 10 psi (69 kPa) {0.7 bar} and final measurement taken when the proof test pressure is reached
- Maximum twist per 50 feet (15 m) length when proof tested not exceeding 7½ turns
  - Final twist is in the direction that tightens the coupling connection.
  - Warp of the hose is no more than 20 inches (508 mm).
  - No rise greater than 7 inches (178 mm) is permitted when tested.

Occupant-use hose is usually preconnected to a small-diameter occupant-use-only standpipe connection or yard hydrant. The hose, complete with a nozzle, is suspended from a swivel pin rack in an accordion fashion within a hose cabinet or stored on a pivoting hose reel **(Figure 1.5, p. 18).** On ships, fire hose is often stowed on a rack or hose reel on deck in the open, making it vulnerable to misuse and damage. A major disadvantage of occupant-use hose is that it becomes unreliable with age deterioration. Because of the reliability question of occupant-use hose, emergency responders should use attack hose that is carried on their pumping apparatus or that maintained by their organization on standpipe systems.

Figure 1.5 Occupant-use hose is designed for use in locations where occupants or incipient fire brigade members might operate from standpipes.

Figure 1.6 Forestry fire hose meets design specifications outlined in NFPA 1961, *Standard on Fire Hose* (2002) and U.S. Department of Agriculture Specification 5100-186C.

Lightweight attack hose (maintained regularly and tested annually) folded into a portable pack or bundle can be carried on the pumping apparatus. These "high-rise" hose packs along with fire and emergency services nozzles and appliances can then be conveniently carried into a structure or on board a marine vessel and connected to the fire and emergency services standpipe system or international shore connection. Hose packs of this type are discussed in Chapter 7, Attack Hose Loads, Finishes, Hose Packs, and Deployment Procedures.

Forestry Fire Hose

Forestry fire hose is a single-jacket, small diameter (1 to 1½ inches [25 mm to 38 mm]) lightweight hose used to combat fires in the forest and other wildland settings. All hose must be lined and composed of a rubber compound, a thermoplastic material, blends of rubber compounds and thermoplastic materials, or a natural rubber-latex-coated fabric. The lining is a uniform thickness throughout.

The small size and lightweight characteristics of forestry fire hose are necessary because it needs to be carried long distances over uneven terrain (often uphill) and used in a way that conserves water. In some regions it can be used with a pumping apparatus in a mobile fire attack. In less accessible terrain, it can be transported by backpacking or airdropping to a remote site where it can be pressurized using a portable pump that drafts water from a stream or other open water source. Forestry fire hose is also used with pumping apparatus when wildland fires occur in areas that are accessible by road and have some type of water supply.

Forestry fire hose design, testing, and performance characteristics differ slightly from those found in the attack hose category. Because forestry hose requires long deployments through rugged terrains, the lightest design weight possible is required. Without sacrificing its strength and maneuverability, forestry fire hose (as a category of fire hose) remains much lighter than attack hose found on municipal pumping apparatus **(Figure 1.6).** Minimum design and manufacturer testing requirements for forestry hose are as follows:

- Maximum weight for a 50-foot (15 m) length of lined, cotton-synthetic woven reinforcement forestry hose of 10.9 pounds (4.9 kg) for 1-inch (25 mm) diameter hose and 14.8 pounds (6.7 kg) for 1½-inch (38 mm) diameter hose
- Maximum weight for a 50-foot (15 m) length of lined, synthetic woven reinforcement forestry hose of 5 pounds (2.3 kg) for 1-inch (25 mm) diameter hose and 8 pounds (3.6 kg) for 1½-inch (38 mm) diameter hose
- Minimum design service test of 300 psi (2 068 kPa) {20.7 bar}
- Withstand a proof test pressure of two times its design service test pressure without movement of couplings, leakage, or breakage of any thread in the reinforcement component of the hose
- Test sample withstands a burst test pressure of three times the design service test pressure without failure
- Full section of forestry hose withstands a kink test conducted using a pressure of 1½ times the design service test pressure without breaking any thread in the reinforcement component of the hose
- No elongation greater than 10 percent of its nominal section length when measured from its initial length when pressurized at 10 psi (69 kPa) {0.7 bar} and final measurement taken when the proof test pressure is reached
- Maximum twists of 12 turns per 50 feet (15 m) when proof tested
  - — Final twist is in the direction that tightens the coupling connection.
  - — Warp of the hose is no more than 25 inches (635 mm).
  - — No rise greater than 8 inches (203 mm) is permitted when tested.

Suction Hose

Suction hose is intake hose that connects a pumping apparatus or portable pump to a water source. Both hard suction and soft sleeve types are available. In reality, the soft sleeve type is generally a short length of supply hose with female couplings on both ends. See Supply Hose section.

Hard suction hose (or *hard sleeve* or simply *suction)* is a rigid, noncollapsible fire hose that draws (or drafts) water from a static (nonpressurized) water source that is below fire-pump level by creating a vacuum within the hose, which draws water into a fire pump and then distributes it to attack or supply hose for fire-suppression purposes **(Figure 1.7).** Hard suction hose is also used for tandem-pumping evolutions (where a pumping apparatus takes water from the supply source and pumps it into a second pumping apparatus that boosts the water pressure higher than the single fire pump was capable of doing) or other specialized applications. This hose is not always found on pumping apparatus in many highly urbanized communities with well-developed municipal water systems. A number of large communities located close to large bodies of water retain the hard suction hose on their pumping apparatus, even though it is seldom used. However, suction hose is vital to rural fire departments in fulfilling their fire flow needs. Suction hose can also draft water from static water sources and siphon water from one portable water tank to another during water tender/mobile water supply vehicle shuttle operations. The efficiency of the operation is maximized when suction hose is used in this way.

Hard suction hose is constructed of a variety of reinforcing materials that withstand the partial vacuum conditions created during drafting operations. It is available in sizes ranging from 2½ to 6 inches (65 mm to 150 mm). Suction hose must meet the following criteria:

Figure 1.7 Hard suction hose is designed to draft water from a static water source into a fire pump or conduct tandem-pumping evolutions.

- Internal diameter of the tube no smaller than the trade size of the suction hose
- Internal surface smooth and free of corrugations
- When coupled, withstands an internal partial vacuum (negative pressure) of 23 inches of mercury (0.78 bar) and sustains it for 10 minutes

  NOTE: Suction hose is not tested by positive-pressure test methodology unless it is specially designed for that use.
- Must withstand a service test pressure of 165 psi (1 138 kPa) {11.4 bar} when attached to pressurized water sources
- Functions as designed in ambient temperature conditions ranging from –30 to 140°F (-34°C to 60°C) while maintaining its longitudinal strength
- Not necessarily designed to attach to pressurized water sources, municipal water mains, or pressurized supply lines from other pumping apparatus
- If designed for use under vacuum only, indelibly marked with the words FOR VACUUM USE ONLY in letters no less than 3⁄16 inch (4.8 mm) high

Booster Fire Hose

Booster fire hose is a preconnected, rubber-covered noncollapsible hose (often carried on pumping apparatus on reels) made of several layers of braided or spiraled rubberized materials or reinforcement. It is used to extinguish small, incipient-stage exterior fires. It is often desirable for performing overhaul work on debris removed from a structure. In some cases, extruded nitrile rubber (elastomeric or thermoplastic tube) has replaced the conventional booster fire hose on reels.

Braided or spiraled booster fire hose is manufactured in ¾-, 1-, and 1½-inch (20 mm, 25 mm, and 38 mm) diameter sizes. Because booster fire hose is in the small diameter hose category, it can be used at high pressures to overcome the friction loss that occurs when producing maximum water flows. Although it has the same minimum design performance characteristics as attack hose, it is not classified as such due to its smaller size and insufficient fire flow capabilities, with the 1½-inch (38 mm) size being an exception. The most successful application of booster hose is for quick deployment for suppression of brush or grass fires. The 1½-inch (38 mm) booster hose, though capable of producing structural fire-suppression flows at reasonable pump pressures, is very heavy with less maneuverability than an equivalently sized fabric hose.

WARNING

Booster hoses of ¾- or 1-inch (20 or 25 mm) diameter sizes are not considered adequate hoselines for any fire beyond the incipient stage, nor should they ever be deployed to the interior of a structure for fire-suppression and loss control purposes.

For rapid deployment, booster fire hose is usually carried on motorized or hand-powered reels mounted inside a compartment on the front or rear of a pumping apparatus **(Figure 1.8a).** Booster-hose reels can be mounted on top of the pumping apparatus as another option that can be employed, although this placement is not as prevalent today because of the use of preconnected 1½-inch (38 mm) or larger attack hoselines carried in transverse (also called *mattydale*) hose beds that lie across the pumping apparatus body at right angles to the main hose bed. They are located in easy-to-reach positions on top of the pumping apparatus **(Figure 1.8b)** (see Chapter 7, Attack Hose Loads, Finishes, Hose Packs, and Deployment Procedures, for more information). Because of its rigid design, booster fire hose can be partially deployed from the hose reel, which allows charging and operating the hose without removing all of it from the reel.

Fire Extinguisher Hose

Fire extinguisher hose (hose attached to portable fire extinguisher units) is designed to withstand high pressure while remaining maneuverable for use. These hoses are used on extinguisher units of various sizes. When coiled on reels or as short sections attached to extinguishers, the hose delivers liquid, gaseous, dry chemical, or dry powder extinguishing agents from the extinguisher to

Figure 1.8a Hose reel mounted on top of the pumping apparatus.

Figure 1.8b The use of transverse hose beds reduces the need to rely on small diameter booster hoselines mounted on reels.

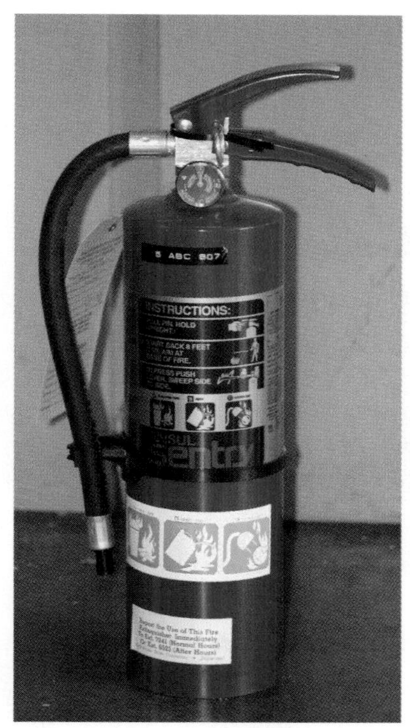

Figure 1.9a A hand-held fire extinguisher intended for use on Class A, Class B, and Class C fires. Construction of extinguisher hose is essentially the same as that of booster hose.

Figure 1.9b A wheeled fire extinguisher unit with hose and nozzle. This hose is a high-pressure type and is rubber-covered, rubber-lined, and fabric-reinforced. *Courtesy of Ansul, Inc., Marinette, Wisconsin.*

the nozzle. Extinguisher units can be stationary, wheeled, or vehicle mounted **(Figures 1.9 a and b).** Although fire extinguisher hose is not included in NFPA 1961, limited mention of its construction and design capabilities are included for basic information. For an in-depth discussion of fire extinguishers and their designs, refer to NFPA 10, *Standard for Portable Fire Extinguishers* (2002).

Two categories of extinguisher hose exist: (1) low-pressure style that is used when discharge pressures are no greater than 400 psi (2 758 kPa) {27.6 bar} and (2) high-pressure style that withstands pressures up to 1,250 psi (8 618 kPa) {86 bar}. Each type of extinguisher hose is constructed similarly to booster fire hose and can be made into twin-agent, dual-agent, or quad-agent hoses. Dual- and quad-agent extinguisher hoses are becoming more common because they offer the ability to disperse several different extinguishing agents separately or in combination.

WARNING

Numerous hoses are available that are used for nonfire applications in industrial, manufacturing, and agriculture settings. These hoses are not necessarily designed for fire suppression. Use only NFPA-compliant hose for fire-suppression activities.

Special Cold-Resistant Hose

Fire and emergency service organizations that are located in extremely cold climates should use special cold-resistant hose. This hose must exhibit all the qualities present in other hose classifications including that it must endure a temperature of -65°F (-54°C).

Fire Hose Construction

NFPA sets precision standards for the construction of fire hose in NFPA 1961, Chapter 5 **(Figure 1.10).** These standards provide for safety and effective use of fire hose at an emergency incident. The primary hose components addressed in the standard are listed and then discussed in the sections that follow:

- Diameter
- Length
- Reinforcement
- Linings and Coverings
- Markings

Diameter

According to NFPA, the internal diameter of a hose should not be less than the trade size of the hose, which means that the diameter of the hose is no less than its actual internal diameter **(Figure 1.11, p. 24).** For example, hose labeled 3 inches (77 mm) in diameter must have an internal diameter of 3 inches (77 mm). Some types of fire hose can expand beyond their actual manufactured internal diameter because of the elastic qualities of modern materials used in hose construction, which result in a larger orifice when the hoseline is pressurized with water and has lower friction loss. Not all fire hose exhibit these characteristics. The performance of a particular hoseline is contingent upon the materials and methods used in its construction.

Length

Each kind and type of fire hose is designed for a specific purpose. Fire hose (either for attack or supply) are typically cut and coupled into lengths of 50 or 100 feet (15 m or 30 m). Regardless of the selected length of the hose, each coupled piece is referred to as a *section.* These lengths allow for convenience and ease of handling. There is no "standard" length that must be followed, thus hose may be cut into shorter or longer lengths as desired by the local organization. However, the traditional length of fire hose in North America is 50 feet (15 m) per section. Even though this hose length is easily rolled and stored, this old standard length is being challenged by fire and emergency service organizations that are purchasing supply fire hose with diameters of 4 inches (100 mm) or greater. The increased use of high-strength, low-weight synthetic hose components allows fire and emergency responders to manage greater lengths with the same relative weight as the older hose. The associated reduced cost of longer fire hose with fewer hose couplings has also contributed to this trend.

Suction supply hose (as defined in NFPA 1901) requires 15 feet (4.6 m) of large soft sleeve hose or 20 feet (6 m) of hard suction hose on a pumping apparatus. Additionally, 800 feet (244 m) of hose that is 2½ inches (65 mm) or larger in diameter and 400 feet (122 m) of 1½-, 1¾-, 2-, or 2½-inch (38 mm, 45 mm, 50 mm, or 65 mm) attack hose must be provided on a pumping apparatus **(Figure 1.12, p. 25).** Local fire and emergency service organizations may increase the lengths of hose available as long as the minimum quantities and sizes are provided for pumping apparatus. Additionally, the Insurance Services Office, Inc. (ISO) also has requirements for hose length and type and number of nozzles and hose appliances that are carried on pumping apparatus for insurance rating purposes.

Fire Hose Construction Features

Hose Types/Diameters	Cutaway Views	Linings, Coverings, and Reinforcements
Booster Hose ¾ or 1 inch (20 mm or 25 mm)		• Rubber Covered • Rubber Lined • Fabric Reinforcement
Woven-Jacket Hose 1 to 6 inches (25 mm to 150 mm)		• One or More (Two Showing) Woven-Fabric Jackets • Rubber Lined
Rubber-Covered Hose 1 to 6 inches (25 mm to 150 mm)		• Rubber Covered • Woven Polyester Tube • Nitrile Rubber • Synthetic Rubber Outer Cover
Impregnated Single-Jacket Hose 1½ to 5 inches (38 mm to 125 mm)		• Woven Polyester Nylon or Combination of Synthetic Fibers Form Tube • Polymer Covered • Polymer Lined
Noncollapsible Intake Hose 2½ to 6 inches (65 mm to 152 mm)		• Rubber Covered • Fabric and Wire (Helix) Reinforcement • Rubber Lined
Flexible Noncollapsible Intake Hose 2½ to 6 inches (65 mm to 150 mm)		• Rubber Covered • Fabric and Plastic (Helix) Reinforcement • Rubber Lined
Flexible Noncollapsible Clear Intake Hose 2½ to 6 inches (65 mm to 150 mm)		• Polyvinyl Tube • Polyvinyl Reinforcement (Helix)

Figure 1.10 Fire hose construction features for various hose types and diameters.

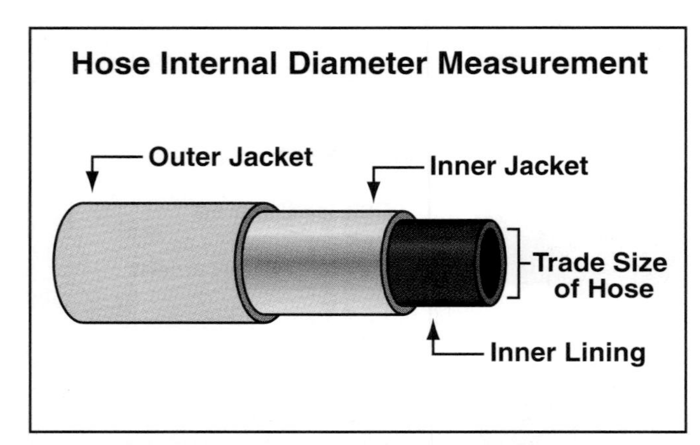

Figure 1.11 Hose internal diameter measurement.

The minimum design requirement for hose lengths is that it be between –2 and +4 percent of its stated length at the time of its acceptance. The manufacturer measures all fire hose from the inside edge of each coupling, along its centerline while it is under a pressure of 10 psi (69 kPa) {0.7 bar}, which is then steadily raised to its proof test pressure where it is held for a minimum of 15 seconds. Then the hose is measured again and checked for movement of the couplings.

Reinforcement

According to NFPA 1961, the *reinforcement* (previously called *jacketing* or *jacket*) of fire hose is the structural support for fire hose in the form of woven yarn and is made from one of the following materials:

- Natural fiber thread
- Synthetic fiber thread
- Combination of natural and synthetic fiber threads

The reinforcement component needs to be evenly and firmly woven and free from defects, dirt, knots, lumps, and other twist irregularities. All knots are tucked under the warp thread. Each fabric layer needs to be seamless and have the filling woven around the hose throughout its length with the warps interwoven with and substantially covering the filling **(Figure 1.13, p. 26).**

Linings and Coverings

According to NFPA 1961, *all* fire hose shall have linings and coverings (also sometimes called *jackets*) that are made from one of the following approved materials:

- Rubber compound (nitrile)
- Thermoplastic material
- Blends of rubber compounds (nitrile) and thermoplastic material
- Natural rubber-latex-coated fabric

The lining needs to be a uniform thickness of standard commercial quality and the waterway surface of the lining free from pitting, irregularities, or other imperfections. When tested, the adhesion between the lining and the reinforcement is such that the rate of separation of a 1½-inch (38 mm) strip of lining from the reinforcement is not greater than 1 inch per minute (25 mm/min) with a weight of 12 pounds (5.5 kg) applied to the reinforcement or lining. If a rubber backing is used between the lining and the reinforcement, the adhesion between the lining and the backing and between the backing and the reinforcement also is such that the rate of separation of a 1½-inch (38 mm) strip is not be greater than 1 inch per minute (25 mm/min) with a weight of 12 pounds (5.5 kg) applied to the reinforcement or lining.

The outer cover (jacket) located on a double-jacketed hose is most often referred to as the *wear jacket.* It provides a barrier from damage caused by use during fire operations. Although it adds some marginal strength to the hose, its value is the protection provided to the inner lining and the reinforcement.

The lining is manufactured by calendaring or extrusion methods. *Calendaring* is a process where the rubber is pressed between opposing rollers to produce a flat sheet of material. A tube is then formed by lapping and bonding the edges of the rubber sheet, which becomes the inner diameter of the hose. The tube is glued, vulcanized, and sealed. This method is rarely used today because of the inherent weakness along the seam. The calendaring method has been replaced by the *extrusion* process, which forms a seamless tube by forcing a heated mass of rubber or thermoplastic through the orifice of an extrusion machine die. This die is the diameter of the final hose product.

Markings

NFPA requires each length of fire hose to be indelibly marked in letters and numbers at least 1 inch (25 mm) high with the manufacturer's identification,

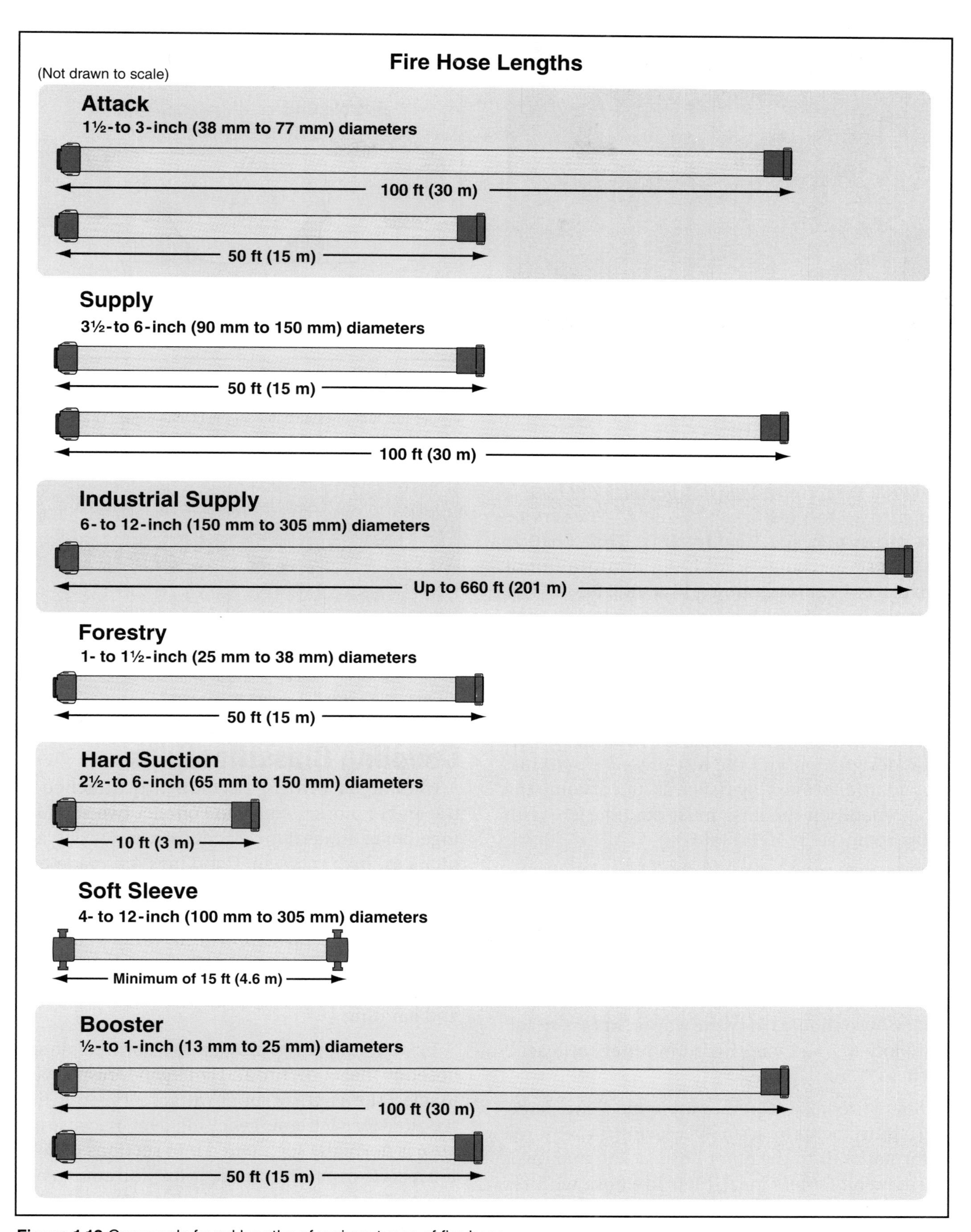

Figure 1.12 Commonly found lengths of various types of fire hose.

Figure 1.13 A circular loam produces a seamless fabric tube for hose reinforcement or abrasion jacket. *Courtesy of Sam Goldwater.*

the month and year of manufacture, and the words "SERVICE TEST TO [*service test pressure the hose is designed to*] BAR [PSI] PER NFPA 1962." The service test pressure is specified in NFPA 1962, Section 7.1. These markings shall be in a minimum of two places beginning approximately 5 feet (1.5 m) from the ends of each hose length or at no less than 12-foot (3.7 m) intervals if continuously printed along the hose length **(Figure 1.14).** No marking referring to pressure, other than the service test pressure, appears on the hose. Identifying marks for the owning organization are found after the hose designation and the test pressure marking. Any additional markings (such as unit or company assignment) on the hose must not interfere with those required by NFPA 1961.

Supply hose that meets the design and test requirements given in the Supply Hose Section needs to be indelibly marked lengthwise along the centerline of the hose in letters at least 2 inches (51 mm) high with the words "SUPPLY HOSE" **(Figure 1.15).** Supply hose intended to also be used as attack hose should carry the words "ATTACK HOSE" indelibly marked lengthwise in letters at least 2 inches (51 mm) high.

Some fire hose manufacturers meet the marking requirements by actually weaving the required information into the outer hose jacket, making it permanent. Others mark their fire hose with reflective material to enhance the visual properties of the hose.

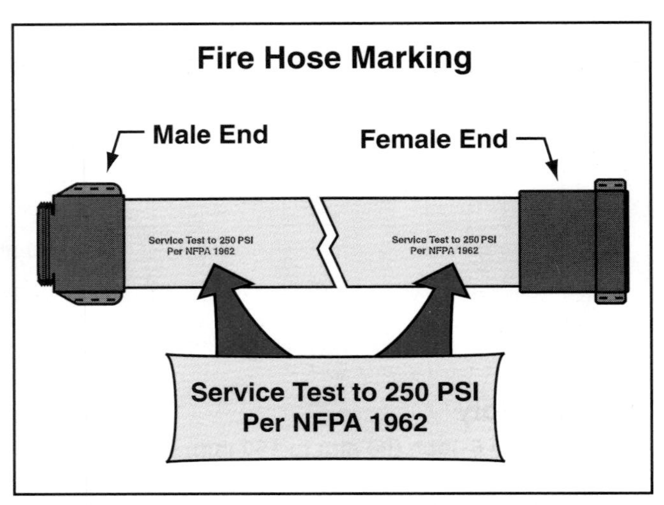

Figure 1.14 The service test pressure of fire hose is marked in 1-inch letters by the manufacturer. In this example the test pressure is 250 psi (1 724 kPa) {17.24 bar}.

Figure 1.15 The manufacturer permanently marks the hose lengthwise along the centerline in letters at least 2 inches (51 mm) high with the words SUPPLY HOSE.

Coupling Classifications

A *coupling* is a fitting permanently attached to the end of a hose length to connect two sections together or attach it to devices such as nozzles, appliances, hydrants, etc. Couplings are made of a variety of materials, with brass or aluminum being the most common. Couplings are divided into two types: *threaded* and *nonthreaded.* Various coupling attachment methods are available such as expansion ring, screw-in expander, collar, tension ring, and banding.

Fire hose couplings provide the vital links along a hoseline that allow fire and emergency responders to select the length of supply, attack, or suction line required for each emergency incident. Because fire hose is normally segmented into sections of 50 or 100 feet (15 m or 30 m), a single fire and emergency responder can usually deploy the amount of fire hose that is needed for a specific task. If a hoseline develops a leak, it can be easily removed and replaced with another section of hose. If necessary, hoselines can be extended to reach further into areas as fire-suppression or hazardous materials liquid spill-control activities progress. Ultimately, fire hose couplings allow for the versatility required for the effective deployment of hoselines.

All fire hose and couplings must be service tested to their service test pressures by the purchasing organization before being placed into service. The purchasing organization has 90 days to conduct the service test after delivery of the new hose/couplings.

Threaded Couplings

Couplings are manufactured by one of the following methods:

- ***Casting*** — Couplings are cast in molds of molten metal. These couplings are very weak and only found on occupant-use fire hose. They often crack if reattachment to the hose is attempted.
- ***Extruding*** — Couplings are produced by extruding metal through a die forming a long tube that is cut to the length of the coupling. They are most often made of aluminum or aluminum alloy, allowing for their lightweight and high strength. They are somewhat stronger than cast couplings.
- ***Drop forging*** — Couplings are made of brass or other malleable metal and shaped in forging die by pounding the metal into the desired shapes. Because of the compression of metal molecules, it is the strongest and most expensive of the three coupling types.

One of the oldest coupling designs involves the casting or machining of a spiral thread into the face of two distinctly different couplings — male and female. A *male coupling thread* is cut on the exterior surface, while a *female coupling thread* is on the interior surface of a free-turning ring called a *swivel.* The swivel permits connecting two sections of hose without twisting the entire hose. Each section of fire hose with threaded couplings has a male coupling at one end and a female coupling at the other. Together, the two couplings are referred to as a *set* (also referred to as a *three-piece coupling*): The male coupling is considered one piece, and the female coupling is a two-piece assembly allowing it to swivel during coupling with the male end.

A threaded coupling has several other parts. The portion of the coupling that serves as a point of attachment to the hose is called the *shank* (also called the *tail piece, bowl,* or *shell*) **(Figure 1.16).** A flattened angle at the end of the threads on the male and female couplings called the *Higbee cut*

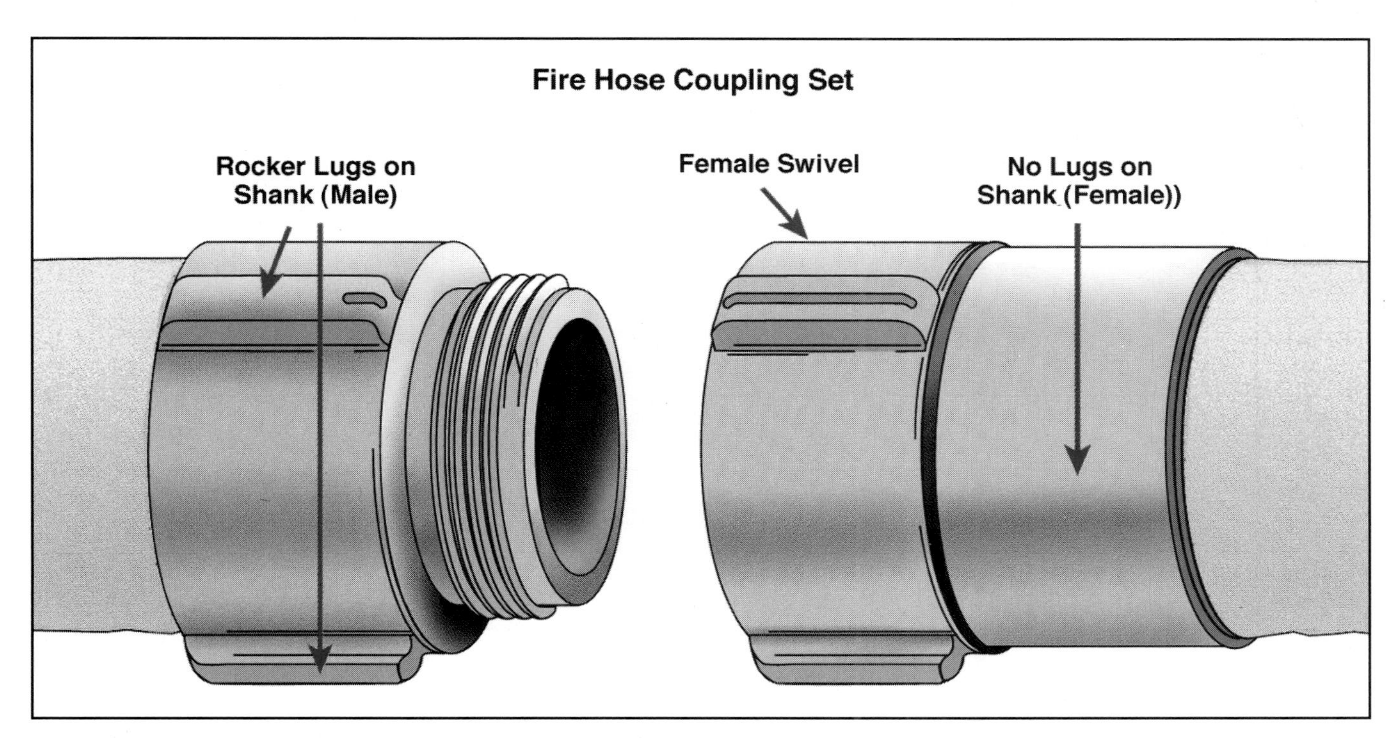

Figure 1.16 The fire hose coupling set is comprised of the male shank, female swivel, and female shank. The male shank has rocker lugs; the female shank does not.

(also called *blunt start*) prevents cross-threading when couplings are connected. The *Higbee indicator* (indentation) marks where the Higbee cut begins. See Chapter 5, Basic Methods of Handling Fire Hose for more information.

Unlike common pipe threads that are relatively fine, fire hose coupling threads are coarse (with wide tolerances), which aid in connecting the couplings quickly. Some manufacturers make the large coupling sizes (3½ inches [90 mm] and above) with either ball bearings or roller bearings under the swivels to ensure their smooth operation. **Table 1.2** provides information about the sizes of threaded couplings available for common fire hose sizes.

Depending on coupling design, if the shank fits inside the fire hose, it reduces the diameter of the waterway. If it fits over the outside of the fire hose, it does not reduce the size of the waterway. Many fire and emergency service organizations that use 3-inch (77 mm) fire hose have 2½-inch (65 mm) couplings installed, which allow the hose to easily connect to fire hydrants and the discharges of the pumping apparatus without the use of adapters. This application reduces friction loss significantly (approximately 60 percent of that experienced with 2½-inch [65 mm] hose).

Drop-forged brass couplings frequently have an embossed ridge on the shank of the male coupling and female swivel. The ridge serves to protect the swivel if a coupling is dropped or when hose is deployed from the pumping apparatus **(Figure 1.17).**

Recent improvements in manufacturing threaded couplings retain standard thread capabilities but allow the couplings to "slide" together for a quick connection using a one-eighth-turn rotation of the female coupling (or nonthreaded/sexless couplings). The threaded speed coupling is machined in a manner that provides gaps in the thread that allow this quick

Table 1.2
Coupling Sizes Available for Hose Sizes

Coupling Size	Hose Size											
	¾ inch (20 mm)	1 inch (25 mm)	1½ inches (38 mm)	1¾ inches (45 mm)	2 inches (50 mm)	2½ inches (65 mm)	3 inches (77 mm)	3½ inches (90 mm)	4 inches (100 mm)	4½ inches (115 mm)	5 inches (125 mm)	6 inches (150 mm)
1 inch (25 mm)	•	•										
1½ inches (38 mm)			•	•	•							
2½ inches (65 mm)						•	•	•				
3 inches (77 mm)							•	•				
3½ inches (90 mm)								•				
4 inches (100 mm)									•	•	•	•
4½ inches (115 mm)									•	•	•	•
5 inches (125 mm)									•	•	•	•
6 inches (150 mm)									•	•	•	•

Figure 1.17 Drop-forged brass couplings frequently have an embossed ridge on the shanks to protect them from damage if they strike the ground during deployment operations.

Figure 1.18 Threaded speed coupling. *Top:* Coupling is machined with gaps in the threads. *Middle:* Higbee indicators allow for accurate aligning of thread ends. *Bottom:* One-eighth rotation completes the quick connection. *Coupling provided by Kochek Co. Inc.*

connection when it is aligned using the Higbee indicator. Regular threaded couplings can be screwed together with these new couplings in the normal manner **(Figure 1.18).**

Threaded couplings are manufactured with either lugs or handles to aid in tightening and loosening connections. Lugs are located on the shank of a male coupling and on the swivel of a female coupling. Lugs aid in grasping the coupling when making and breaking coupling connections. Connections may be made by hand or with *spanner wrenches* (special wrenches that fit against the lugs; often referred to as *spanners*). The following three types of lugs are typically found today **(Figure 1.19, p. 30):**

- ***Pin lugs*** — Resemble small pegs; usually found on couplings of old fire hose. Although still available, pin-lug couplings are not commonly ordered with new fire hose because of their tendency to catch when hose is dragged over objects or deployed from the hose bed of a pumping apparatus.
- ***Recessed lugs*** — Are simply shallow holes drilled into the coupling; booster fire hose normally has couplings with recessed lugs. This lug design prevents abrasion that would occur if the hose had protruding lugs and was wound onto reels. Recessed lugs require using a pin-lug spanner wrench (which inserts into the recessed lug holes) to tighten or loosen the coupling.
- ***Rocker lugs*** — Have rounded shapes (unlike pin lugs) that help prevent snagging the hose; are found on modern threaded couplings. On the

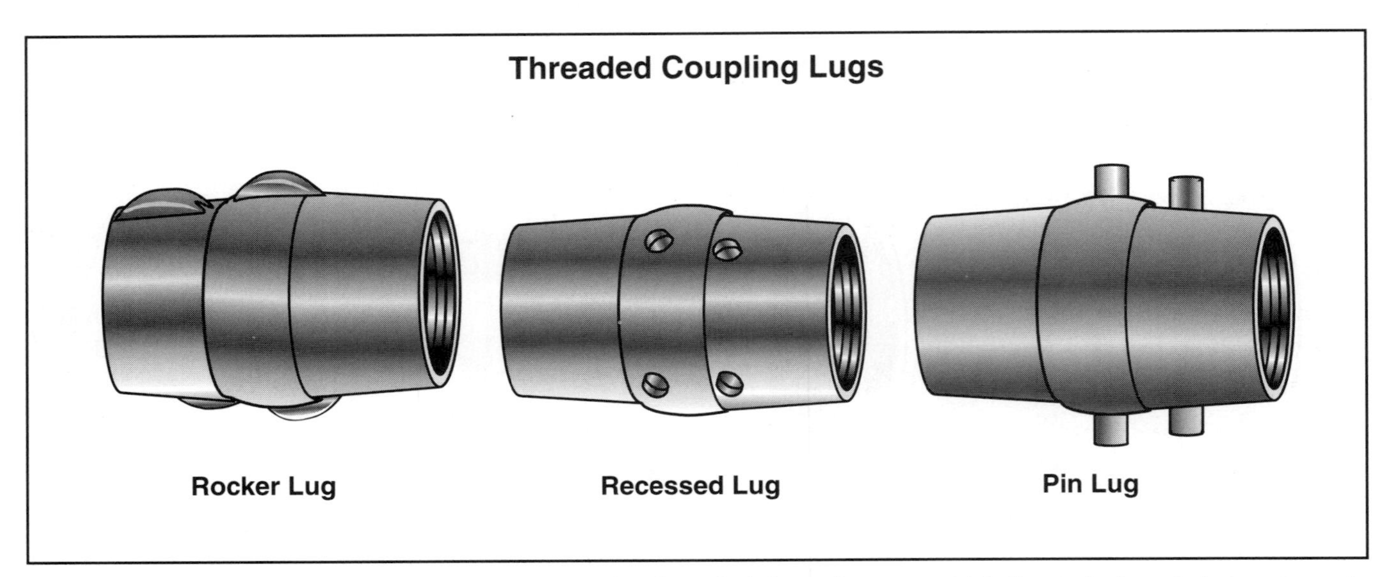

Figure 1.19 Three types of threaded coupling lugs: rocker (rounded shapes), recessed (shallow holes), and pin (small pegs).

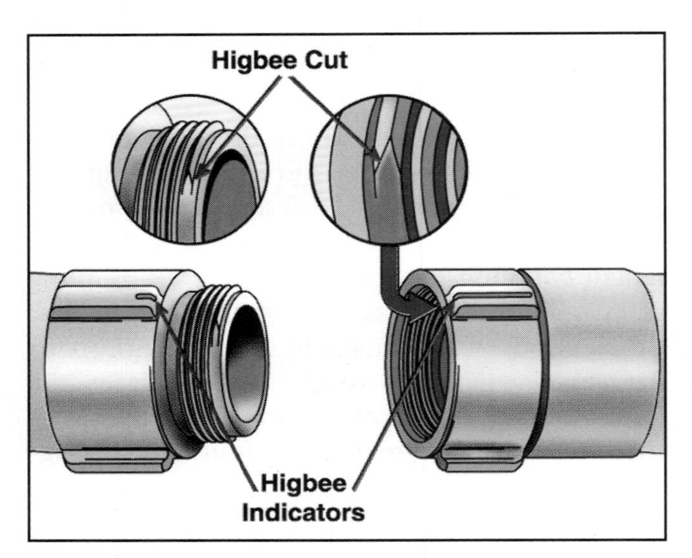

Figure 1.20 The Higbee indicator and Higbee cut assist fire and emergency service responders when making screw thread connections.

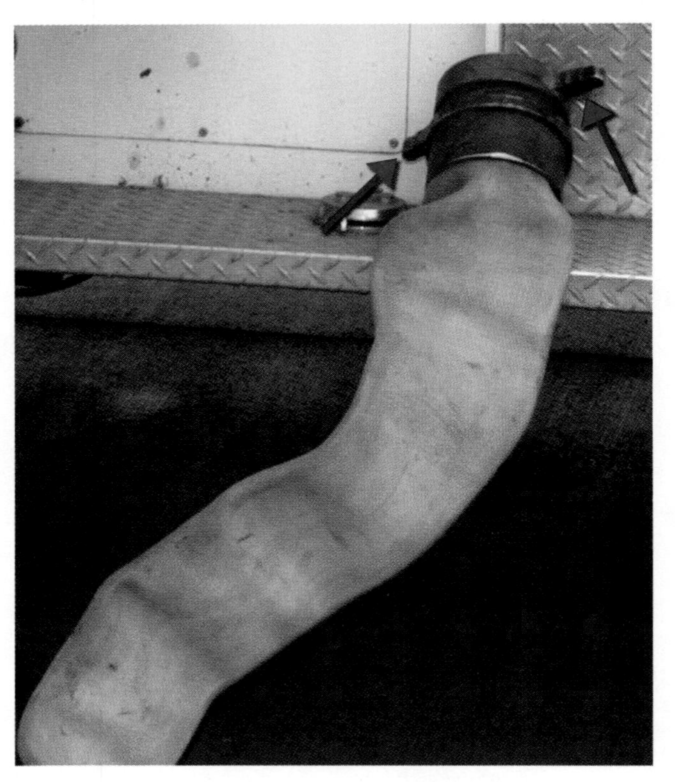
Figure 1.21 Extended lugs on an intake hose coupling.

couplings, one of the rocker lugs on the swivel is scalloped with a shallow indentation (the Higbee indicator) to mark where the Higbee cut begins. This indicator aids in matching the male coupling thread to the female coupling thread during low-light situations or when the threaded end of the coupling is not readily visible **(Figure 1.20).**

Handles or extended lugs are located only on the swivels and used primarily on large intake supply hose or suction hose **(Figure 1.21).** They aid in tightening the large coupling by hand when connecting the hose to a pump intake valve. Striking the handles with a rubber mallet can further tighten the coupling.

CAUTION

Connect couplings hand tight to avoid damage to the coupling and gasket.

Nonthreaded and Sexless Couplings

Nonthreaded fire hose couplings have been used in the North American fire and emergency services since the early 1900s. With this type of coupling, the mating of two couplings is achieved with locks or cams without the use of screw threads (see NFPA 1963, Chapter 4).

One of these nonthreaded coupling designs, although not in widespread use, is a *snap coupling set* that has both a male and female component. The female coupling has a shallow bowl that fits over the end of the male coupling. When a connection is made, two spring-loaded hooks on the female coupling engage a raised ring around the shank of the male coupling. The primary advantage with snap couplings is that hose can be connected more quickly than with threaded couplings. A disadvantage, however, is that the coupling hooks tend to catch on curbs and other obstacles when the hose is pulled along the ground **(Figure 1.22).**

Another type of nonthreaded coupling that has received widespread acceptance in North America is the *sexless coupling set.* It has no distinct male and female components so both couplings are identical. Two kinds of sexless couplings exist: quarter-turn and Storz.

The *quarter-turn coupling* has two hooklike lugs on each coupling. The lugs, which are grooved on the underside, extend beyond a raised lip or ring on the open end of the coupling. When the couplings are mated, the lug of one coupling slips over the ring of the opposite coupling and then rotates 90 degrees clockwise to lock. A gasket on the face of each coupling seals the connection to prevent leakage **(Figure 1.23).**

Figure 1.22 Snap couplings interlock when the spring-loaded hooks located on the female coupling engage the ring flange on the shank of the male coupling.

The *Storz couplings,* commonly found on large diameter hose, are similar to quarter-turn couplings in that they are connected by joining and rotating until locked into place, but they have interlocking grooved lugs instead of hooklike lugs **(Figure 1.24).** Locking components consist of grooved lugs and inset rings built into the face of each coupling swivel. When mated, the lugs of each coupling fit into recesses in the opposing coupling ring, and then slide into locking position behind the ring with a one-third-turn rotation. External lugs at the rear of the swivel provide leverage for connecting and disconnecting couplings. They also indicate complete connection by coming into alignment when the couplings lock into place **(Figure 1.25, p. 32).**

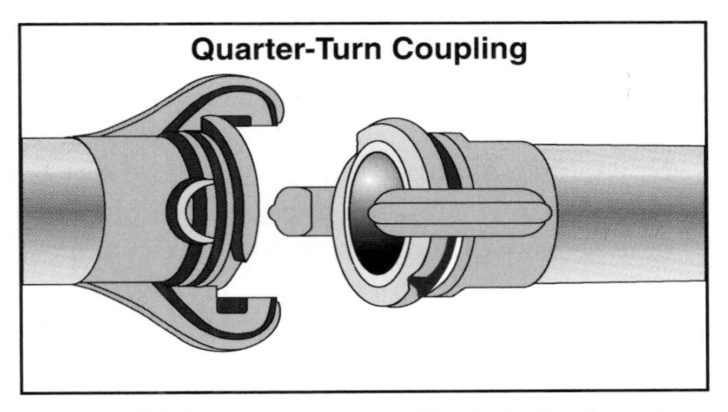

Figure 1.23 The quarter-turn coupling (not often found in municipal fire service use) can be found in industrial applications.

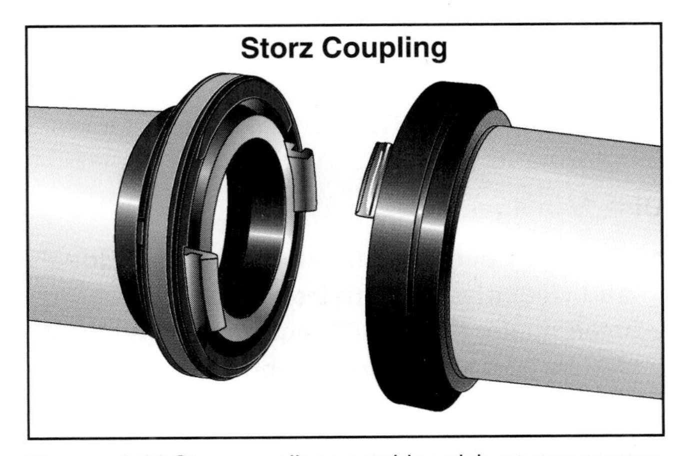

Figures 1.24 Storz couplings provide quick, secure connections for supply hose.

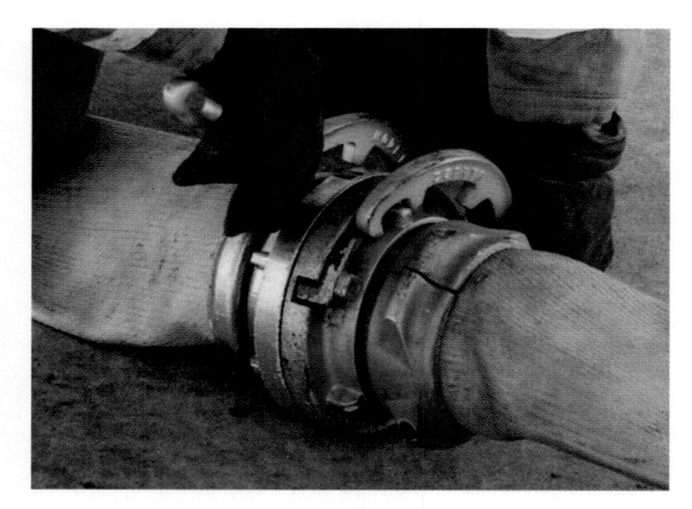

Figure 1.25 The Storz couplings are securely connected by using Storz spanner wrenches.

Several advantages as well as disadvantages to using hose with nonthreaded couplings are as follows:

Advantages:

- Fire hose can be quickly connected. However, spanner wrenches should be used to ensure a complete connection, which slows the connecting operation somewhat. According to NFPA 1963, spanner wrenches should not be needed; however, they may be with old hose or hose from different manufacturers. On most manufacturers' couplings, the lugs align to give a visual indicator of a connected coupling.
- The possibility of cross-threading is eliminated.
- Double male or double female adapters (adapters connecting two threaded couplings of the same thread type, size, and sex) are not needed, thus hose can be deployed from the hose bed regardless of hose load type. See Chapter 5, Basic Methods of Handling Fire Hose, for more information.

Disadvantages:

- Hose can become uncoupled, often suddenly and violently, if a complete connection has not been made. NFPA 1963 requires that a locking system be installed on all nonthreaded coupling systems; however, many old couplings may not be so equipped.
- An adapter is required at the hydrant, which lengthens the hydrant connection and hose deployment time if it is not preconnected.

  NOTE: Although not widely seen, permanent adapters for fire hydrant connections are used by some fire and emergency service organizations that have adopted sexless couplings for their supply hose operations **(Figure 1.26a).** Storz adapters can also be attached to sprinkler and standpipe connections **(Figure 1.26b).**
- Due to deep grooves where locks and cams travel during the connecting process, dirt and other large debris can become lodged inside the coupling, giving the impression that a tight seal has been made when in fact, the hose is not connected **(Figure 1.27).**

Figure 1.26a Storz adapters can be fitted on fire hydrants, facilitating the connection of these couplings to a water source. *Courtesy of Sam Goldwater.*

Figure 1.26b Storz adapters can be attached to sprinkler and standpipe connections. *Courtesy of Sam Goldwater.*

Figure 1.27 Inspect the couplings' grooves for dirt or debris that could cause coupling separation after connection.

Coupling Attachment Methods

A primary requirement of coupling design is that the coupling be firmly attached in a manner that resists detachment when the fire hose is pressurized and pulled or dragged. Five different components for attaching couplings to fire hose are listed and the methods for each are described in the sections that follow:

- Expansion rings
- Screw-in expanders
- Collars
- Tension rings
- Bands

Expansion-Ring Method

One of the oldest ways to attach a threaded coupling to a fire hose involves the use of a malleable metal band called an *expansion ring.* The expansion ring, which is slightly smaller than the hose, is placed inside the hose flush with the hose end. The hose is pushed into the coupling hose bowl, and then the ring is expanded against the hose with a manual or hydraulic expanding device **(Figures 1.28 a and b).** This procedure compresses the hose tightly against the inner surface of the coupling. The expanded ring is the same diameter as the hose liner so that it does not obstruct the waterway. This method is not only used at the factory where hose is manufactured but also in the maintenance facilities of fire and emergency service organizations to repair and recouple hose.

Screw-In Expander Method

Some types of hose, particularly rubber-jacket booster hose, have threaded couplings attached with expanders that are screwed into place. Unlike the metal expansion ring described in the previous section, the *screw-in expander* is actually an integral component of the coupling, which is made of two pieces: shell and expander. The inside surface of the shell is serrated to prevent slippage of the attached hose. One end of the expander, rather than the shell, contains the coupling threads. The opposite end of the expander is also threaded so that it may be screwed into the coupling shell.

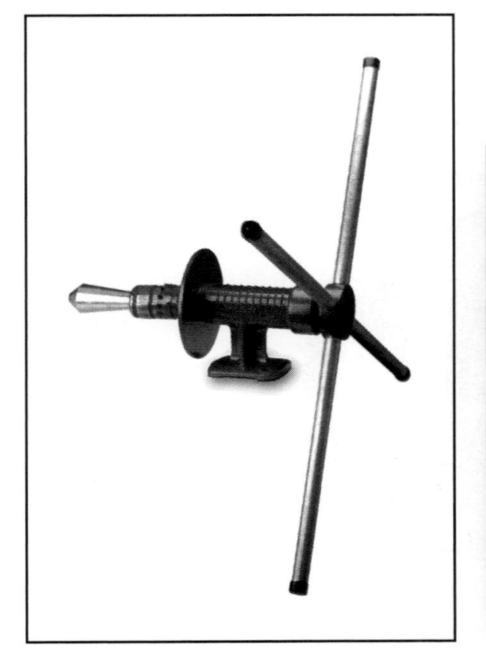

Figures 1.28 a and b Expanders are used to attach hose couplings by expanding a brass ring inside the shank to secure the hose to the coupling. *Left* (a) Manual expander. *Courtesy of Akron Brass Company. Right* (b) Hydraulic expander.

Figure 1.29 A screw-in expander coupling is composed of a shell and an expander.

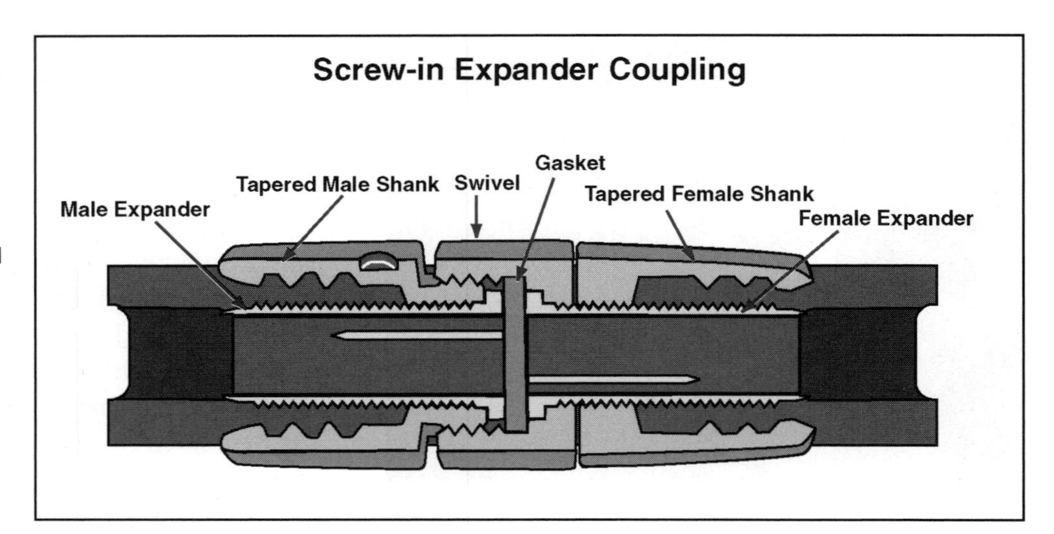

The coupling is attached to the hose by placing the shell over the hose end and then screwing the expander into the hose until it seats against the face of the shell. This procedure compresses the hose tightly against the serrations inside the shell **(Figure 1.29).**

Collar Method

Another coupling designed with a shank that fits inside the hose is fastened to the hose with a *collar* that bolts into place. This method is one of the simplest ways to attach a coupling because no equipment except a wrench is required. The collar coupling is most often used with rubber- or synthetic-covered large diameter supply hose using Storz couplings. With this method, the hose is slipped over the coupling shank (which is serrated), and then a two- or three-piece collar is fastened in such a way that it compresses the hose against the shank **(Figure 1.30).**

Tension-Ring Method

A method similar to the collar method for attaching couplings uses a flange ring, tension ring, and clamp ring (also known as *contractual sleeve*). In this case, the coupling shank is made with two grooves around its outer circumference. As with the collared coupling, this shank fits inside the hose end. A nylon sleeve with inside ridges that correspond to the grooves on the shank is placed on the hose directly over the shank grooves. The *tension ring* is then placed over the nylon sleeve and tightened with Allen-head bolts. As the bolts are tightened, the ridges on the inside of the nylon sleeve compress the hose material against the grooves on the coupling shank, making a tight-fitting attachment **(Figure 1.31).**

Banding Method

The *banding method* is used with couplings found most commonly in industrial applications where special conditions challenge the use of other connecting methods or for fire hoses with very large diameters. The method is rarely used in the fire and emergency services. The method attaches the coupling to fire hose with steel bands or tightly wound bands of narrow-gauge wire. In this case,

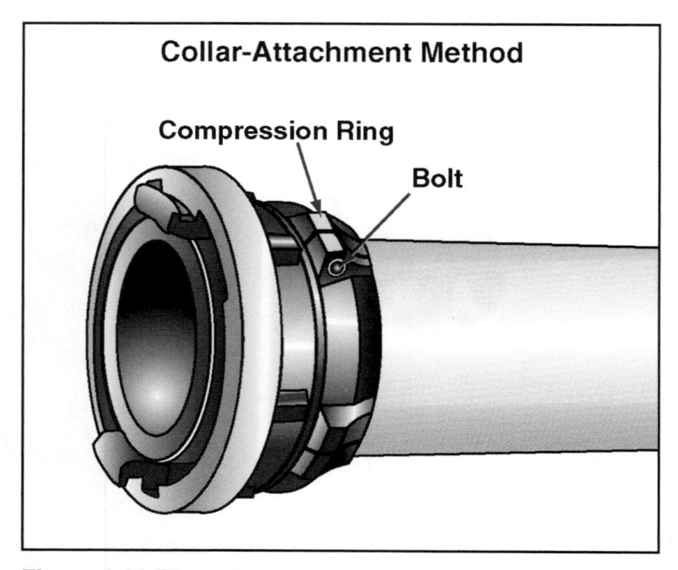

Figure 1.30 The collar-attachment method uses a grooved shank with a nylon sleeve that has ridges corresponding to the grooves. A metal ring is tightened over this assembly, compressing the hose to the shank.

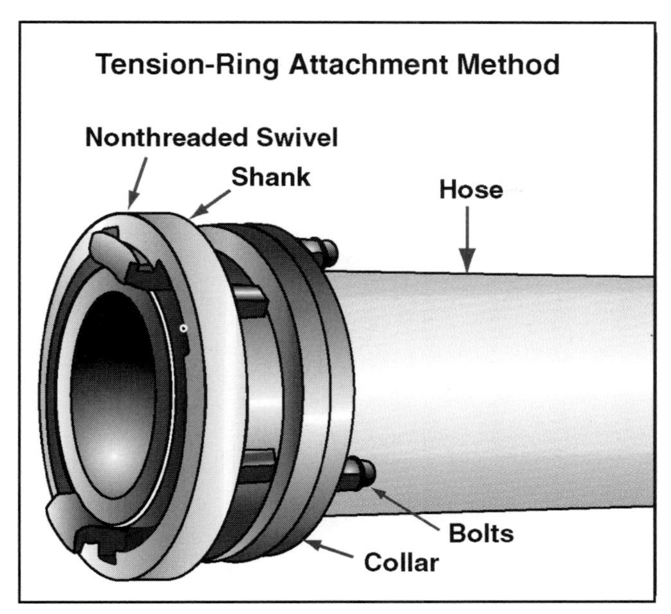

Figure 1.31 The shank of the tension-ring coupling fits inside the hose. A collar is fastened with bolts, holding the hose to the shank.

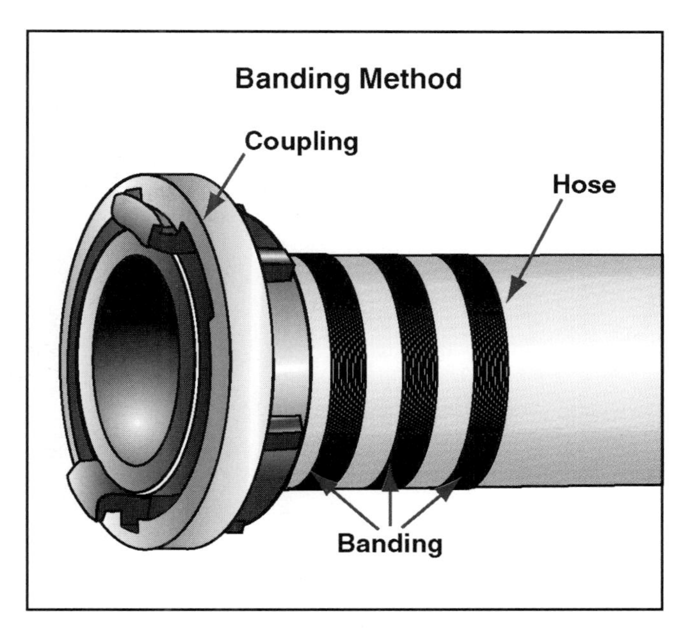

Figure 1.32 The banding method attaches the coupling by tightly binding narrow-gauge wire or steel bands over the coupling shank.

the coupling shank fits inside the end of the hose rather than over the outside like most other types of couplings. The coupling shank is made with ridges and grooves so that the wire or steel band sets into the grooves. The ridges prevent the coupling from detaching under pressure **(Figure 1.32).**

Summary

Regardless of its outwardly simple appearance, fire hose is actually part of a very complicated water-delivery system comprised of numerous design and material alternatives. Each of these design and material alternatives allows a fire and emergency services organization to select the fire hose and coupling system that is best suited to its needs.

Standardization of coupling threads provides a community with a method to ensure conformity to a local thread size. Standardization helps guarantee that fixed fire-suppression systems can be supported by the fire and emergency services organization without special devices or adapters. Additionally, should assistance from neighboring communities be necessary, their equipment will readily fit that of the local fire and emergency services organization.

The century-long program of implementing complete standardization of hose threads has been partially successful. Most old metropolitan areas have established thread standards that may or may not comply with the national standard. Many young communities have established the national hose thread standard and have successfully formed mutual and automatic aid agreements with neighboring fire and emergency service organizations that have also implemented this standard.

Attack, supply, occupant-use, forestry, suction, and booster fire hose, although often similar in appearance, have vastly different qualities — each serving a distinct fire and emergency services mission. Hard suction hose, actually the most dissimilar of all of the hoses studied, functions in a completely opposite manner than the others discussed — not under pressure but in a vacuum. Fire extinguisher hose also delivers a variety of extinguishing agents to a fire from a contained source and often under very high pressures. The commonality of fire hose lies in the fundamental concept that water or other extinguishing agent must be delivered from a source to the location where it can be effectively applied to protect lives and prevent loss of property.

Chapter 2
Fire Hose Care, Maintenance, and Service Testing

Chapter 2
Fire Hose Care, Maintenance, and Service Testing

Fire hose, like other fire-fighting equipment, must be used properly during fire-fighting operations as well as during training and other activities. When fire hose and couplings are not in use, they must be well maintained to ensure that they can be relied upon to function without failure. It is essential that a comprehensive maintenance schedule for the cleaning, inspection, storage, and testing of fire hose be a part of a fire and emergency services organization's standard operating procedures (SOPs). An invaluable aid to accomplishing this maintenance function is outlined in a standard from the National Fire Protection Association (NFPA): NFPA 1962, *Standard for the Inspection, Care, and Use of Fire Hose, Couplings, and Nozzles and the Service Testing of Fire Hose* (2003). Much of the information presented in this chapter is based on this standard.

This chapter describes the types of damage that can occur to fire hose and fire hose couplings and those procedures that will guide fire and emergency service responders who maintain fire hose to prevent this damage. Procedures for inspection, cleaning, and storage of fire hose and couplings are also outlined with several different procedures depicted in the skill-sheet section at the end of the chapter. Fire hose records and the information necessary for these records are described, and several examples are provided. Fire hose service testing procedures are also discussed. The repair of couplings and their reattachment to fire hose is presented. Each type of coupling used by the fire and emergency services is presented with descriptive skill sheets at the end of the chapter clearly defining the methods of coupling reattachment.

Fire Hose Damage Causes and Prevention

During initial fire attack, little effort is usually given to protecting fire hose from injury. It is essential that fire and emergency service responders (through training and practice) fully understand the limitations of their fire hose systems and learn how to minimize those factors that could contribute to a catastrophic failure during fire-suppression and other emergency activities. After a fire is contained, however, there is little reason not to take preventive measures to protect this essential equipment from damage. During loss control operations and when cleaning up after fires, particular care should be taken when moving hose through fire debris. Fire hose can be quickly damaged by contact with embers, broken glass, or hazardous products of combustion or other hazardous materials.

As described in Chapter 1, Fire Hose and Couplings, most fire hose is constructed of a woven material and a rubberized (thermoplastic) coating or vulcanized rubber liner. These components can be damaged in a number of ways: abrasion from or impact with objects (mechanical damage), exposure to high heat or freezing cold (thermal/cold damage), attack by mold and mildew (organic damage), chemical contact/exposure, and age deterioration. Hose-damage factors and ways to prevent or reduce their effects are discussed in the sections that follow.

Mechanical Damage

Mechanical damage occurs when an object contacts the hose somewhere along its length and cuts, abrades, tears, or stresses the reinforcement

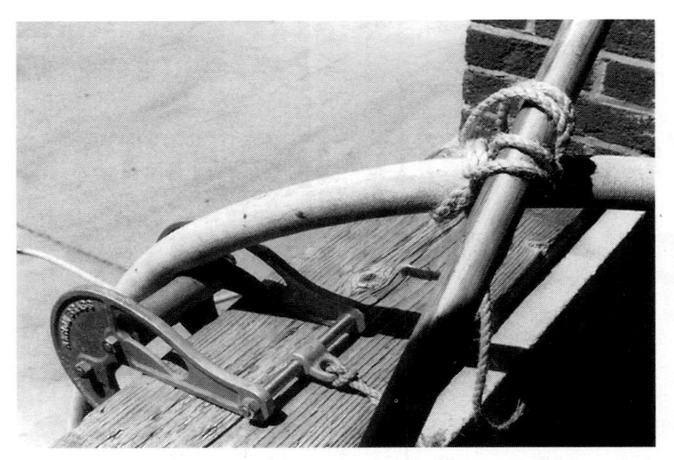

Figure 2.1 A hose roller prevents hose from being damaged when dragging it over rough or sharp edges.

jacket and underlying material. Fire hose can be damaged when it is pulled over sharp edges such as corners, cornices, parapets, and windowsills. Avoid this damage by using a hose roller or improvised padding such as a bundled salvage cover **(Figure 2.1)**. While these procedures may not always be practical during the initial stages of fire-fighting operations, they should become routine during activities such as loss control operations or training evolutions.

Fire hose may also be damaged when it is advanced through window openings containing broken-glass fragments. It only takes a few seconds to completely break glass fragments from window openings. Use a spanner wrench or axe to sweep the window sash clear of glass shards **(Figure 2.2)**. This action not only prevents glass from cutting the hose jacket but it ensures the safety of emergency responders climbing through the opening.

CAUTION

Glass shards and other sharp debris may become embedded in the hose during fire-suppression operations. To avoid injury, ensure that all emergency responders wear hand protection during hose and cleanup operations.

Mechanical damage can occur when fire hose is dragged through debris. Debris usually contains sharp-edged objects such as glass, nails, and metal,

Figure 2.2 Clear glass fragments with an axe or spanner wrench before advancing a hoseline through a window.

each of which can cause cuts or abrasions to the hose jacket. Hot spots are also found in debris piles. Although not recognized as an immediate threat to the fire hose, close or direct contact with hot spots can cause severe damage to the reinforcement jackets and the liner. Route hoselines around debris, move debris from areas where hose must be moved, or cover the debris with a protective material to prevent damage to hoselines.

Tools or equipment carried on top of pumping apparatus hose beds may damage hose. Store tools and equipment, even temporarily, in places other than on top of fire hose. While the tools may cause little damage from simply lying on a hose bed, they can do great damage when fire hose is pulled from the hose bed. Equipment mounting devices are available to attach to the rear steps of pumping apparatus, which allow fire and emergency service responders to have ready accessibility to equipment during deployment operations. This practice not only helps to protect the fire hose from damage, but emergency responder safety is also enhanced by eliminating the possibility of a tool "flying" from the rear of the pumping apparatus during hose deployment **(Figure 2.3)**.

Fire hose may be abraded when emergency responders drag it over rough surfaces or against bolts that protrude through and into the hose bed. This type of damage actually occurs most often during nonemergency activities such as training

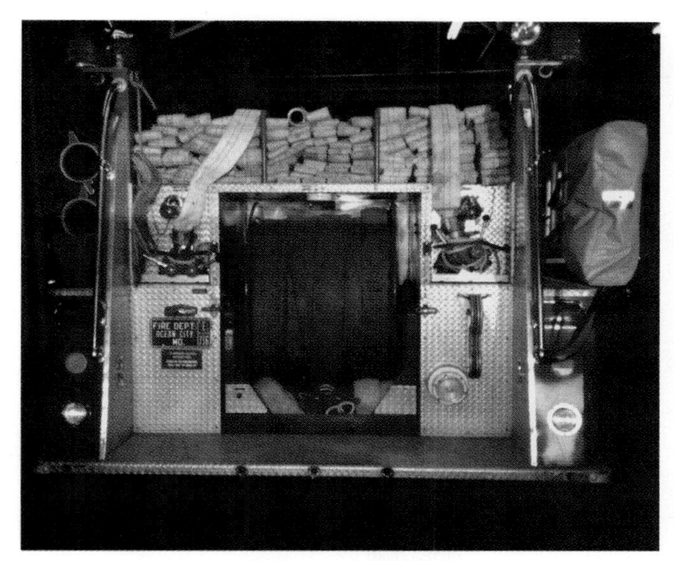

Figure 2.3 Equipment is mounted in a safe manner on the exterior of this vehicle so that it is convenient for use but does not interfere with hose deployment. *Courtesy of Sam Goldwater.*

and hose-maintenance procedures. When hose is dragged over rough surfaces such as asphalt or concrete pavement, the outer jacket receives a mild abrasion. If this situation happens repeatedly, the outer cover or jacket may significantly weaken. To avoid this damage, emergency responders should carry the hose in rolls or loose folds rather than dragging it **(Figure 2.4).** If hose must be dragged, avoid surfaces such as graveled roadways and rocky grounds that are exceptionally abrasive. Also drag hose with the flat side down, rather than on its edge. Hose edges are particularly susceptible to abrasive damage when hose is dragged. Fire hose can also suffer damage when it is dragged on the roadway behind moving apparatus. Both hose jackets and couplings suffer from this kind of treatment.

Damage can occur when hose catches in the bed during hose-deployment operations or when a section of hose vibrates loose from the hose load when a pumping apparatus is in motion. Prevent snags from occurring by not loading the hose so tightly that couplings lodge against adjoining couplings or against hose bed rails **(Figure 2.5).** Prevent hose from accidentally dislodging during travel by firmly securing the end coupling when loading the hose bed.

Damage occurs when vehicles drive over fire hose even though damage may not be immediately recognizable. In this case, the inner liner and inner

Figure 2.4 Do not drag hose, but carry it in rolls or loose folds over the shoulder.

Figure 2.5 Hose must be loaded in the hose bed in a manner that limits the possibility of couplings lodging against the hose bed rails. *Courtesy of Sam Goldwater.*

reinforcement, which are bonded together, may become separated because of stresses caused by the weight of the vehicle. Uncharged hoselines suffer more damage than hose that have been fully charged with water, but it is recommended that vehicles never be allowed to cross any hoseline. The liner may also become cracked at the sharp bends on the edges. This damage is usually not apparent until the hose is pressure tested. Prevent this damage by laying hose to one side of the street/road near a curb or shoulder berm so that vehicles cannot drive over it while allowing pumping apparatus and other emergency vehicles access to the incident scene. Care must be taken, however, to avoid hose and coupling contact with oil, grease, or other petroleum products that may have collected in these locations **(Figure 2.6)**.

Provide traffic control at fire scenes to prohibit nonemergency vehicles from entering an area where fire hose is deployed. If it becomes necessary for emergency vehicles to drive where hose has been deployed, attempt to move the hoseline to the side of the street/road away from the travel lane. A large diameter supply hose charged with water is extremely difficult to move, requiring that it be deployed correctly from the hose bed onto the street/road where it will not need to be moved once it is filled with water. Place hose bridges at locations where the supply line cannot be relocated and vehicles must repeatedly drive over it **(Figures 2.7 a – c)**. Hose bridges redistribute a vehicle's weight to the bridges and not to the hose. Consider the assignment of emergency responders at all roadway intersections to ensure that vehicles are directed over the hose bridge.

Figure 2.6 Do not deploy hose where it will contact oil, grease, or other petroleum products. Avoid placing hose in street gutters or near curbs.

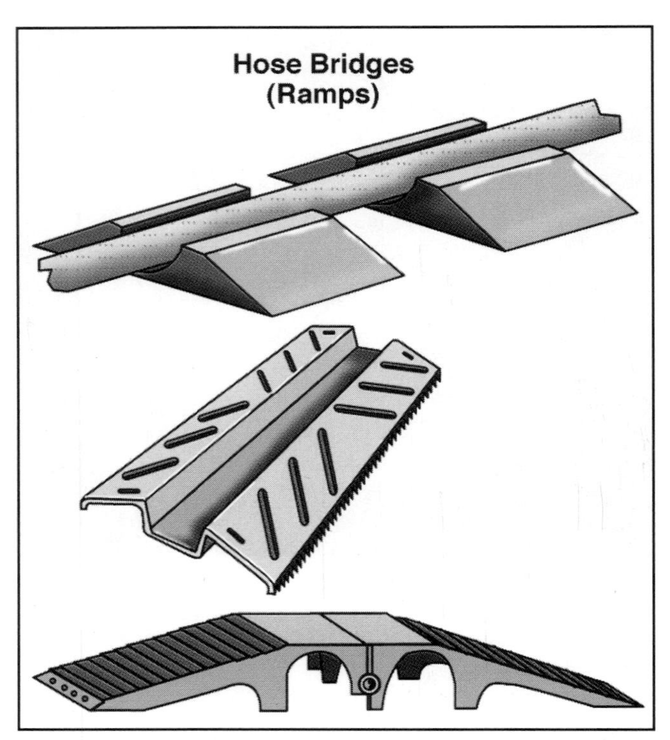

Figure 2.7a Various types of hose bridges and ramps (constructed of wood, rubber, or metal).

Figure 2.7b Large diameter hose ramp constructed of plywood. *Courtesy of Sam Goldwater.*

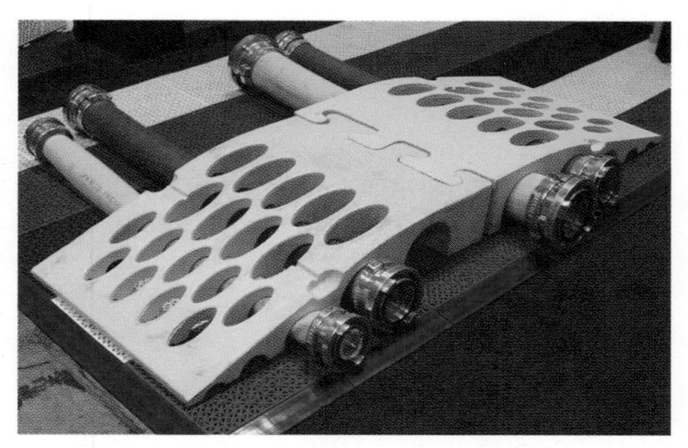

Figure 2.7c Commercial hose ramp for multiple hoselines.

Vibration from the pumping apparatus may chafe fire hose. Some pumping apparatus can vibrate so much that the hose connected to it (particularly the intake hose) chafes where it touches the street/road surface. Use a chafing block at the point where the hose contacts the ground **(Figure 2.8).** During supply relay operations, protect supply hoses from chafing anytime the bend of the hose leading from or into the pump allows the hose to vibrate and move while in contact with a hard surface. When connecting a pumping apparatus to a hydrant, bend the soft sleeve supply hose slightly to avoid kinks when the water is turned on. When preconnected hard suction hose is carried, provide padding at contact points between the hose and supporting brackets to prevent chafing.

Reloading dirty fire hose on the pumping apparatus may cause damage to the hose. Dirt and grit on the hose abrades the cover jacket fibers much like sandpaper. This action is increased by apparatus vibration when the pumping apparatus is moving. To prevent this damage, clean dirty hose and dry it immediately upon returning from an emergency incident. Refer to the manufacturer's recommendations regarding the level of cleaning and drying required for its hose before returning it to service.

Water hammer (sudden increase in pressure caused by closing nozzles or valves too quickly) can also damage fire hose **(Figure 2.9, p. 44).** This pressure can cause hose to burst at the couplings or at weak points. Prevent pressure damage by closing nozzles and valves slowly. Pressure-control devices such as governors and pressure-relief valves can also prevent surges in hoseline pressures. Use an inline pressure-relief valve to protect supply and attack hose against water hammer. The use of these devices is described in detail in the IFSTA **Pumping Apparatus Driver/Operator Handbook.**

CAUTION

Water hammer is generally the result of closing a pump valve, hose appliance, or nozzle too quickly. Burst hoselines, damaged fire pumps, and serious injuries to emergency responders can result.

Thermal\Cold Damage

Heat damage to fire hose occurs when it contacts fire or hot objects such as those found in fire debris. This damage causes charring, melting, or weakening of the covering or the reinforcement fibers. Heat also dries the lining, which promotes cracking. Fire hose can also suffer severe damage from heat sources that are not directly related to the fire such as when hose is near vehicle engine exhaust pipes. Placing hose away from these high heat sources and other power equipment during emergency operations can prevent this damage.

Hose can also become damaged when exposed to extreme ambient temperatures for prolonged periods such as when it is left in a hose dryer or in direct sunlight on an extremely hot day, exposing

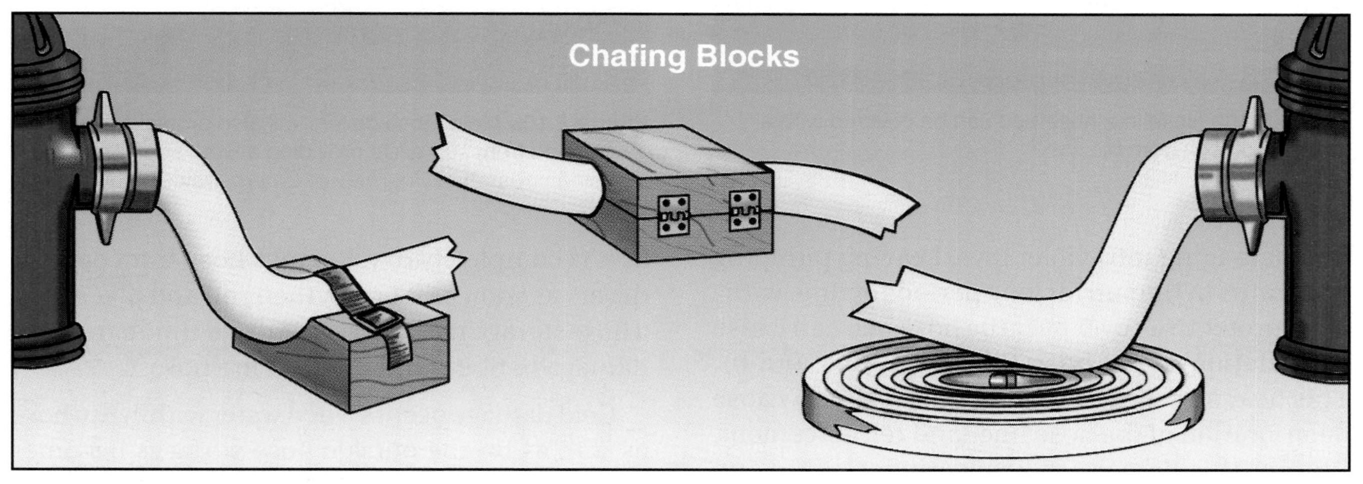

Figure 2.8 Chafing blocks prevent vibration wear damage to intake hose by keeping it from rubbing against a rough surface.

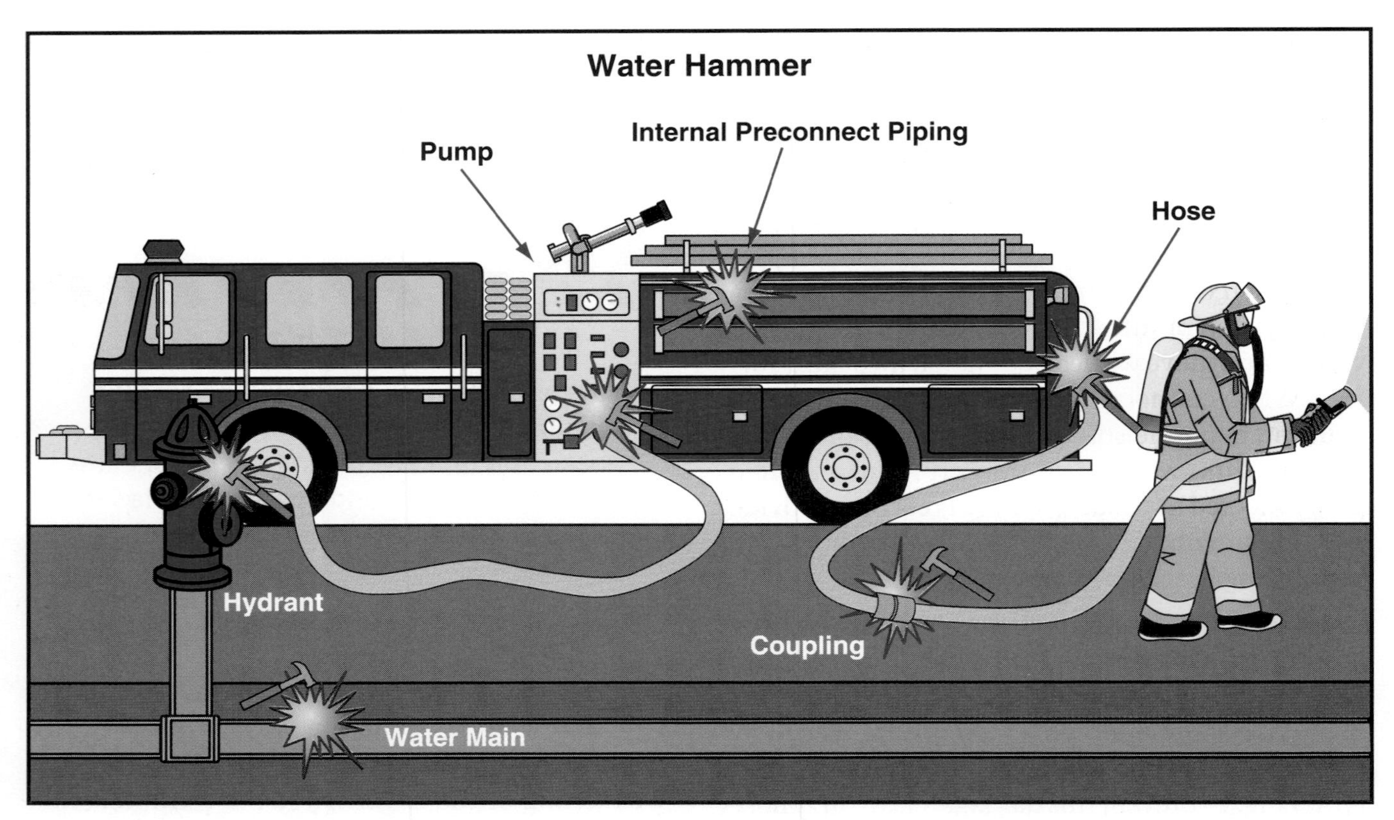

Figure 2.9 Water hammer can cause damage to all parts of the water system and fire equipment. Open and close nozzles and valves slowly.

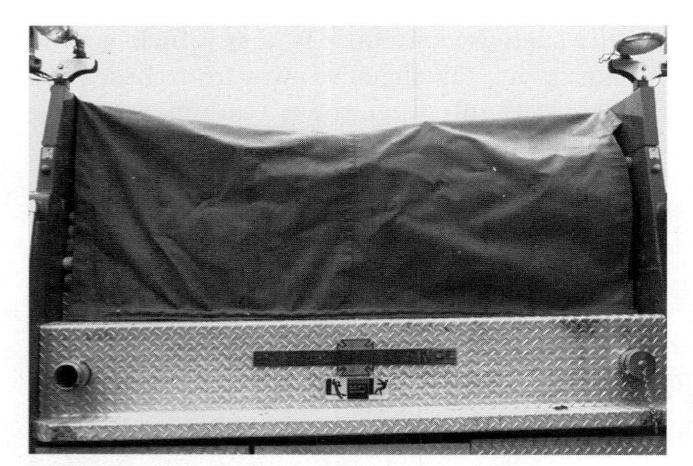

Figure 2.10a Hose in a hose bed can be covered with a tarp to protect it from the sun.

Figure 2.10b Metal hose covers will also protect the hose in a hose bed from sun while providing a work surface on the pumping apparatus. *Courtesy of Emergency One, Inc.*

it to damaging ultraviolet rays. Leaving pumping apparatus in the sun for long periods of time without a protective cover over the hose bed can cause degradation of the hose **(Figures 2.10 a and b).** This heat and ultraviolet-ray exposure can cause deterioration of the hose liner and reinforcement. Prevent this damage by inspecting, rolling, and storing hose in an appropriate location as soon as it is completely dry. Remove hose from cabinet dryers as soon as it is dry, then roll and place it on a hose storage rack. Excessive dryer time can cause damage to fibers and liners of fire hose.

Cold damage occurs when water within fire hose as well as on the outside hose surfaces becomes frozen. If a fire and emergency services organization is located in a climate that suffers severely cold temperatures, special cold-resistant hose (as described in NFPA 1961, *Standard on Fire Hose,* 2002) should be used. This type of hose is designed for use at temperatures down to -65°F (-54°C) with an expectation that it will perform with the same reliability as regular fire hose while withstanding the rigors of freezing and thawing. Although this hose is available, it is still necessary to prevent internal freezing in hose during intermittent use by allowing some water to flow through the nozzle at all times. Extremely cold temperatures require greater water flows. Maintain water flow in intake hose by circulating water from a hydrant through the fire pump, discharging it through a drain-off hose that routes water down a gutter or to a place away from the pumping apparatus. Immediately drain and roll hose that is no longer needed for firefighting purposes. Prevent couplings from leaking and thus freezing by tightening all connections. In extremely cold conditions, some manufacturers recommend applying an approved cold-weather lubricant with an antifreeze agent on the swivel and gasket portions of the couplings to help limit freezing.

When fire hose becomes frozen in ice, there are three ways to remove it: (1) melt the ice with a steam-generating device, (2) chop the hose loose with axes, or (3) leave the hose until the weather warms enough to melt the ice. When chopping hose out of ice, make all cuts in the ice well away from the hose to reduce the chance of the axe blade glancing into the hose fabric. Avoid using exhaust manifold heat from the pumping apparatus because it can be very hot and poses a carbon-monoxide hazard to fire and emergency service responders working with the hose. Do not fold frozen hose because folding will damage the structure of the hose, weakening it in a manner not always apparent during visual inspections. If hose sections can be uncoupled, carefully load hose onto a flatbed vehicle and transport to a location where it can be thawed and protected from damage. Perform a service test before placing thawed hose back in service to ensure that no damage has occurred. See Fire Hose Service Testing section.

Organic Damage

When fire hose reinforcements are woven from organic fibers, they are susceptible to attack from fungus (a parasite that feeds on dead organic matter). This fungus is commonly referred to as *mildew* or *mold.* Mildew weakens the hose reinforcement as the fungus consumes the fibers. The ideal condition for mildew to form on hose is when moisture is in the reinforcement jacket and evaporation is inhibited by a lack of airflow. These situations can occur when hose is in a pumping apparatus hose bed, compartment, or storage rack. Organic hose damage can be avoided by taking the following actions:

- Ensure that cotton or cotton-blend fire hose is completely dry before storing or loading **(Figure 2.11).**
- Cover hose beds with water-repellent covers to keep hose loads dry during inclement weather.
- Remove wet woven-jacket fire hose from the apparatus and dry it thoroughly to prevent mildew from forming.
- Inspect fire hose in storage racks and hose beds periodically; also remove and rotate hose periodically **(Figure 2.12, p. 46).** Even if mildew is not visible, a musty smell is often an indicator that it is hidden somewhere within the hose.
- Ventilate all areas where fire hose is kept, including pumping apparatus hose beds and compartments.

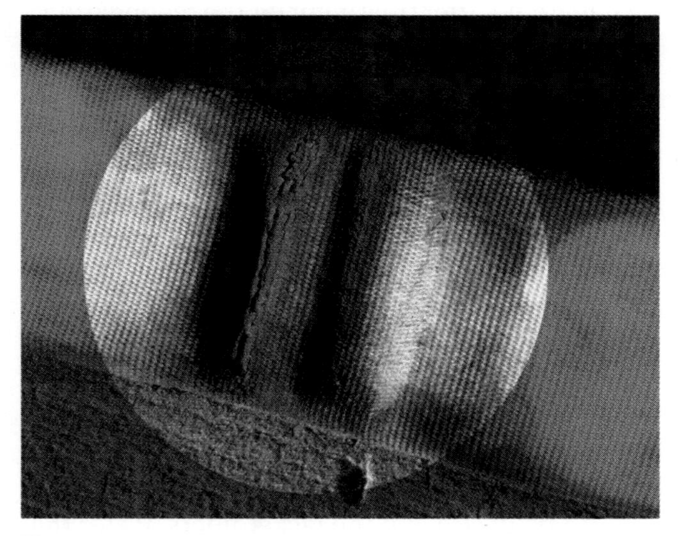

Figure 2.11 Mold or mildew can weaken the jacket of woven-jacket hose.

Figure 2.12 Remove and reload hose periodically.

If mildew is discovered on hose, immediately wash it. Scrub the cover jacket with a very mild soap or bleach solution (5 percent in water) and dry completely. Inspect the hose section within the next few days after treatment for the reappearance of mildew.

Chemical Exposure

Many chemicals (in dry, liquid, or gaseous forms) are injurious to fire hose. Motor oil, found in some quantity on most streets and highways where hose is laid, is an example of a petrochemical that will penetrate the woven cover and produce a solvent action that separates the rubber lining from the reinforcement or deteriorates the synthetic material in the reinforcement. This action is even more drastic with gasoline contact. Battery acid is another chemical that causes significant damage to hose by destroying the jacket fibers. Water that is not drained completely from the hose can form sulfuric acid, which weakens or destroys the liner. The following recommended practices help prevent chemical damage:

- Avoid laying fire hose directly against curbs where oil, gasoline, and battery acid may accumulate or pool from parked automobiles. Hose laid in gutters may also come in contact with fire-fighting runoff water that could contain harmful chemicals. Even though some types of fire hose are constructed from materials that are not affected by chemical exposure, other hose types are still used. Procedures:
  - — Place the hose 2 to 4 feet (0.6 m to 1.2 m) out into the street/road, but away from vehicle travel lanes.
  - — Move the hose, if possible, onto a sidewalk or into a median to avoid vehicle and contamination damage.
- Avoid exposing fire hose to hazardous material spills where it might be damaged by the materials.
- Avoid exposing fire hose to spills of foam concentrate, which is mildly corrosive and can deteriorate the hose lining or cover material if it remains on the hose.
- Scrub fire hose suspected of having contacted acid or other caustic chemical thoroughly with a solution of bicarbonate of soda and water **(Figure 2.13).** Remove the hose from service and contact the manufacturer for further maintenance procedures.
- Service test and inspect all inactive fire hose periodically. Inspect regularly (about every 90 days), and conduct service testing on an annual basis.
- Service test fire hose if there is any suspicion of exposure to chemical or other damage.

Age Deterioration

Fire hose may become worn if it is loaded on edge continuously, particularly with horseshoe or accordion loads. Most fire hose manufacturers recommend loading hose in a flat load (see hose load descriptions in Chapter 6, Supply Hose Loads and Deployment Procedures).

Damage from deterioration and cracking also occurs if sharp folds are left in tightly loaded fire

Figure 2.13 Scrubbing hose that has been exposed to acid with bicarbonate of soda and water will neutralize any acid on the outer cover.

Figure 2.14 When reloading hose, attempt to place the folds in different locations from previous loads.

hose for long periods. NFPA 1962 recommends rotating hose loads four times a year if they are not used. Fire and emergency service organizations should include these hose rotations in their standard operating procedures. Fire and emergency service organizations frequently realize that they seldom deploy all of the fire hose on their pumping apparatus during normal fire-suppression operations. The hose near the top of the bed is often deployed, but lower layers may seldom be removed. When reloading the hose, refrain from packing it tightly, and relocate folds at previously unfolded places in the hose **(Figure 2.14)**.

Age deterioration can also occur if fire hose is left hanging in a hose tower for excessive periods of time. The inner lining of hose can become weakened at the point where it hangs over the support peg. Reinforced jacketed fabric hose may suffer a separation of the rubber or plastic lining from the inner reinforcement, reducing the strength of the hose at that point. Prevent this damage by removing the hose from the tower as soon as it dries. If fire hose must remain in a tower for prolonged periods, change the hose/peg contact point periodically.

Coupling Damage Causes and Prevention

Fire hose couplings can be damaged in many of the same ways that fire hose is damaged. As with fire hose, care and normal maintenance procedures can minimize or eliminate common coupling maladies. Pay special attention to the hose coupling during use, and protect it from damage due to careless handling or exposure to other degrading environments. Because couplings are constructed of brass or aluminum, damage due to abuse is often overlooked during routine inspections. When damaging exposure to chemicals, heat, or abuse goes unnoticed, the cumulative effect of these damages can result in the complete failure of a hose coupling system. The following four general categories of coupling damage are possible:

- Mechanical damage
- Thermal/cold damage
- Corrosion
- Chemical exposure

Mechanical Damage

Mechanical damage is the most prevalent form of coupling injury and failure. Couplings usually become damaged through rough handling such as when they are dropped or vehicles drive over them. Dropping a coupling, especially one made of brass, may cause the coupling to become misshapen or out-of-round, which results in an inability to screw/unscrew the connection. This type of damage is more likely to occur when couplings have been disconnected from hoselines, although it can occur when couplings are still connected. Swivels and male threads are particularly susceptible to damage when couplings are dropped during hose deployment. Prevent this damage by simply taking care when handling fire hose.

The pumping apparatus driver/operator can greatly limit this type of damage by deploying the hose at low speeds (between 5 and 10 miles per hour (mph) [8 kmph and 16 kmph]) during hose-deployment operations, rather than a higher speed that allows the hose to "whip out" of the hose bed **(Figure 2.15).**

When hose is deployed from a hose roll (in rack storage or any other hose-roll situation), care must be taken to protect the male coupling threads. These threads can be damaged if the coupling is allowed to drop on a hard surface during the operation.

Figure 2.15 Drive the pumping apparatus between 5 and 10 mph (8 kmph and 16 kmph) so that the couplings clear the tailboard as the hose comes from the hose bed.

Few couplings can withstand the weight of a vehicle **(Figure 2.16).** When a tire rolls over a coupling connection, the usual result is that the couplings are slightly flattened, becoming out-of-round. Although this damage is not always visible, it can be easily detected because the swivel on the female coupling will no longer spin freely and the male coupling will not accept a female coupling. If vehicle tires pass over the hose immediately adjacent to a coupling, the hose can be pulled partially from the coupling shank. When this pulling action happens, the coupling will appear to be cocked at an angle on the hose. This situation is extremely dangerous because the integrity of the coupling attachment has been compromised. A catastrophic failure of the coupling could result when the hose is pressurized. Prevent this damage by deploying hose and couplings out of the path of vehicular traffic or prohibiting vehicle access where hose has been deployed.

Figure 2.16 Lay hose to one side of the roadway so that following emergency vehicles are not forced to drive over it or the couplings.

CAUTION

Deploy hose out of the path of vehicles to minimize damage to hose and couplings.

Excessive fire pump pressure can also cause damage to a coupling. When the design limitations of the connection are exceeded, the expansion ring and hose can slip out of the shank of the coupling (normally referred to as a *slipped coupling*). Water hammer is the most common explanation when this type of damage occurs. Reduce this type of damage with proper training in pumping apparatus fire pump techniques as well as pump-valve, pressure-relief valve, and nozzle operations. The slow opening and closing of ***all*** nozzles and valves provides a much safer environment on the emergency scene.

WARNING

Excess fire pump pressure is an extremely dangerous situation that can result in serious injury or death to emergency responders working around charged hoselines. Ensure that pressure-relief valves are in place and operating during hoseline operations.

When a section of hose is dragged with couplings trailing behind, there is always a chance of damaging a coupling in addition to damaging the hose cover, especially on rough surfaces. Male threads can wear or become misshapen in these situations. If it becomes necessary to manually drag hose, fold it in half and carry both couplings or fold the ends of the hose back over itself so that the couplings lie on top, away from the rough surface. Dragging couplings behind a moving pumping apparatus is a particular concern; severe damage can result to couplings, hose, and cover jackets.

Screw threads on couplings can be damaged if they are cross-threaded or mismatched with couplings of another thread type. Hose couplings manufactured with a Higbee indicator help avoid cross-threading the male and female threads of the coupling and ease the connecting process by allowing fire and emergency service responders to correctly align the two couplings. See Chapter 5, Basic Methods of Handling Fire Hose. If it is difficult to mate the threads, turn the swivel counterclockwise against the male thread until a distinct "click" is heard, then turn the swivel clockwise to begin attaching the couplings together **(Figure 2.17).** If it becomes difficult to turn the swivel when connecting couplings, couplings may be out-of-round, threads may be damaged, or threads may not match. Never force threads; take action to correct the problem. When brass threads have been damaged, they can sometimes be repaired with a thread tap or die tool of the proper size **(Figure 2.18, p. 50).** Check for irregularities on the thread surface and, if needed, remove burrs and abrasions with a fine three-cornered file **(Figures 2.19 a and b, p. 50).**

Figure 2.17 Use the Higbee cuts and Higbee indicators to connect the male and female threaded couplings without cross-threading or mismatching.

Figure 2.18 Standard tap *(bottom)* and die *(top)* are used to ensure consistency during coupling manufacturing or repair process. *Courtesy of Akron Brass Company.*

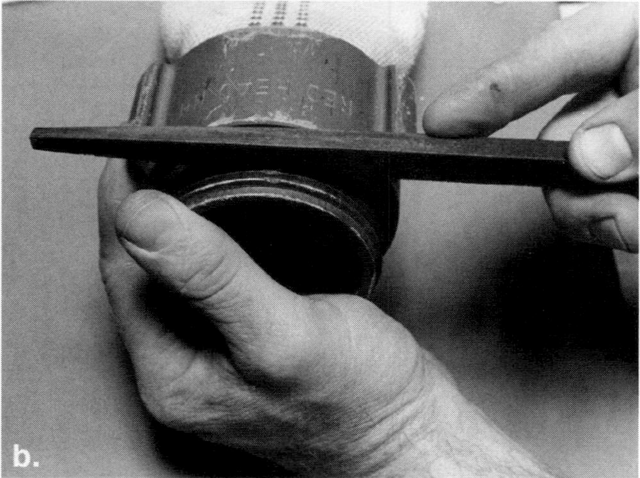

Figure 2.19 (a) Visually check threads for irregularities. (b) Remove burrs and abrasions on the thread surface with a three-cornered file.

When a coupling is damaged or out-of-round, the hose must be removed from service and the coupling replaced. If the damage is less severe, however, the coupling can sometimes be restored to near its original shape. This restoration depends on a number of factors that include the degree and place of distortion, the type of metal in the coupling, and the coupling design. A brass screw-thread coupling, for example, can be gently hammered back into shape if it is the type that attaches with an expansion ring. The coupling can be placed on the mandrel of an expansion machine and tapped while under slight expansion pressure. If done properly, the coupling can be restored to almost its original shape (see Fire Hose Recoupling section). Only a factory-trained technician should attempt this procedure, and repair will not always be possible with metals other than brass. The recoupling equipment (mandrel and/or expansion machine) can receive serious damage during this procedure if it is not done properly.

CAUTION

ONLY a factory-trained technician should perform coupling repair.

Thermal/Cold Damage

The effects of extreme heat and cold can also damage hose couplings. Severe damage in the form of metal fatigue and distortion can occur to the coupling, although this damage is not as visible as damage to the fire hose itself. The expansion ring or other attachment method may become loosened or completely separated from the hose. The coupling gaskets may be damaged beyond the point of being able to seal the couplings. These circumstances can occur when the coupling is subject to the direct heat of a fire. When that is the case, remove the coupling/hose from service until full inspection, maintenance, and service testing can be performed.

Often overlooked is similar damage that can occur when couplings are subjected to extreme cold and from the effect of ice on the coupling. As water freezes, it expands. Any water trapped in a space around a gasket or within the fibers of the hose often stresses the coupling to the point of failure. Brass couplings, for example, will deform. Aluminum and other aluminum alloy couplings, while having a higher strength than brass, are more friable and tend to crack in this situation. A common indication of freezing is the failure of the female coupling swivel to operate. Wrap couplings in towels and pour hot water over them to melt the ice binding the threads and swivels or place them near a heat source that will slowly and evenly warm them until they thaw. Using a propane torch to thaw couplings is not recommended because overheating can damage the hose and gaskets.

CAUTION

Do not try to force frozen couplings loose with hammers or torches; severe coupling damage could occur.

Corrosion

Corrosion is a chemical process in which a metal is attacked by some substance in its environment and converted to an unwanted compound that gradually weakens or destroys the metal. The most common fire coupling metals are made of brass and aluminum, and each of these metals possesses a high resistance to corrosion but each will suffer some deterioration when exposed to certain conditions. NFPA 1925, *Standard on Marine Fire Fighting Vessels* (2004), NFPA 1964, *Standard for Spray Nozzles* (2003), and NFPA 1962 require that all couplings, nozzles, and hose appliances be resistant to corrosion caused by the effects of saltwater exposure or that anticorrosion lubricants be applied to lessen the effects of corrosion.

Brass is an alloy of copper, zinc, and lead, and it is highly resistant to corrosion. Over time, however, the metal darkens and turns green as copper oxides are formed. This process is most prevalent when the coupling is allowed to contact moist organic material or earth. Although these copper oxides are usually found only on exposed surfaces, they can form on the interior of female swivels or inside surfaces of nozzles, reducing the ease of the device's operation. Normal cleaning removes most of the surface corrosion; however, the only way to free the swivels or operating mechanisms is to lubricate moving parts on a regular basis. Follow the manufacturer's recommendations.

Aluminum couplings develop a layer of corrosion (aluminum oxide) that in effect "seals" the metal against further oxidation. This protective layer can be scratched or abraded during normal use, resulting in a new layer being formed.

Chemical Exposure

Fire hose couplings are often exposed to chemicals during fire-suppression operations. As the mission of the fire and emergency services has expanded to include responding to hazardous materials incidents, couplings are subjected to a much higher risk of chemical attack and possible damage. Corrosive chemicals (either acids or bases) aggressively react with the metals in the couplings and nozzles if they are exposed. In concentration, these acids and bases can destroy the coupling or nozzle and degrade its parts to the point of failure. Aluminum, a reactive metal, deteriorates in the presence of strong acids and bases, which may result in coupling damage and failure in extreme cases. This situation must be considered during hazardous materials incidents. Couplings that have been exposed to these materials must be completely neutralized, inspected, and service tested before returning them to service. See recommendations from chemical manufacturers for procedures on how to neutralize various chemicals.

Fire Hose and Coupling Care

Fire hose is one of the most essential types of equipment carried on a fire-fighting apparatus. It must be maintained on a regular basis to be dependable in every emergency. Fire hose rarely stays clean during fire-fighting operations. It often suffers great abuse, especially in intense fire situations when it becomes difficult to follow ideal hose-care rules. Cleaning the hose, therefore, is very important to prolonging the life of each hose section. Equally important, the coupling must also be protected from damage by incorporating regular inspection and maintenance procedures. When couplings are stored, they must be afforded the same level of care given to the hose. It is easy to damage couplings while rolling or unrolling hose when it is improperly stored or carelessly loaded into hose beds. Maintain accurate and up-to-date fire hose and coupling records and include any maintenance or preventative work that has been performed.

Maintenance

All fire hose and couplings require the employment of a comprehensive maintenance and repair program. Station personnel can conduct most initial maintenance. Trained technicians, however, must perform many equipment repairs. Although fire hose and coupling category types vary, the fundamental procedures for maintenance are somewhat similar for each. Wet cotton fabric hose promotes mildew growth; therefore, dry this hose type thoroughly. Care must also be taken when loading synthetic fabric fire hose. Although moisture and the possibility of mold and mildew growth does not affect this hose type, complications arising from corrosion of the coupling and coupling attachment must be considered. Load this fire hose type on pumping apparatus in such a way that air can circulate under the hose load to eliminate or reduce the growth of mildew in the hose jackets and corrosion in the hose couplings.

Rubber-coated and synthetic hose, including booster and suction hose, are susceptible to damage when they are exposed to petroleum products. Normal fire apparatus maintenance may inadvertently cause this type of exposure if fuel, oil, transmission fluid, or hydraulic fluids are not carefully handled during these procedures. Additionally, when pouring foam concentrates from their containers, some can accidentally spill and damage hose/couplings if it is not immediately removed. Coupling gaskets can be contaminated if fire and emergency service responders who have completed fire apparatus maintenance procedures involving petroleum products handle them. When inspecting and handling any hose or coupling, ensure that gloves or cleaning rags are free of contaminants.

Remove fire hose from the pumping apparatus and reload with sufficient frequency so that the folds occur at different positions to prevent damage that can occur at the folds. When fire hose is used infrequently, it weakens at these fold locations and may fail when deployed and pressurized. Also repack soft sleeve supply hose in a different position after each use if it is stored in a folded load. Rolling and then storing soft sleeve hose is the best way to avoid permanently creasing the hose **(Figure 2.20)**.

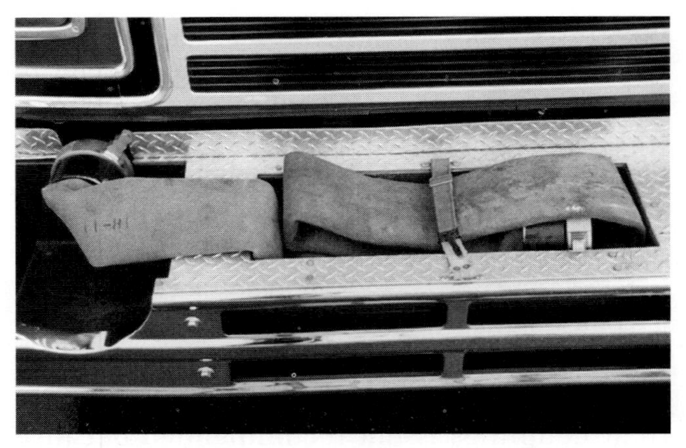

Figure 2.20 Soft sleeve supply hose can be tied or clamped onto the exterior of the pumping apparatus or placed in a compartment.

Inspection

According to NFPA 1962, hose should be inspected and service-tested within 90 days before being placed in service for the first time and at least annually thereafter. See **Skill Sheet 2-1** for general procedures for inspecting fire hose periodically.

Each time a section of hose is used, whether for emergency incidents or training, it needs to be inspected to ensure that it is free of visible soil or damage. Check couplings for ease of operation, any deformations, or other visible damage. While gathering equipment and rolling fire hose immediately following emergency incidents, conduct a postincident inspection. This quick inspection allows fire and emergency responders to identify and mark possible damaged hose locations and couplings.

Before fire hose and couplings are stored or placed back in service after use or inspection, correct any of the following deficiencies:

- Evidence of dirt or debris on the hose cover jacket or couplings
- Damage to the hose jacket
- Coupling loosened from the hose
- Damage to male or female threads

- Obstructed operation of the swivel
- Absence of a well-fitting gasket in the swivel

Routine Care Tasks

Hose washing is a laborious, time-consuming job that can be simplified by using such devices as a commercial hose washing machine or a jet-spray washer. A cabinet-type hose washing machine washes, rinses, and drains fire hose **(Figure 2.21).** This automated washer can be operated by one person and used with or without detergent. A jet-spray washer can be attached to a section of hose or attached directly to a hydrant **(Figure 2.22).** Water under high pressure is directed through small jets within the washer housing so that as hose is manually guided through the washer, dirt and debris are washed away.

If these washing devices are not available, an adequate job of cleaning can be done by hand. Simply brushing off accumulations of dirt, leaves, and other debris with dry brooms, however, does not adequately clean hose. Although this practice leaves the hose looking clean, tiny particles of grit and sand remain within the cover jacket among the threads where they can cause wear to individual fibers. To remove this grit, wash hose with clear water as soon as possible after use. Scrub the hose cover jacket with brushes or brooms along with a high-pressure water stream **(Figure 2.23, p. 54).** See **Skill Sheet 2-2** for hand washing fire hose procedures. Use a mild soap solution if the hose has been exposed to oil. Soap helps dissolve the oil absorbed by the jacket fibers. Rinse hose thoroughly after a soap washing to remove all traces of contaminants and soap.

Figure 2.22 A jet-spray washer cleans the hose cover with a high-pressure water stream that surrounds the hose.

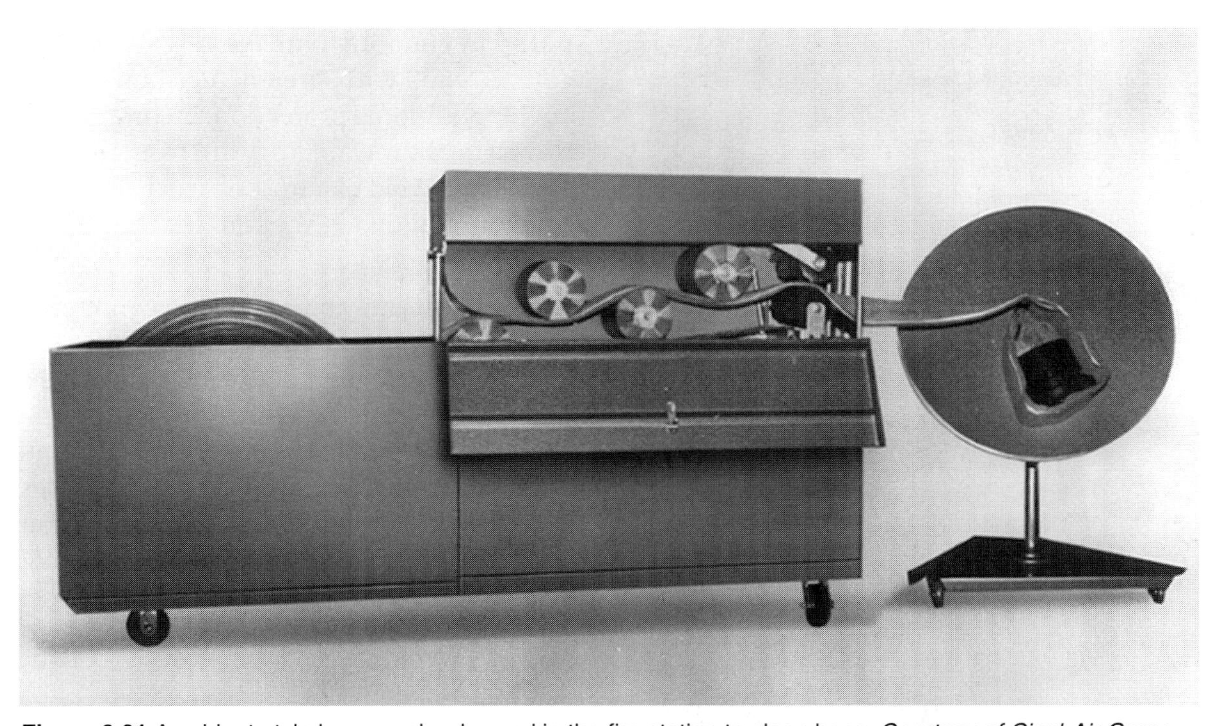

Figure 2.21 A cabinet-style hose washer is used in the fire station to clean hose. *Courtesy of Cicul-Air Corporation.*

Figure 2.23 Use a high-pressure water stream to rinse hose and dislodge dirt and other debris on the outside covering.

After fire hose has been thoroughly washed, hang it in an interior or exterior hose tower, place it on an inclined drying rack, or place it in a cabinet-type hose dryer **(Figures 2.24 a and b).** For drying procedures, see **Skill Sheet 2-3** for drying fire hose using a hose tower and **Skill Sheet 2-4** for using a cabinet-type drier. An alternate drying method is to lay hose on edge on the floor of a large room such as an apparatus room/bay **(Figure 2.25).** The disadvantage of drying hose in this manner is that it is difficult to completely drain residual water from inside the hose. Use this method only as a "last resort" when all other drying methods are not available.

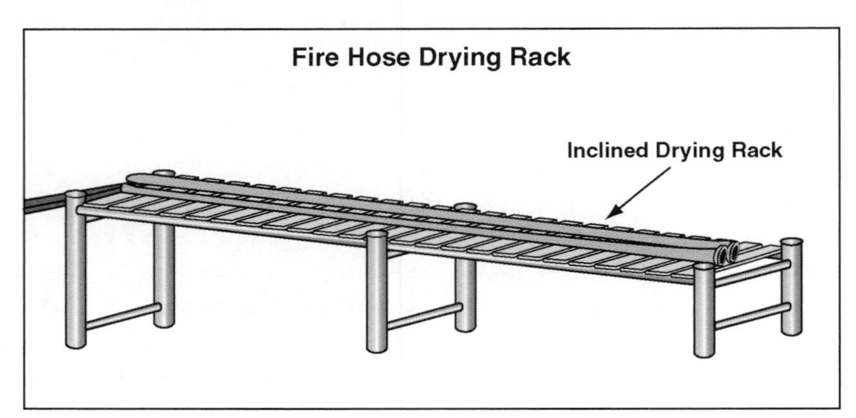

Figure 2.24a Fire hose drying rack.

Adequately ventilate and protect hose towers and drying racks so that fire hose is not exposed to excessive temperatures or direct sunlight. Remove hose from exterior hose towers as soon as it is dry to protect it from damage by the sun's ultraviolet rays. Secure hose hung in outside drying towers by lashing/tying the coupling ends to prevent them from swinging in the wind. Such movement could cause couplings to bang against each other or against tower supports, resulting in coupling damage **(Figure 2.26).** Cover male threads with precut sections of tubing to provide additional protection. Incline drying racks enough to allow water to drain from the hose during drying. Avoid placing hose sections too closely together or allowing them to touch, thus slowing the drying process.

Hose dryers may be purchased commercially or built into a building. If a hose dryer is built into the building, ensure that fans of adequate size are installed to circulate the necessary air to dry the hose and expel moist air to the outside. Incorporate hose racks with strong rack supports and slides designed to support the weight of the wet hose. These racks should be a size that two persons can manage. Additionally, include commercial grade temperature control units and timers to protect the hose during the drying process.

Figure 2.24b This cabinet-type hose dryer safety dries hose without radiant heat and also gives extra storage space.

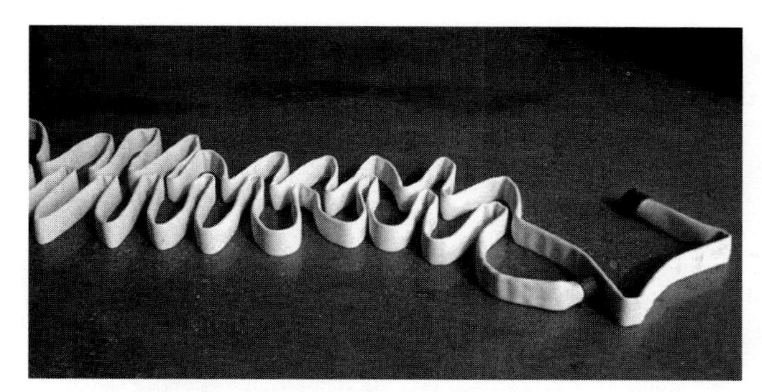

Figure 2.25 Hose can be laid on an apparatus room/bay floor to dry.

Figure 2.26 Secure hoses in outside hose towers to prevent them from swinging in the wind and possibly damaging the couplings.

Figure 2.27a Clean coupling threads thoroughly with a stiff bristle brush.

Figure 2.27b Use a wire brush to clean male threads that are clogged with foreign materials.

Basic coupling maintenance is essential for the continued useful service of fire hose. When a coupling swivel becomes stiff or sluggish with dirt or other foreign matter, remove the gasket and put the coupling in a container of warm, soapy water. Work the swivel back and forth to help loosen accumulations of dirt between the swivel collar and coupling body. Clean coupling threads with a stiff bristle brush **(Figure 2.27a).** If the threads are occluded with tar, asphalt, or other foreign matter, use a wire brush to loosen stubborn material **(Figure 2.27b).** Rinse the coupling thoroughly in clear water after washing, and lubricate it according to the manufacturer's instructions. Should the coupling attachment to the hose become loose or its integrity is questionable, remove the hose from

Figure 2.28 If the swivel gasket has become hardened and inflexible, replace it to avoid leaks.

service until a new coupling is attached. Lubricants such as graphite or silicone are usually all that is needed to maintain swivels so that they spin freely. If a gasket is cracked, scored, or has become hardened and inflexible, replace it **(Figure 2.28).**

Storage

After fire hose has been washed and dried, roll and store it in racks **(Figure 2.29).** Pack cotton fabric hose loosely so that air circulates around it. Synthetic and rubber-jacketed hose can be stored in tight rolls following normal cleaning procedures. Locate hose racks in a clean, well-ventilated room that is easily accessible to the pumping apparatus. Hose that is stored in the fire apparatus room/bay may be exposed to cleaning solvents, lubricants, oils, diesel fumes, and other airborne contaminants.

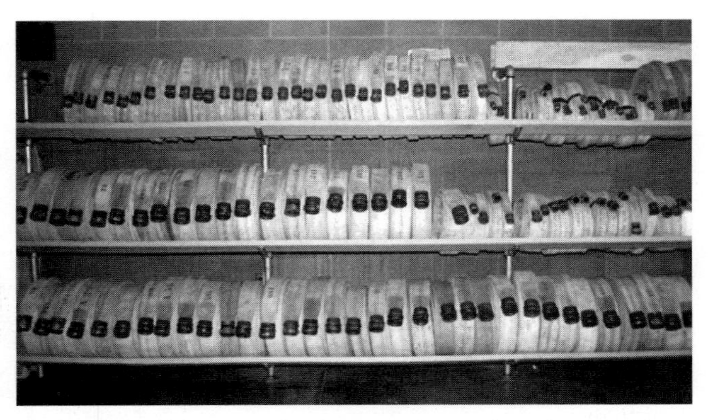

Figure 2.29 Roll clean dry hose and store it on hose racks.

If hose must be stored in the fire apparatus room/bay, inspect and clean it more frequently than fire hose that is stored in a separate space.

CAUTION

Never store solvents, petroleum products, or other chemicals in close proximity to fire hose and couplings.

Also protect fire hose from exposure to direct sunlight. The sun's ultraviolet rays break down the natural or synthetic fabric of the hose, reducing its expected service life. Mount racks permanently on the wall or stand them free on the floor. Mobile hose racks can be used to both store and move hose from storage rooms to pumping apparatus for loading **(Figure 2.30).**

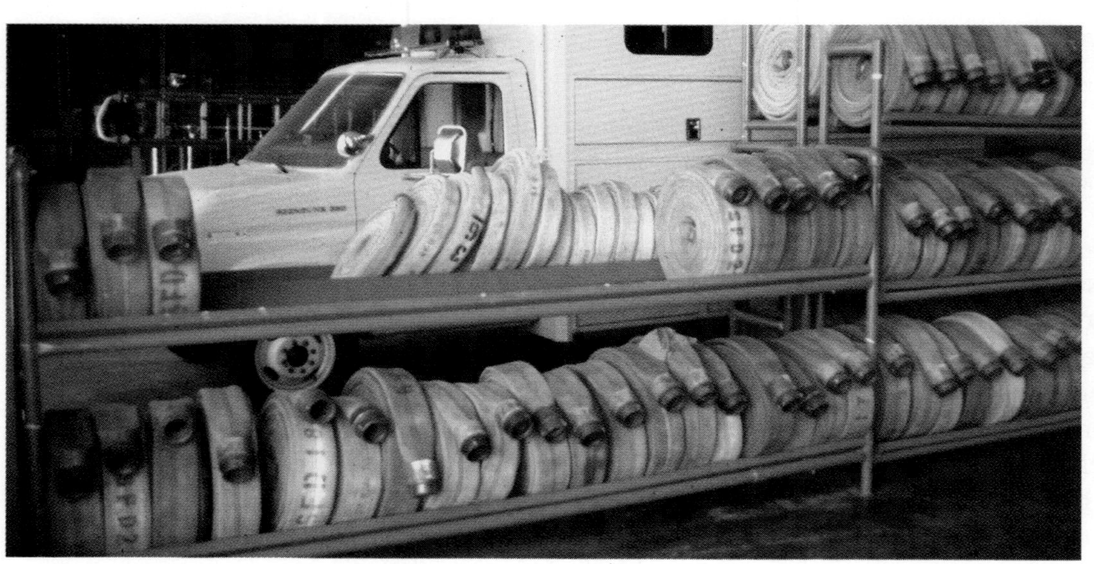

Figure 2.30 Store hose on free-standing (shown), wall-mounted, or portable racks. Many free-standing racks are close to the pumping apparatus for quick loading. *Courtesy of Sam Goldwater.*

When storing hose and couplings, place them in a storage rack in a manner that protects them from damage. Do not locate them in a rack in a way that allows them to hang into a walkway where they could be struck by other equipment or cause injury to fire and emergency responders. Protect the male coupling threads by rolling the hose with the male end inside the roll. When it is necessary to store fire hose with the male coupling on the outside of the roll, protect the exposed threads with a cap or other protective device. Place sexless couplings in a storage rack in a way that prevents dirt or other foreign objects from collecting in their ramp grooves. These contaminates would interfere with their ability to securely lock into place.

NOTE: Some fire and emergency service organizations, depending upon the hose loads that they use, may store hose with the male coupling on the outside of the hose roll.

CAUTION

Never stack equipment on top of fire hose that is stored in a hose storage rack.

Fire Hose Records

A *fire hose record* is a case history of each section of hose (including the attached couplings) from the time it is purchased until it is taken out of service. These records are required by NFPA 1962 and NFPA 1500, *Standard on Fire Department Occupational Safety and Health Program* (2002). To accurately keep records of hose tests, repairs, and inventory, it is necessary to permanently mark each section of hose with an identification number. Mark hose/couplings by using one or more of the following methods:

- Die-stamp the section identification number and fire and emergency service organization's initials on the bowl or swivel of the female coupling **(Figure 2.31a)**.
- Stencil the identification number and fire and emergency service organization's initials on the hose cover jacket near each end using indelible ink **(Figure 2.31b)**. Avoid marking the fire hose in locations where required NFPA markings are located.
- Color-code the coupling shank, swivel, or lugs to indicate the pumping apparatus or station to which the hose is permanently assigned. Avoid paint or colored tape near the swivel or other locations that might hinder its operation. If the coupling is reassigned to another section of hose, modify the marking and record to reflect this change.
- Attach a permanent bar code number to track couplings and hose. The code carries all of the significant data associated with a particular section of hose (similar to those used in merchandising). The bar code is generated with all pertinent information and placed on one of the couplings with an adhesive. The codes can then be scanned (read) and modified as needed. Each hose section is scanned after each use, which greatly enhances the ability to track every hose in inventory. This system requires a bar code reader for each station.

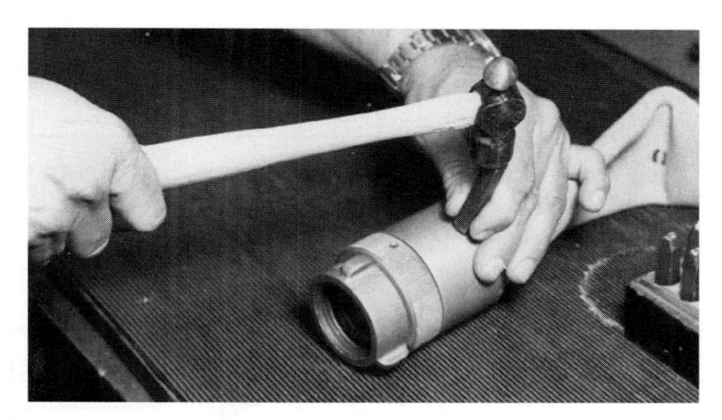

Figure 2.31a One hose-marking method is to die-stamp the section identification number on the coupling.

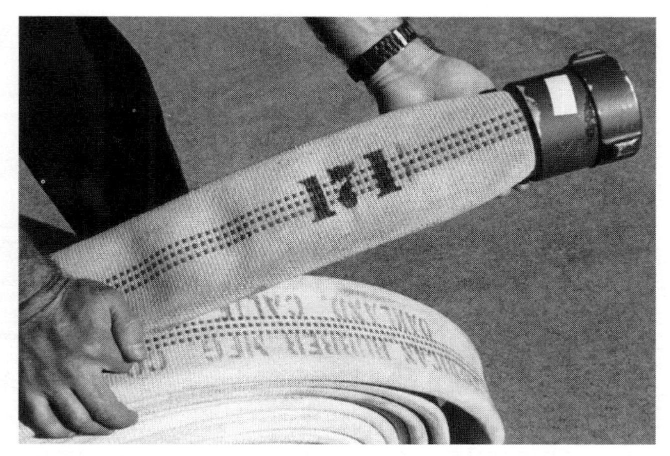

Figure 2.31b Another marking method is to stencil the identification number on the hose jacket with indelible ink.

Records can also be kept on cards, log sheets, or computers **(Figure 2.32).** Include such information as the date of purchase, the dates and results of annual testing, remarks concerning testing, the dates of required maintenance, unusual features, and causes of failure if any (complete bulleted list follows). These records are kept as part of the fire and emergency service organization's or individual company's equipment inventory. These records indicate the disposition of the hose and its assigned location, pumping apparatus, apparatus compartment, rack storage, etc. These records are also vital when the fire and emergency service organization is evaluated for insurance rating purposes. See **Appendix B,** Fire Hose Record Forms, for sample forms. The following information is included in the hose record for each section of hose:

- Assigned identification number
- Name of manufacturer and part number
- Name of vendor
- Trade size of hose
- Hose length
- Hose type
- Hose construction type
- Date received
- Date put in service
- Date of each service test and the service test pressure (see Fire Hose Service Testing section)
- Dates of required maintenance
- Date of actual damage
- Date and type of repairs made to section and new length if shortened
- Date of adverse exposure events (heat, freezing, chemical, etc.)
- Date and reason hose was removed from service
- Date and reason hose was condemned
- Date and indication that hose has been removed from service or condemned within the warranty period because of an in-warranty failure

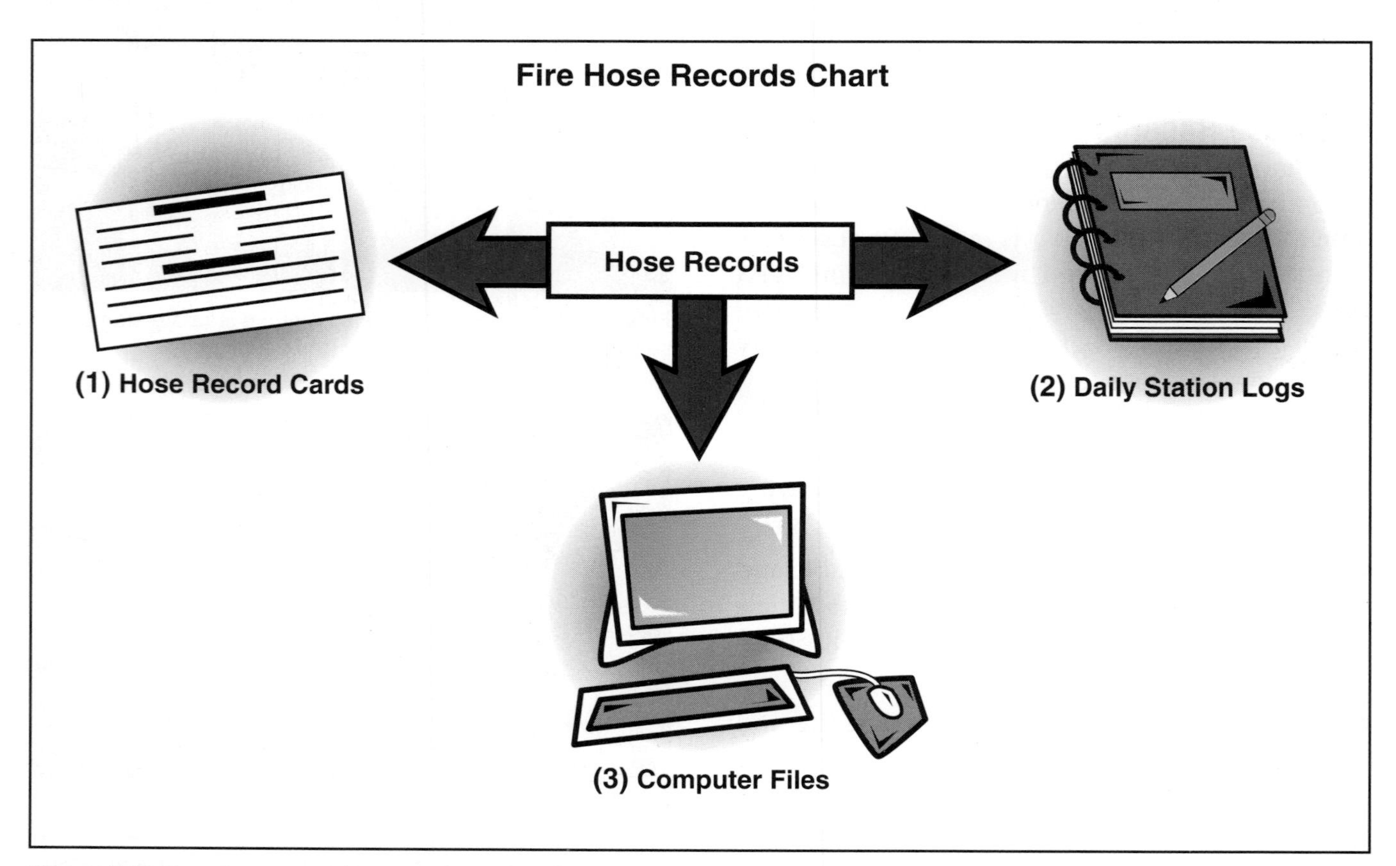

Figure 2.32 Keep hose records on cards, on log sheets, or in computer files.

Fire Hose Service Testing

There are two types of tests for fire hose: (1) proof test (formerly referred to as acceptance test) conducted by the manufacturer to certify the fire hose and (2) service test performed periodically by the fire and emergency services organization. *Proof testing* (as outlined in NFPA 1961) is a rigorous test during which the hose is subjected to pressures two times the service test pressure. This type of testing is ***not*** conducted by the fire and emergency services organization. *Service testing* of in-service hose confirms that it is still able to function as designed during fire-fighting or other emergency operations.

Testing Fire Hose Manufactured Before July, 1987

NFPA 1961, *Standard on Fire Hose* (2002), was completely revised in 1987, with major changes in how fire hose is labeled and tested. Before 1987, the proof test (or acceptance test) pressure was stenciled on each section of hose with the words *TESTED TO ___ PSI*. This pressure is not used, however, when service testing this hose. NFPA 1962 now requires that the annual service test pressure be labeled on each section of hose as follows: *SERVICE TEST TO ___ PSI PER NFPA 1962* or *SERVICE TEST TO ___ BAR PER NFPA 1962*. **Table 2.1** (adapted from Table 7.1.1.1, NFPA 1962) lists pressures that are used for service testing purposes for hose manufactured before 1987. Each service test requires a minimum duration of 3 minutes once the desired service test pressure is attained.

Table 2.1
Service Test Pressures for Hose Manufactured Before July 1987

Trade Size inches (mm)	Jackets	New Hose Rated Acceptance Test Pressure psi (KPa)	Service Test Pressure psi (kPa)
Lined industrial, standpipe, and fire department			
1½–2½ (38–65)	Single	300 (2 070)	150 (1 030)
1½–4½ (38–114)	Single	400 (2 760)	250 (1 720)
1½–2½ (38–65)	Single	500 (3 450)	250 (1 720)
1½–4 (38–100)	Multiple	400 (2 760)	250 (1 720)
1½–4 (38–100)	Multiple	600 (4 140)	250 (1 720)
Lined forestry			
1 and 1½ (25 and 38)	Single	450 (3 100)	250 (1 720)
Relay supply			
3½–5 (90–125)	Single	400 (2 760)	200 (1 380)
6 (150)	Single	300 (2 070)	150 (1 030)
Pump supply (soft suction)			
4–6 (100–150)	Multiple	400 (2 760)	200 (1 380)

Service testing of attack, supply, forestry, booster, and suction hoses must be performed within 90 days of purchase and before being placed in service. An annual test is required thereafter. Occupant-use hose is service tested within 90 days of being placed in service and then at 5 years. Three-year testing intervals are then required following the 5-year test. Normally the owner of the structure where occupant-use hose is installed is responsible for conducting these tests. Fire hose manufactured after 1987 is tested to the labeled test pressure marked on the hose. The test is conducted for 3 minutes after the test pressure is achieved.

For an effective and safe fire hose service test, it is essential to thoroughly prepare the testing site **(Figure 2.33).** The test location must be convenient to the fire and emergency services organization while being removed from areas where it would interfere with traffic or pose a risk to the public or personnel. Ensure test site safety and follow the testing procedure outlined in NFPA 1962 plus any fire and emergency services organization safety guidelines. Additionally, apply the testing procedures described in NFPA 1962 to the category of hose couplings being tested. The sections that follow describe the procedures to successfully accomplish the hose and coupling service tests. Unlined fire hose no longer requires testing and is being removed from service.

Test Site Preparation

Test fire hose in a place that has adequate room to deploy the hose in straight rows, free of kinks or twists. Isolate the site from traffic and ensure that it is well lighted if testing is done at night. The site surface should be smooth and free of dirt and debris. A water source sufficient for filling the hose is also necessary. The following equipment is needed to service test hose:

- Hose-testing machine, portable pump, stationary pump or pumping apparatus that is equipped with gauges certified as accurate within 30 days before testing **(Figure 2.34)**
- Fire hose test gate valve for use during tests conducted with a stationary pump or pumping apparatus

Figure 2.33 Prepare the test site by providing a large, clear working area and excluding all spectators from the area.

Figure 2.34 Example of a hose-testing machine.

- Fire hose test report sheet to record the hose identification numbers and test results
- Tags or other means to identify hose sections that fail
- Nozzles with shutoff valves

Test Site Safety

As is the case when working with any equipment, exercise care when working with fire hose, especially when it is under pressure. Air is compressible (but water is not), and the sudden release of energy from expanding air when a pressurized hose fails during testing can result in a serious injury or fatality. Pressurized hose is potentially dangerous because of its tendency to "whip" back and forth if a break occurs. The water pressure remains constant throughout the length of the hose during the test once the air expels. The hose will not move as violently as long as the volume of water is limited. To prevent this type of accident, use a specially designed hose test gate valve or restricting disk to reduce the quantity of water available to the hose during the test **(Figure 2.35)**. The test gate valve has a ¼-inch (6 mm) hole drilled through the gate, which permits pressurizing the hose but will not allow water to surge through the hose if it fails. The same effect can be achieved by nearly closing down the pumping apparatus outlet valve after the desired pressure has been reached. Even when using the test valve, only stand or walk near the pressurized hose as necessary.

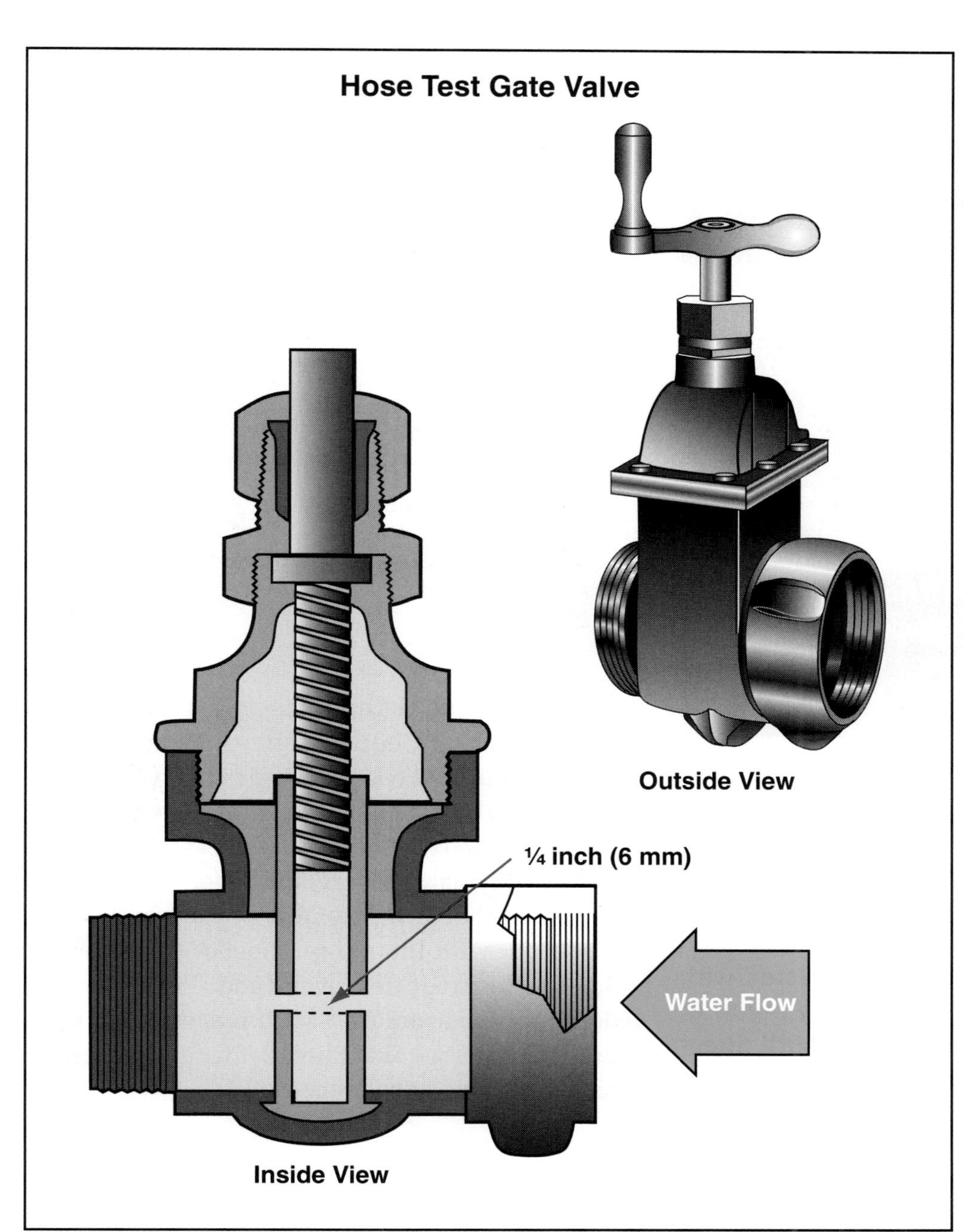

Figure 2.35 A hose test gate valve showing the location of the ¼-inch (6 mm) hole.

Open and close all valves slowly to prevent water hammer in the hose and pump. Test lengths of hose should not exceed 300 feet (91 m) in length (longer lengths are more difficult to purge of air).

Lay hose flat on the ground before charging to prevent unnecessary wear at the edges. Do not stand in front of the discharge valve connection when charging because some types of hose, especially fabric-reinforced hose, tend to twist when filled with water and pressurized. Although hose is designed to limit the possibility of uncoupling, this twisting could cause the connection to come loose.

Keep the hose testing area free of water when filling and discharging air from the fire hoses. This procedure aids in detecting minor leaks around couplings during testing.

WARNING

While visually inspecting hoselines and couplings for leaks when the hose is at the service test pressure, walk the test layout and remain 15 feet (4.6 m) to the left side (facing the free end from the pressure source) of the nearest hoseline. *NEVER* stand in front of the free end of the hose, stand on the right side of the hose, stand closer than 15 feet (4.6 m) on the left side of the hose, or straddle a hose in the test layout during the test. Check more closely for coupling slippage once the pressure has been removed from the hose.

Service Test Procedures

Fire hose service testing procedures are very similar for all types and categories of hose. Service test pressures vary for each hose type, but the fundamental process can be used for each type (see information box). Pressurized tests may be conducted with a hose-testing machine or a stationary pump or pumping apparatus fire pump. See **Skill Sheet 2-5** for annual service testing procedures using a hose-testing machine. See **Skill Sheet 2-6** for annual testing procedures using a stationary pump or pumping apparatus fire pump.

NFPA Service Test Pressure Specifications

- ***Attack hose*** — Minimum service test pressure: 300 psi (2 068 kPa) {20.7 bar} (or as marked on hose cover by manufacturer)
- ***Supply hose*** — Minimum service test pressure: 200 psi (1 379 kPa) {13.8 bar} (or as marked on hose cover by manufacturer)
- ***Occupant-use hose*** — Minimum service test pressure: 150 psi (1 034 kPa) {10 bar} (or as marked on hose cover by manufacturer)
- ***Forestry hose*** — Minimum service test pressure: 300 psi (2 068 kPa) {20.7 bar} (or as marked on hose cover by manufacturer)
- ***Suction hose (hard sleeve or hard suction)*** — Dry-vacuum test at 22 inches mercury (74.5 kPa) {0.75 bar} and hold for 10 minutes; if used on pressurized systems, minimum service test pressure is 165 psi (1 138 kPa {11 bar} (or as marked on hose cover by manufacturer)
- ***Soft sleeve hose (soft suction)*** — Test to 165 psi (1 138 kPa) {11 bar} when used to supply water to pumping apparatus fire pump intake from a pressurized water source
- ***Booster hose*** — Minimum service test pressure: 110 percent of its maximum working pressure or as designated by the manufacturer

Source: NFPA 1962, *Standard for the Inspection, Care, and Use of Fire Hose, Couplings, and Nozzles and the Service Testing of Fire Hose*, 2003 edition.

The preferred method of testing is to use a hose-testing machine. The use of other fire-rated pumps, although authorized, can cause excessive wear and strain on these devices. The moderate-to-high pressures used with very little water flow can cause overheating as well as recirculation cavitation (a condition that forms vacuum pockets and causes vibration) with the pump impeller (see sidebar). Suction hose requires several additional tools and devices to measure the vacuum developed in the drafting tube. See **Skill Sheet 2-7** for annual service testing suction hose procedures.

Recirculation Cavitation

This condition is the result of the formation of low-pressure vacuum pockets within the liquid. These pockets violently collapse at the ends of the pump impeller, causing shock waves to form that cause vibration that can damage the pump shaft. Additionally, it can cause pitting (erosion) on the face of the impeller vane, which seriously limits the efficiency of the pump.

All attack, supply, forestry, occupant-use, booster, and suction (if used on pressurized water sources) fire hose is tested at its appropriate test pressure. If a hose-testing machine is used, the desired test pressure must be held for 3 minutes once it is attained. Should a stationary fire pump or pumping apparatus fire pump be employed, the test pressure must be maintained for 5 minutes.

Unlined Fire Hose Testing

The fire and emergency services have used unlined fire hose for many years. Because it is very lightweight and relatively inexpensive, it found favor for use in hose racks for occupant use and with some forest service applications. However, the strength of this hose is very limited, making it subject to failure during aggressive fire-suppression operations. Therefore, NFPA 1962 requires that old unlined fire hose be replaced at its next annual service test date. Because this requirement was first established in 1998 and repeated again in 2003 by NFPA 1962, all unlined fire hose should be removed from service and replaced with lined hose by 2004.

Fire Hose Recoupling

Most damage to fire hose is repairable, and fire and emergency services personnel can make these repairs. Some repair work is simple such as that previously described for threads and couplings. More complex repair, such as recoupling, requires special tools but can still be easily accomplished in the fire and emergency services maintenance shop by factory-trained technicians.

If a hose bursts near a coupling, remove the portion of hose between the hose and the coupling and recouple the remaining hose, now slightly shortened. If a hole appears too far away from the coupling, however, take the hose out of service. The determining factor in deciding whether to recouple a hose or take it out of service is usually a matter of the fire and emergency services organization's policy. An example of such a rule is the following: *"If fire-fighting hose requires recoupling, it shall not be shortened to less than 90 percent of its original length."*

Some organizations find that short sections can improve their on-scene setup time, or a special operation can be more efficiently accomplished by using a nonstandard length of fire hose. It is very important to ensure that if a short section of hose is used by a fire and emergency service organization during its operations, the length must be clearly marked in a manner that is easily seen during all emergency operations.

Each procedure described in the skill sheets at the end of the chapter for expansion ring, screw-in expander, bolted-on collar, and tension-ring coupling attachments requires that the hose end be cut squarely so that it fits tightly to the coupling. Use a straightedge hose cutter or similar device to square cut the frayed edge of the hose before recoupling **(Figure 2.36).**

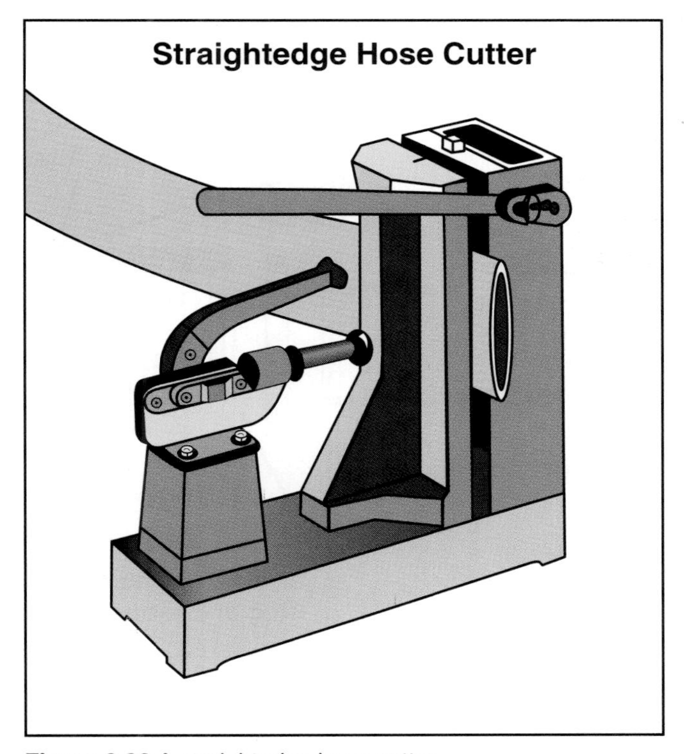

Figure 2.36 A straightedge hose cutter.

Expansion Ring

Couplings designed for attachment with expansion rings are manufactured with serrations on the inside of the shank. The components and assembly of an expansion ring coupling assembly for a set of threaded couplings are shown in **Figure 2.37** as given in NFPA 1963. The attachment procedure involves placing an expansion ring, which is slightly smaller than the fire hose, inside the end of the hose. The hose and ring are pushed into the coupling shank against a gasket, and then the ring is expanded against the hose to compress it against the coupling shank, thus affixing it permanently to the hose.

The attachment procedure requires a machine called an *expander.* Three types of expanders are available: manually operated, hand-hydraulic, and power. All expanders contain a mandrel assembly that fits inside the expansion ring and spreads to expand the ring against the hose and coupling. Each size coupling requires a different size mandrel **(Figure 2.38).**

Expansion rings vary not only in diameter but also in length **(Figure 2.39).** Because fire hose is made to withstand a higher internal pressure than non-fire hose, it requires a longer expansion ring. This length gives it more surface area to bind the hose

Figure 2.38 An example of a mandrel attached to a hydraulic expander.

Figure 2.39 Expansion rings vary in diameter and length. *Courtesy of Niedner, Ltd.*

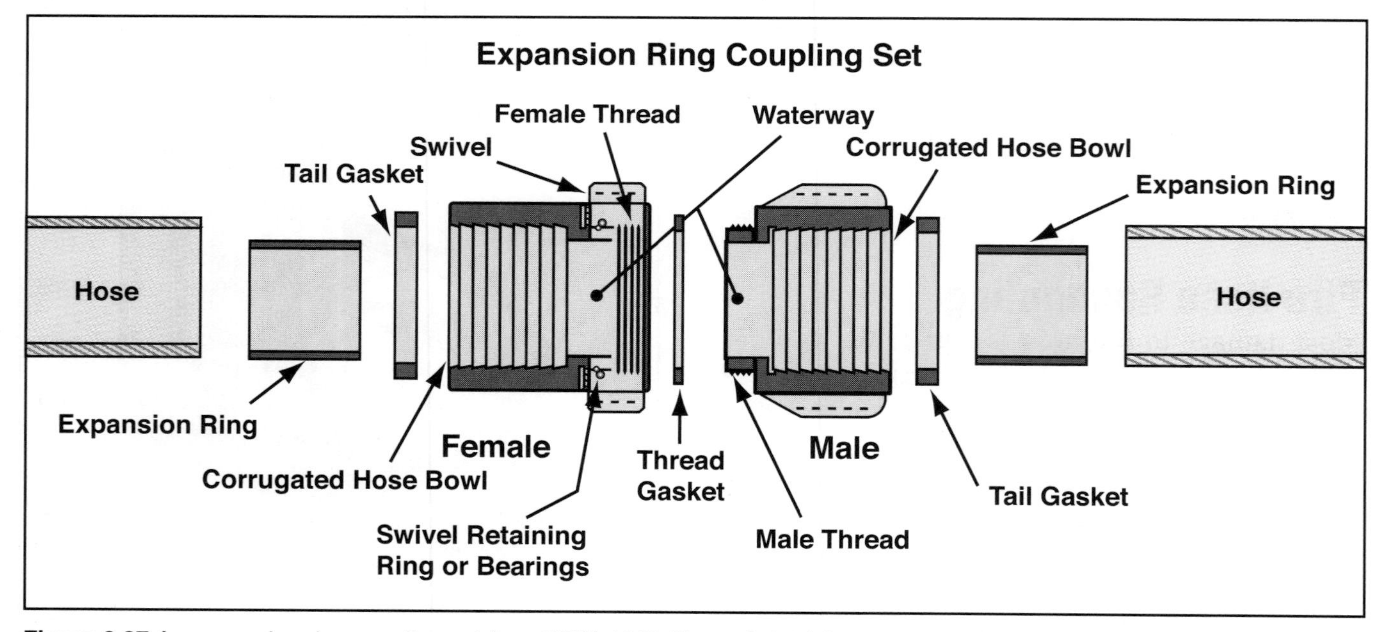

Figure 2.37 An expansion ring coupling set (see NFPA 1963 Figure A.4.1 (a), page 1963-22). *Reprinted with permission from NFPA 1963, Standard for Fire Hose Connections, Copyright © 2003, National Fire Protection Association, Quincy, MA 00269. This printed material is not the complete and official position of the National Fire Protection Association on the referenced subject, which is represented only by the standard in its entirety.*

Figure 2.40 An expansion-ring gasket *(right)* has a smaller surface area than a standard coupling gasket *(left). Courtesy of Niedner, Ltd.*

to the coupling shank. It is important, therefore, to purchase the proper type of expansion rings for re-coupling fire hose. The technician must know the maximum and minimum ring pressures to ensure a tight fit and not damage the coupling; refer to the manufacturer of the coupling for this information.

Expansion-ring gaskets that prevent seepage around the end of the hose fit between the end of the hose and the coupling. Although similar to a standard coupling gasket, the expansion-ring gasket has a smaller surface area **(Figure 2.40).** See **Skill Sheet 2-8** for attaching couplings (both male and female) to fire hose with expansion rings.

Screw-in Expander

Fire hose such as booster hose that is subjected to high pressures requires a coupling that is attached very securely. A coupling with a screw-in expander (often referred to as a *bar-way* coupling) is capable of withstanding extremely high pressure because of its design. These couplings are usually found on booster or other high-pressure hoselines. As illustrated in **Figure 2.41, p. 66**, the screw-in expander is a separate but integral component of the coupling body. It is threaded in two ways: (1) it has a fine thread along most of its length so that it may be screwed into the coupling shell and (2) it has standard threads so that the complete coupling assembly can be connected to another coupling. **Skill Sheet 2-9** gives the procedures for attaching male type couplings with screw-in expanders.

The procedure for attaching the female coupling is essentially the same as for the male coupling, except that the expander has a female-threaded swivel nut rather than male threads. Screw the expander into the coupling shell and hose until the expander reaches the bottom of the shell, and then reverse the expander until the swivel nut turns freely. The final step in the female coupling attachment process is to place a gasket in the swivel.

Bolted-On Collar

Possibly the simplest method of attaching a coupling to the hose is by using a bolted-on coupling. These couplings are designed with a shank that fits inside the hose and securely fastens with a collar that bolts into place **(Figure 2.42, p. 67).** Couplings with bolted-on collars require only an Allen wrench for attachment. See **Skill Sheet 2-10** for steps in attaching couplings with bolted-on collars.

Tension Ring

Couplings attached with tension rings (also known as *contractual sleeve couplings*) have several components: coupling body, flange ring, tension ring, and clamp ring (contractual sleeve) **(Figure 2.43, p. 67).** These couplings require a special installation tool kit. They also require a special disassembly key to remove them from the hose because prying the components apart with screwdrivers or other similar tools could result in damage. See **Skill Sheet 2-11** for the procedures for attaching couplings with tension rings.

Summary

The care and maintenance of fire hose and couplings are essential components in guaranteeing reliability during fire and emergency operations. Failure of the hose or coupling not only risks the loss of property but also the lives of fire and emergency service responders and civilians. By simple inspection, proper cleaning, and proper storage and repair of fire hose, the fire and emergency services organization can have a reasonable expectation that an uninterrupted water supply for fire suppression will be available.

Following the provisions of NFPA 1961 and 1962 (as described in this chapter) allows the fire and emergency services organization to implement standard operating procedures for a comprehensive fire hose and coupling maintenance program. Proper hose and coupling records are also vital not only to record when testing was conducted but also to track the performance of this equipment to help assist the fire and emergency services organization when new hose and coupling purchases are made. As with many other aspects of the fire and emergency services, constant vigilance and care are necessary to provide a consistent high level of service to the community.

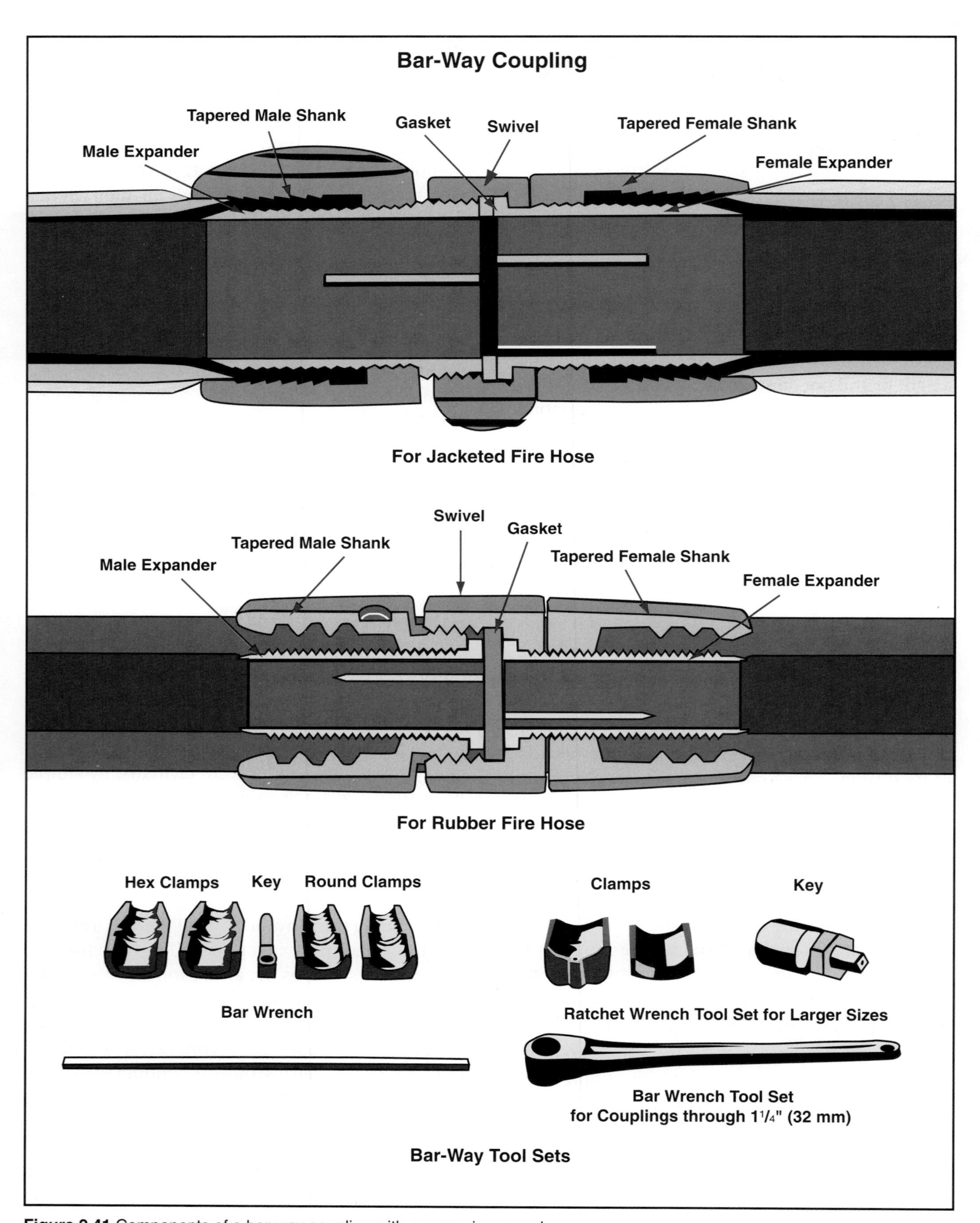

Figure 2.41 Components of a bar-way coupling with a screw-in expander.

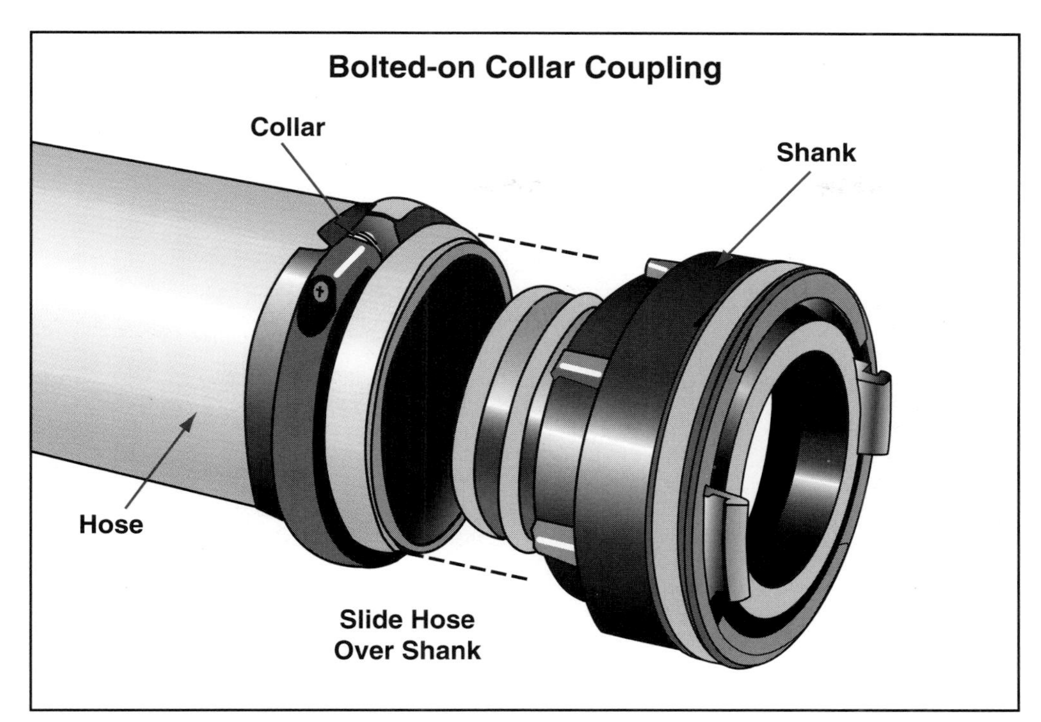

Figure 2.42 Components of a bolted-on collar coupling.

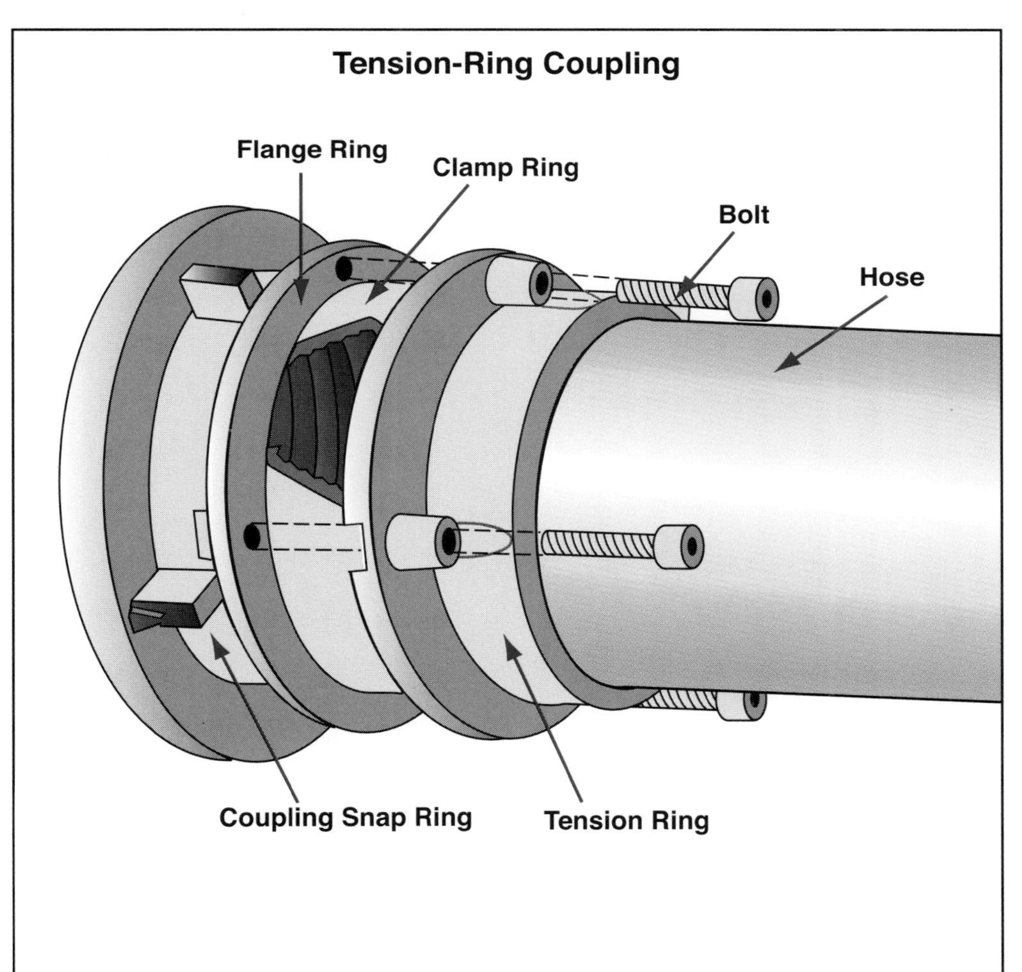

Figure 2.43 Components of a tension-ring coupling.

Inspecting Fire Hose

Skill Sheet 2-1

Step 1: Stretch fire hose to its full length on a flat, clean, and dry surface. Ensure that grease or other chemicals cannot come in contact with the fire hose. Attempt to unroll the hose rather than drag it.

Step 2: Begin at the male coupling and check the condition of the threads, the attachment to the hose, and the lugs.

2-1 Skill Sheet

Step 3: Carefully place the male coupling back on the surface, protecting the threads from impact with the ground.

Step 4: Slowly walk along the section of fire hose, visually inspecting the surface for abrasion, burns, or other damage.

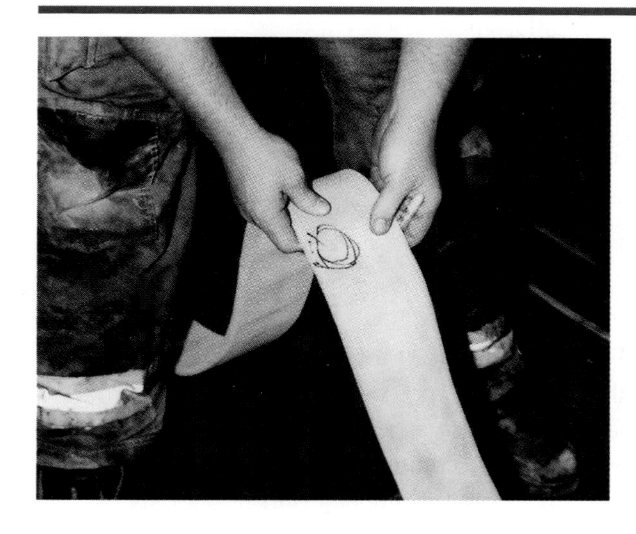

Step 5: Circle any damaged spots with a piece of chalk or other nonpermanent marker. Mark both sides of the fire hose for easy location when the hose is reversed.

Step 6: Inspect the threads, swivel, and hose coupling attachment upon reaching the female hose coupling. Also inspect the fire hose coupling swivel gasket. It should be pliable with no cracks or deep depressions caused by over-tightening.

Step 7: Turn the fire hose over to inspect the bottom side. Following the same procedure, inspect the hose back to the male coupling. Pay particular attention at locations that were marked on the other side of the hose.

2-1 Skill Sheet

Step 8: Roll the hose from the male coupling (with the male coupling on the inside) back to the female coupling once the hose is completely inspected. Roll the hose with the opposite side "out" from the way it had previously been rolled.

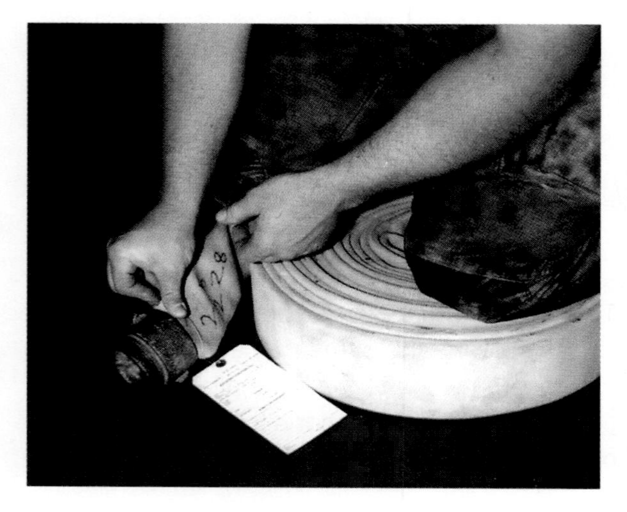

Step 9: Notice general inspection results and update the fire hose service log, noting the date of the inspection and the disposition of the individual sections:

- If the hose is damaged or has other defects noted during the inspection, tag with an out-of-service tag and remove from service until repaired and tested.
- If the hose is free of damage, return it to the pumping apparatus, storage rack, or other location for use.

Cleaning Fire Hose — Hand

Skill Sheet 2-2

Step 1: Unroll dirty hose on a clean, level, and well-drained surface. Ensure that the male coupling threads are protected. If more than one section is to be cleaned, place several sections of fire hose side by side. Allow spacing to facilitate drainage of water.

Step 2: Rinse hose with clean water. For fire hose types constructed with a rubber outer cover, rinsing may be sufficient for cleaning and return to service.

2-2 Skill Sheet

Step 3: Apply a mixture of mild soap and water.

Step 4: Scrub fire hose with a long-handled broom with medium-to-heavy bristles (see photo).

Step 5: Turn fire hose to opposite side and repeat Steps 2 through 4.

Step 6: Rinse soap from hose and prepare it for drying.

2-3 Skill Sheet

Drying Fire Hose

Hose Tower

Step 1: ***Firefighter 1 at Bottom of Tower:*** Connect end of hose to hoist system. Protect the male coupling threads.

Alternate Method:
Double the hose back upon itself. Protect the male coupling threads. Place the center of the section of hose over the hose hoist hook.

Step 2: ***Firefighter 1 at Bottom of Tower:*** Hoist hose to top of tower using a rope or electric chain hoist system.

Firefighter 2 at Bottom of Tower: Do not allow hose to drag on floor. Support couplings to protect them.

Step 3: ***Firefighter 3 at Top of Tower:*** Transfer hose from hoist hook to stationary hook rack at the top of hose tower.

Step 4: ***Firefighter 3:*** Disconnect hose from hoist system.

Step 5: ***Firefighter 3:*** Separate hose when all sections have been placed in tower to ensure even air circulation around all sections (see photo).

Step 6: Record the date and time that the hose was hung to dry.

2-4 Skill Sheet

Drying Fire Hose

Cabinet Dryer

Step 1: Place loosely rolled fire hose on wire-frame drying racks so that air can easily pass over all hose surfaces.

Step 2: ***Two Firefighters:*** Lift the sides of a single rack of wet hose and place it in the hose dryer.

Safety Tip

Lift with knees, watch hand placement, and wear gloves.

Step 3: Carefully slide the rack into the dryer, ensuring that the hose remains separated and does not "bunch-up."

Step 4: Close and secure the door. Set timer (see manufacturer's recommendations for hose).

Step 5: Record time and date hose was placed in dryer. The location of this record may vary from organization to organization.

2-5 Skill Sheet — Service Testing Fire Hose (Annual) Hose-Testing Machine

Step 1: Check the condition of the hose-testing machine. Perform this inspection each day before starting tests.

NOTE: If damage is discovered, repair damaged components before use.

> **WARNING**
> **Give all personnel a safety briefing regarding test-site safety before the test begins.**

Step 2: Perform a pressure leak integrity test as follows:

- Close or cap the fire hose outlet connection(s).
- Pressurize the machine using the integral pump to a level 10% higher than the highest test pressure required of the hose type being tested.
- With the pump turned off, maintain pressure for 3 minutes.

NOTE: If leaks are detected, remove hose-testing machine from service until repaired.

Step 3: Connect the hoseline to be tested to the outlet side of the hose-testing machine (see photo).

Step 4: Attach a test cap with a bleeder valve at the far end of each hoseline (use a nozzle with a non-twist-type shutoff if a test cap is not available).

Step 5: Secure the hose directly behind the test cap or nozzle with a rope/hose tool or other securing device.

Step 6: Slowly charge the hoseline with the test cap bleeder valve or nozzle slightly open to allow trapped air to escape. Bring pressure to 45 psi (310 kPa) {3.1 bar} ± 5 psi (34 kPa) {0.34 bar}. Once all of the air is discharged from the hose test array, close the bleeder valve(s) or nozzle(s).

WARNING

Purge all air from the hoselines in the test layout to prevent catastrophic failure that may result in personnel injury or death.

Step 7: Mark each coupling with an indelible marker directly behind the base of the coupling.

Step 8: Check each coupling for leakage with hose at 45 psi (310 kPa) {3.1 bar} ± 5 psi (34 kPa) {0.34 bar)}. Tighten with a spanner wrench when necessary.

2-5 Skill Sheet

Step 9: Secure the test area of all unnecessary personnel.

Step 10: Raise the pressure in the hoselines at a rate less than 15 psi (103 kPa) {1.03 bar} per second until the service test pressure as marked on the hose is reached (see NFPA 1962). Allow the pressure to stabilize for 1 minute per 100 feet (31 m) of hose, raising the pressure to maintain the service test pressure.

Step 11: Monitor and record the gauge pressure on the testing machine during this procedure (see photo). Ensure that the hose retains the prescribed test pressure without further pump assistance for 3 minutes following the stabilization period.

Step 12: Inspect hose for leaks. Remain 15 feet (4.6 m) from the hose at all times during the test. Stay on the left side of the hose as viewed looking towards the closed end of the test layout (see NFPA 1962).

Step 13: Terminate the test if the test layout does not hold test pressure for 3 minutes. Fail the leaking hose sections.

Step 14: Tag and mark failed hose.

Step 15: Open the bleeder valve or the nozzle after 3 minutes at the service test pressure and drain the test layout. Drain water away from dry fabric hose (see photo).

Step 16: Record the test results on a fire hose test record sheet. Place hose that passed the test in a separate location from failed hose.

2-6 Skill Sheet — Service Testing Fire Hose (Annual) Stationary Pump or Pumping Apparatus Pump

Step 1: Test-calibrate the test gauge that will be used within 30 days before service testing.

WARNING
Give all personnel a safety briefing regarding test-site safety before the test begins.

Step 2: Place a hose test valve, consisting of a gate valve with a ¼-inch (6 mm) orifice drilled through the gate, between the pump and the hose test layout.

Step 3: Connect the test layout to the hose test valve.

WARNING
Do not attach the hose test valve to a pump discharge outlet at or adjacent to the driver/operator's position.

Step 4: Secure the test valve end of the hose with a rope/hose tool or other securing device within 10 to 15 inches (254 to 381 mm) of the coupling.

Step 5: Attach a test cap with a bleeder valve at the far end of each hoseline (use a nozzle with a non-twist-type shutoff if a test cap is not available).

Step 6: Continue with Steps 5 through 16 of Skill Sheet 2-5.

2-7 Skill Sheet

Service Testing Fire Hose (Annual) Suction Hose

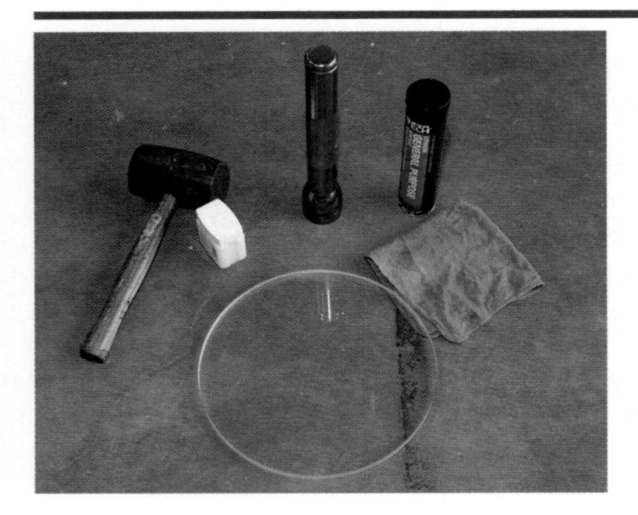

Step 1: Gather equipment required to conduct the suction hose test:

- Equipment used for testing lined hose
- Vacuum pump (if the lined hose-testing equipment is incapable of creating a vacuum)
- Vacuum gauge
- Transparent plastic disk large enough to cover the end of the hard suction hose
- Flashlight

Step 2: Close the valve between the tank and the pump, and then completely drain the pump. Cap and tighten all discharge and intake openings to prevent a vacuum leak.

Step 3: Check the gaskets to ensure that they are in place and free of dirt.

Step 4: Place a lighted flashlight inside the barrel of the pump intake with the light directed outward (see photo).

NOTE: The intake barrel should have an internal screen.

Step 5: Connect the hard suction hose to the intake and provide support so that the hose is held horizontal from the pump intake.

Skill Sheet 2-7

Step 6: Apply a small amount of heavy lubricating grease to the edge of the hard suction coupling to help seal the transparent disk to the coupling when creating a vacuum.

Step 7: Place the transparent disk over the male end of the hose.

Step 8: Prime the pump until at least 22 inches (559 mm) of mercury read on the vacuum gauge.

Step 9: Discontinue priming, turn off the engine of the apparatus, and listen for vacuum leaks. Monitor the vacuum gauge. The hose is to hold the vacuum for 10 minutes.

2-7 Skill Sheet

Step 10: Look through the transparent disk into the hose. Examine the lining for any indication that it is coming loose and protruding into the waterway (see photo).

Step 11: Open the pump drain slowly after 10 minutes to allow air to reenter the pump. Disconnect the hard suction hose and remove the flash-light. Replace the intake cap and prime the pump to prepare it for normal service.

Step 12: Record the test results for each section of hard suction hose.

Attaching Couplings
Expansion Ring

Skill Sheet 2-8

Step 1: Install the segment assembly into the nosepiece of the expander machine. The tapered drawbar threads as well as the segment holder must reach the bottom of the nosepiece.

Step 2: Select the suggested pressure from the pressure chart provided by the manufacturer and preset the designated pressure by using the pressure regulator knob. Pressure requirements vary according to the type of hose as well as whether the couplings are made of brass or aluminum.

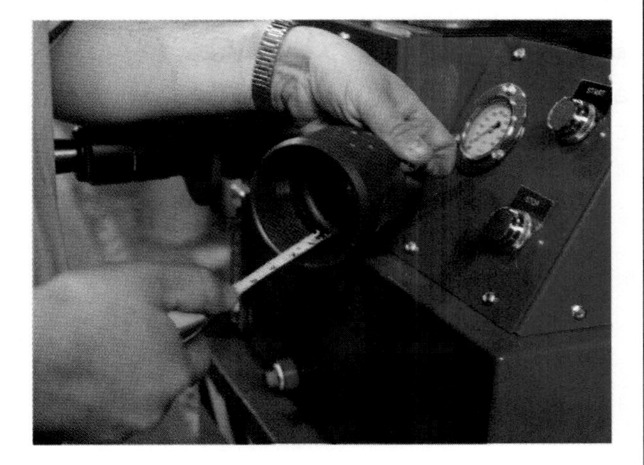

Step 3: Measure the length of the coupling waterway, and then add 1⁄16 of an inch (1.6 mm) to the measurement.

2-8 Skill Sheet

Step 4: Rotate the adjusting collar to set the measured distance on the machine. Measure to the back of the adjusting collar for female couplings; measure to the lip of the adjusting collar for male couplings.

Step 5: Place a coupling over the segment assembly so that the coupling face fits squarely against the adjusting collar. Expand the segments against the inside of the coupling to check that the measurement is correct. The segments should contact only the inside of the bowl and not touch the waterway.

Step 6: Place the expansion ring over the segment assembly so that the ring sets flush against the segment holder.

Step 7: Insert a backup gasket into the coupling.

Step 8: Push the hose into the coupling bowl. Make sure that the hose end is tight against the gasket.

Step 9: Place the hose and coupling over the segment assembly and expansion ring so that the coupling face fits squarely against the adjusting collar (see photo).

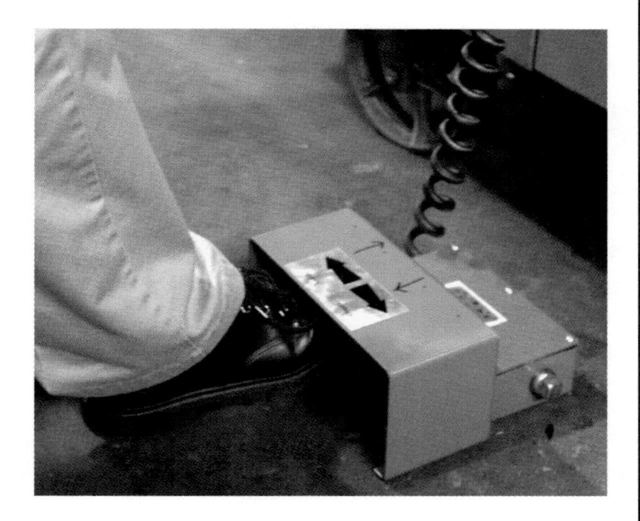

Step 10: Engage the hydraulic expansion control and operate until fully expanded (see photo).

Step 11: Engage the hydraulic retraction control and reverse the mandrel until the coupling can rotate (the expanders retract). Rotate the coupling a quarter-turn.

Step 12: Engage the expansion control and expand again. This procedure ensures a snug, tight fit.

Step 13: Retract the expander until the coupled hose can be removed.

2-8 Skill Sheet

Step 14: Inspect the expansion ring to be sure it is slightly indented with the outline of the mandrel segments and that the rubber behind the expansion ring indicates some rolling effect due to compression (see photo).

Step 15: Check the waterway of the coupling to be sure that the segments did not indent the waterway. If indenting is visible, the locating plate was not adjusted properly.

Step 16: Look for hairline fractures, a rippling effect, or distortion on the outside of the coupling. Such damage indicates that too high a pressure was used.

Step 17: Test the recoupled hose section at the recommended service test pressure to confirm that a satisfactory attachment has been made.

Attaching Couplings
Screw-in Expander

Skill Sheet 2-9

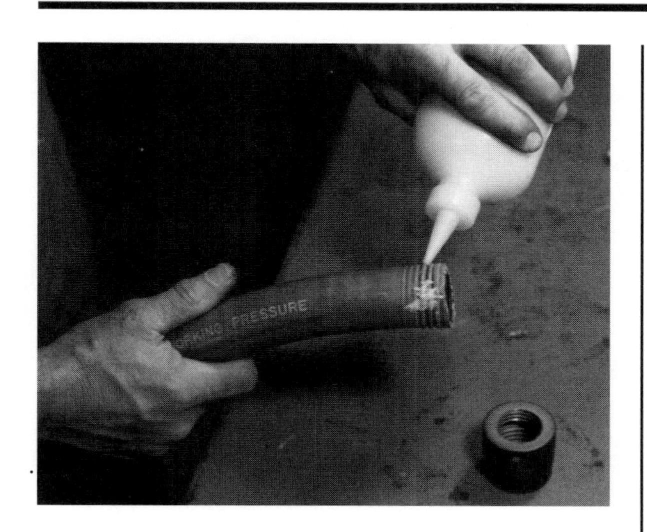

Step 1: Lubricate the shell and hose end with a soapy water solution or lubricant. Use an inert oil-based or silicone-based lubricant as recommended by the coupling manufacturer.

Step 2: Screw the end of the hose into the shell until the hose is seated in the shell.

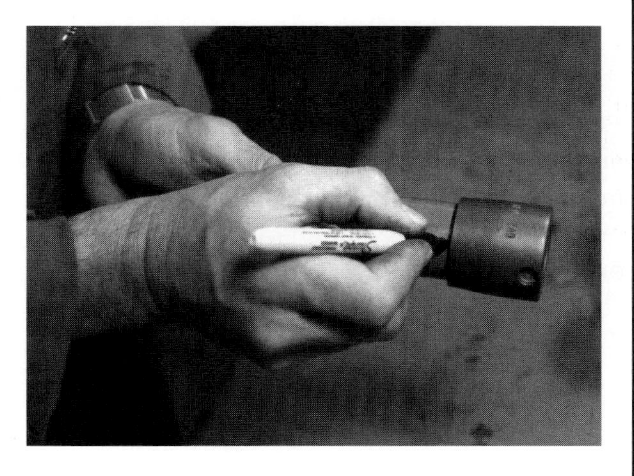

Step 3: Mark the hose at the end of the shell with indelible ink. This mark will be used later to check for slippage.

2-9 Skill Sheet

Step 4: Place the clamps around the shell.

Step 5: Put the clamps, shell, and hose assembly into a vise. Tighten the vise jaws against the clamps until the shell cannot turn.

Step 6: Place the expander on the key so that the key engages the interior grooves and protrudes from the leading end of the expander. This protrusion serves as an entrance guide for the expander.

Step 7: Lubricate the threaded exterior of the expander and the interior of the hose end.

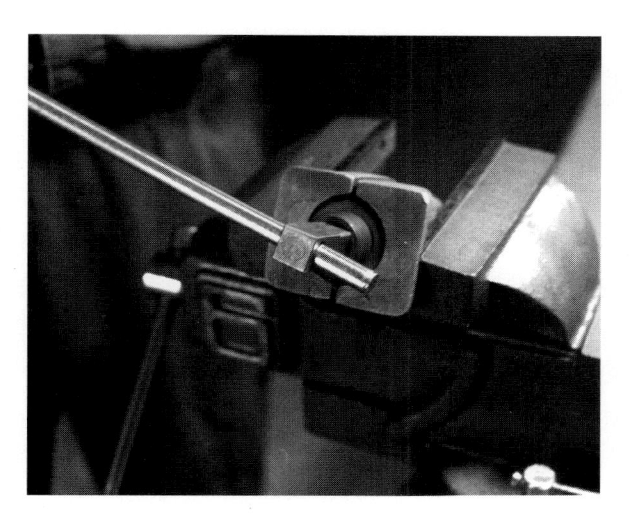

Step 8: Screw the expander into the shell and hose until the expander reaches the bottom of the shell. When driving the expander into the hose and shell, ***do not***

— Interrupt the rotations once they have started.

— Permit the hose to rotate.

— Allow the key to push up.

Step 9: Release compression on the vise, remove the clamps, and inspect the coupling for slippage. The coupling should not have moved away from the scribe mark.

All Skill Sheet 2-9 photographs except 2-9.9 are courtesy of Weis American Fire Equipment, Oklahoma City, OK.

2-10 Skill Sheet Attaching Couplings Bolted-on Collar

Step 1: Slide the hose over the coupling shank.

Step 2: Fit the collar over the hose and the shank.

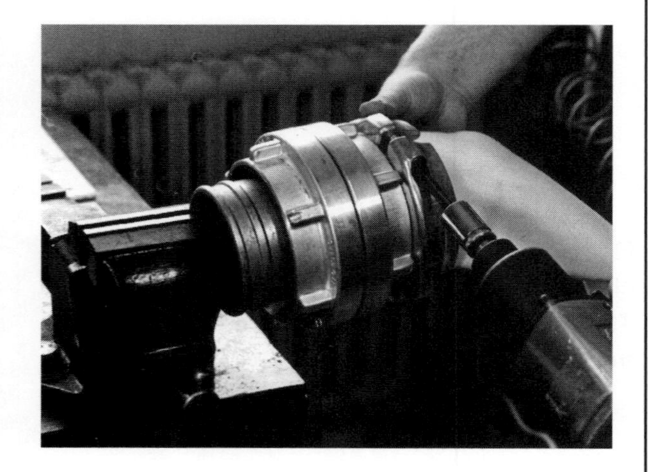

Step 3: Insert the bolts and torque using a small torque wrench with approximately 40 foot-pounds or 54.4 Newtons/meter (joules) of torque.

All Skill Sheet 2-10 photographs are courtesy of Neidner, Ltd.

Attaching Couplings
Tension Ring

Step 1: Slip the tension ring and flange ring over the end of the hose, and then slide the hose onto the coupling shank so that it butts against the coupling snap ring.

Step 2: Move the flange ring up against the snap ring, and then place the lubricated nylon clamp ring over the hose behind the flange ring (making sure the tapered side of the clamp ring is toward the tension ring). Push the tension ring onto the clamp ring as far as it will go.

Step 3: Install the two starter bolts, and then tighten the bolts evenly to draw the tension ring up to within ¼ inch (6 mm) of the flange ring.

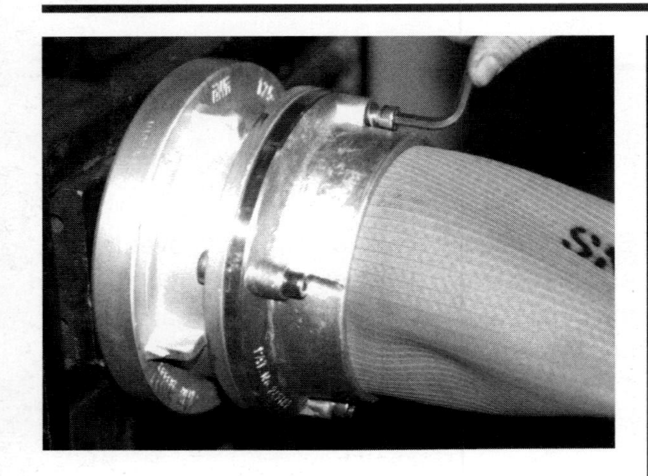

Step 4: Remove the starter bolts and install the four tension ringbolts. Tighten the bolts evenly to draw the tension ring as far as it will go toward the flange ring.

NOTE: The tension ring may not move completely flush against the flange ring. If the flange ring starts to bend, ***do not*** continue to tighten the bolts.

All Skill Sheet 2-11 photographs are courtesy of Snap-tite, Inc.

Chapter 3
Fire Hose Nozzles

Chapter 3
Fire Hose Nozzles

A *nozzle* is a device (hose appliance) that directs water, foam, or other extinguishing agent from a hose to a point required for fire extinguishment or exposure protection. It forms the extinguishing agent into a fire stream and controls the agent so that it is applied in the most efficient manner (that is, by using a minimum amount of extinguishing agent with the least amount of collateral damage resulting from its application). The nozzle, under pressure, also forms the stream pattern and regulates the quantity of water or extinguishing agent passing through it.

NFPA 1964, *Standard for Spray Nozzles* (2003), describes the minimum performance requirements for the construction of fire and emergency services fog nozzles. Smoothbore nozzles are not included in this standard because they are not complicated design nozzles. They are, however, presented within the context of a class of nozzles vital to numerous fire-suppression activities.

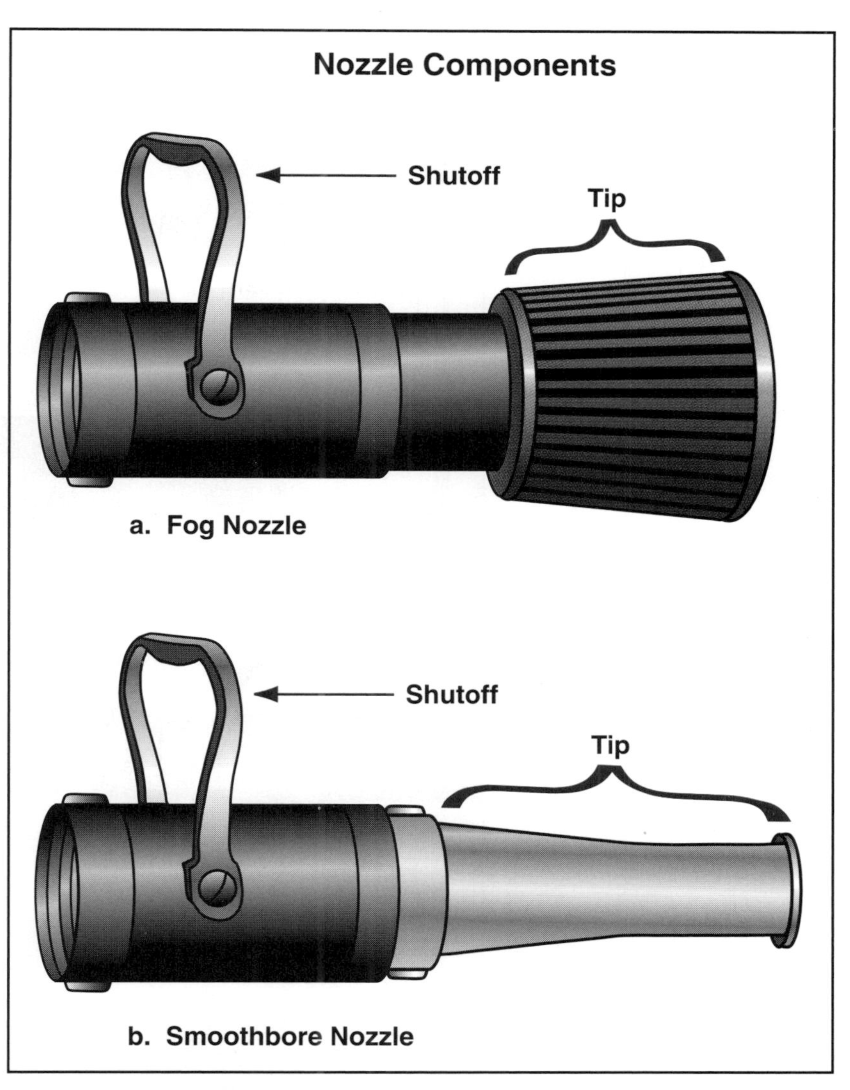

Figure 3.1 Two basic nozzle components: shutoff valve and tip.

A nozzle generally consists of two major components: *shutoff valve* and *tip* **(Figure 3.1).** The shutoff valve provides a means of not only opening and closing the nozzle (turning the extinguishing agent on and off), but in some cases it is a means of controlling the quantity of agent that flows through the nozzle tip. The shutoff valve can also be used to select the type of stream that will be produced by the nozzle. The tip is the component that forms the stream and creates its pattern. A nozzle is a precisely engineered device that directs the water or extinguishing agent to the area of application, much the same as a rifle barrel directs a bullet to its target. In some basic nozzle designs, the shutoff valve and the tip are distinctly separate; in others, they are combined as inherent parts of the nozzle.

Four general categories of fire and emergency services nozzles exist — *solid stream, fog stream, foam,* and *multipurpose* **(Figure 3.2).** Each is designed to produce a particular type of fire stream or stream pattern. The marine all-purpose nozzle (also know as *Naval All-Purpose (NAP)* or *Rockwood nozzle*) can produce two of the stream types (solid and fog) with one nozzle design. The Akron® SaberJet™ nozzle is in the combination nozzle category and the only one that can produce the solid and fog streams separately or simultaneously **(Figure 3.3).** All other nozzles deliver only one of these stream types. For example, a fog nozzle can deliver a straight pattern stream, but the straight pattern is not classified as a solid stream. A solid stream nozzle can deflect a fire stream from a ceil-

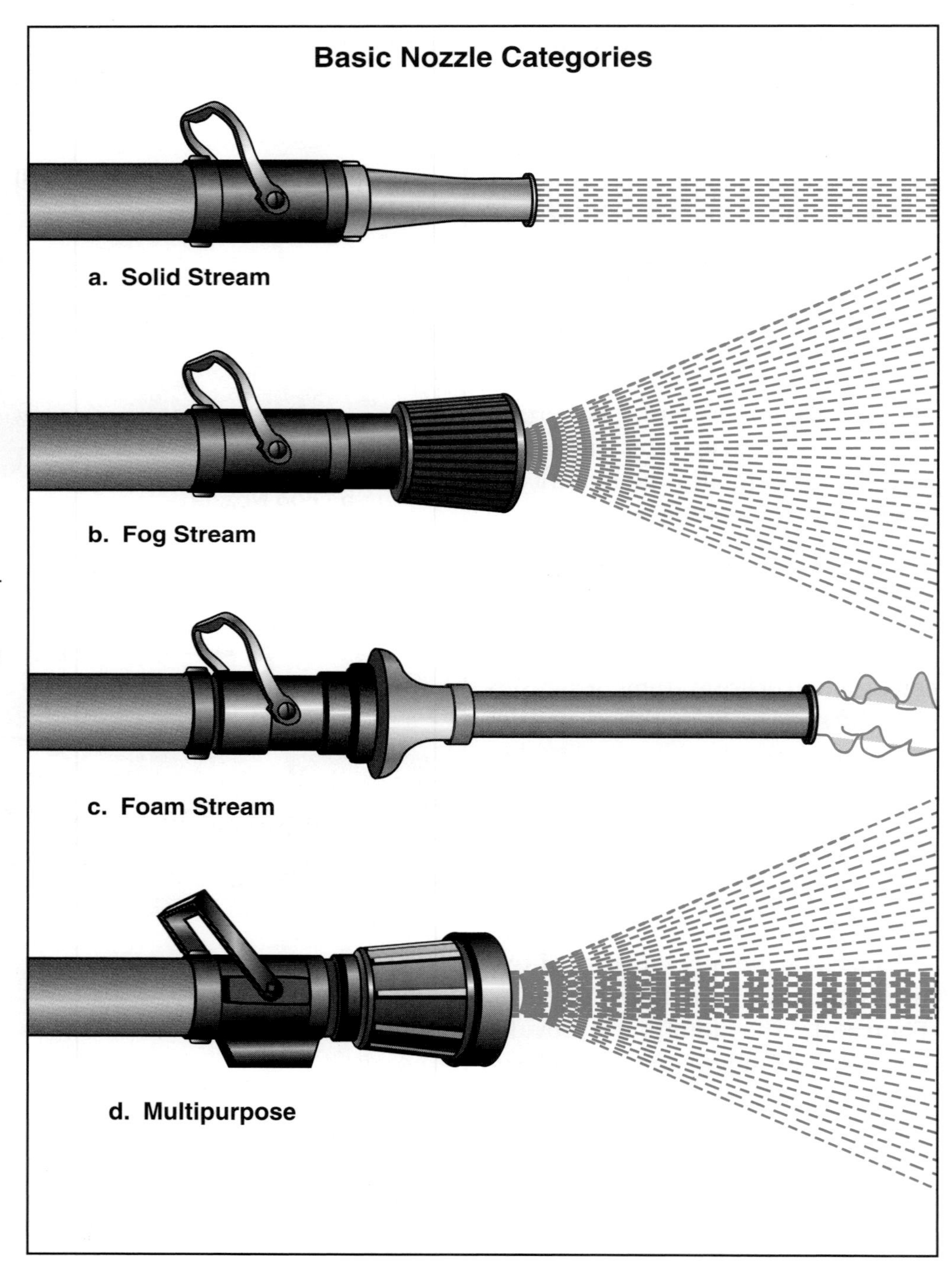

Figure 3.2 Four basic categories of fire service nozzles: (a) Solid stream, (b) Fog stream, (c) Foam stream, and (d) Multipurpose.

Figure 3.3 The Akron® SaberJet™ nozzle can deliver both solid and fog streams at the same time or independently. *Courtesy of Akron Brass Company.*

ing or other object to create a foglike effect, and the nozzle operator can rotate a broken stream nozzle in a way so that it simulates fog hoseline qualities. Each of these nozzle categories retains its own unique design qualities and, in some instances, limitations.

Nozzles are also classified by the quantity of extinguishing agent they are designed to deliver. The three stream flow classifications include *low-volume, handline,* and *master streams* (see information box). The nozzle design and the pressure of the water or other extinguishing agent as it flows from the nozzle determine the rate of discharge **(Figure 3.4).**

Fire Streams

- ***Low-volume stream*** — Fire stream that discharges less than 95 gpm (360 L/min). Nozzles producing these streams include those fed by booster hoselines.
- ***Handline stream*** — Fire stream that is supplied by a hoseline sized from 1½ to 3 inches (38 mm to 77 mm) with a fire flow ranging from 95 to 350 gpm (360 L/min to 1 325 L/min). Many fire and emergency service organizations have elected to "standardize" the 1¾-inch (45 mm) handline as the minimum size for deployment during fire-suppression operations.
- ***Master stream*** — Large-volume fire stream that discharges more than 350 gpm (1 325 L/min) and is fed by multiple 2½- or 3-inch (65 mm or 77 mm) or larger supply hoselines connected to a master stream nozzle.

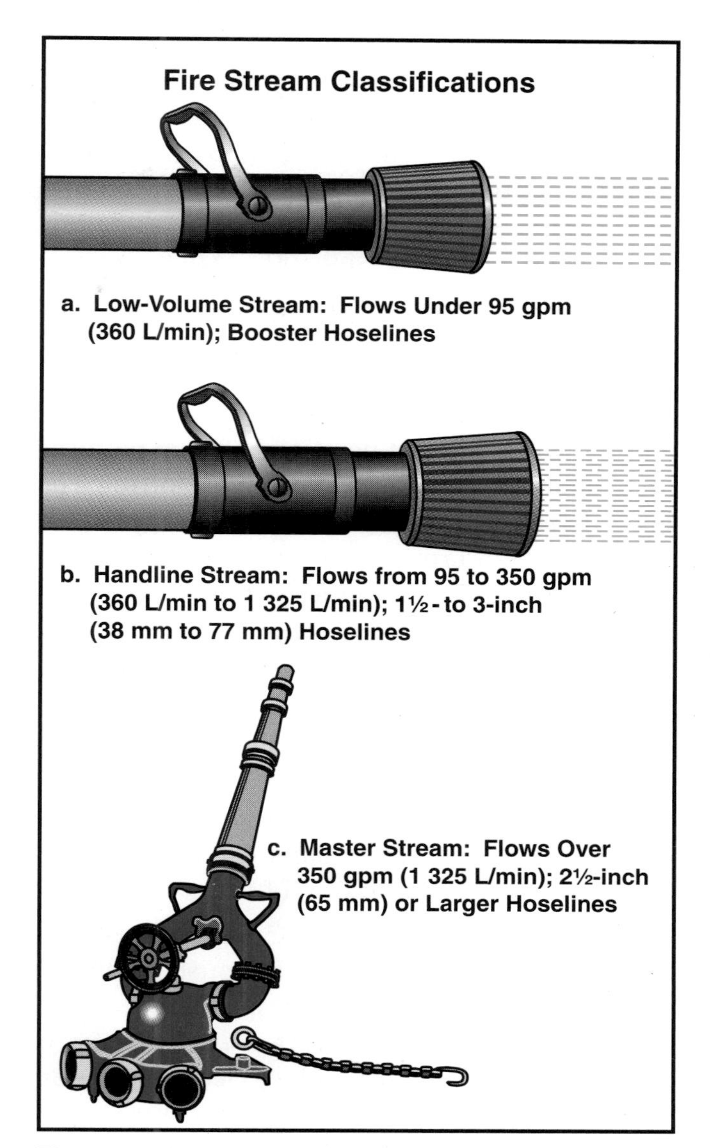

Figure 3.4 Fire stream flow classifications: (a) low-volume stream, (b) handline stream, and (c) master stream.

Low-volume streams (below 95 gpm [360 L/min]) do not deliver fire streams that are adequate or safe to use during interior structural fire attack. Low-volume streams are limited to applications that give fire and emergency service responders an easy-to-maneuver hoseline while not being relied upon to provide life-safety protection during fire-suppression activities.

Booster hoselines with interior diameters of ¾ inch (20 mm) or 1 inch (25 mm) equipped with a fog nozzle can produce 15, 30, or 45 gpm (57 L/min, 114 L/min, or 170 L/min) at 100 psi (689 kPa) {6.89 bar}. The same is true for a 1-inch (25 mm) forestry nozzle operating at 100 psi (689 kPa)

{6.89 bar}, which can produce up to 60 gpm (227 L/min). Solid stream nozzle tips are also available that produce low-volume flows. Standard sizes include ⅜-, ½-, and ⅝-inch (10 mm, 13 mm, 16 mm) nozzle tips, producing stream flows under 100 gpm (379 L/min). These sizes are found on 1½- to 1¾-inch (38 mm to 45 mm) shut-off nozzles.

Handline nozzles and hoselines are deployed by fire-suppression crews and maneuvered by hand to positions where the streams are directed at a fire or placed in a position to stop the advance of a fire. A wide range of flows and stream patterns are available, including solid, straight, and fog. The flows delivered (ranging from 95 to 350 gpm [360 L/min to 1 325 L/min]) are capable of rapid fire extinguishment when properly applied.

Master stream nozzles produce large-caliber streams (more than 350 gpm [1 325 L/min]) that are generally not moved once deployed. These streams deliver large quantities of water often in a defensive manner. Applied from deluge sets, master stream devices (also known as *multiversals*), monitors, or elevated devices, these streams are capable of delivering up to 2,000 gpm (7 571 L/m) of flow.

As water or extinguishing agent discharges from any nozzle, a force generates that pushes back on the fire and emergency service responders handling the hoseline. This counterforce or *back pressure* is known as *nozzle reaction* and is a clear illustration of Newton's Third Law of Motion: *"For every action there is an equal and opposite reaction."* The greater the nozzle pressure (the measured velocity of the water or other agent passing through the nozzle orifice) and the quantity (volume or mass) of water being discharged, the greater the resulting nozzle reaction will be **(Figure 3.5).**

Figure 3.5 Nozzle reaction is a result of the velocity and quantity of the fire steam being flowed through the nozzle and limits the emergency responder's ability to maneuver the hoseline.

This chapter describes those hose appliances identified as fire hose nozzles, their categories, and their uses. This information is a guide to fire and emergency service responders when selecting the appropriate nozzle for a given situation. Characteristics of each nozzle and how they interact with the various hose types are presented, defining the positive and negative aspects of each. Descriptions of the flows and nozzle reaction forces that can be expected with solid stream and fog nozzles are also given. Foam nozzles and the various types of foam streams and finished foam they produce are discussed, matching the expected performance of each type of nozzle with the foam solution being discharged. Finally, multipurpose nozzles are described, and their individual performance characteristics are defined.

Solid Stream Nozzles

A solid stream nozzle is the oldest type of nozzle in the fire and emergency services. The simplest type of solid stream nozzle consists of a shutoff valve and tip (known as a *smoothbore nozzle*). The fire stream emerges from the nozzle orifice in a compact "solid stream" or solid cylinder of water or other extinguishing agent with little spray or showering effect **(Figures 3.6 a and b).** The shape of the stream is gradually reduced once it leaves the fire hose and travels through the nozzle.

One of the most common variations of the nozzle has a shutoff valve with two or more sizes of tips "stacked" in descending order **(Figure 3.7).** This stacking feature permits the nozzle operator to remove one or more tips to achieve the desired rate of fire flow. Master stream smoothbore nozzles are constructed in a similar manner (see Master Stream Solid Stream Nozzle section).

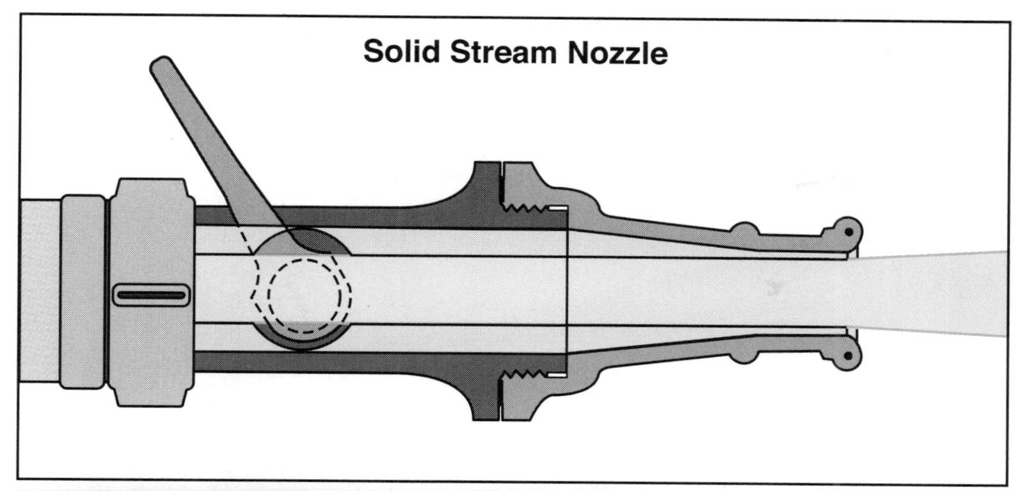

Figures 3.6 a and b *Top* (a) Basic solid stream nozzle design: a shutoff with a tapered bore to shape the stream. *Bottom* (b) A solid stream nozzle delivers a solid cylinder of water or extinguishing agent that is shaped by a tapered nozzle.

The inside diameter of the tip corresponds to a specific flow in gallons per minute (gpm) or liters per minute (L/min) at a nozzle pressure of 50 psi (345 kPa) {3.45 bar}. Nozzle pressure depends on the type of device used. Typically, smoothbore tips on handline nozzles are designed to deliver an optimum stream at 50 psi (345 kPa) {3.45 bar}, and master streams operate best at 80 psi (552 kPa) {5.5 bar}. Some fire and emergency service organizations have selected 70 psi (483 kPa) {4.8 bar} for use with their master stream devices.

An optimum solid stream is one that remains intact, without fragmentation for a specific distance. As a stream loses its forward velocity, referred to as the *breakover point,* its effectiveness is lost **(Figure 3.8, p. 106).** Air and wind resistance overcome the continuity of the stream, breaking it into showers of spray. This distance varies with the size of the tip and nozzle pressure. It is difficult to calculate just exactly where a solid stream will begin to lose its forward velocity at the time of use. As long as the stream maintains its original shape without breaking into spray and mist, it is considered an effective solid stream **(Figure 3.9, p. 106).**

Figure 3.7 To vary the possible rate of flow through a solid stream nozzle, several tip sizes are "stacked" on a shutoff.

An advantage of a solid stream nozzle and the stream produced by it is that the stream can penetrate a mass of burning material of a well-developed fire. Another advantage is that it forms a stream of water capable of reaching a greater distance than that produced by other nozzle

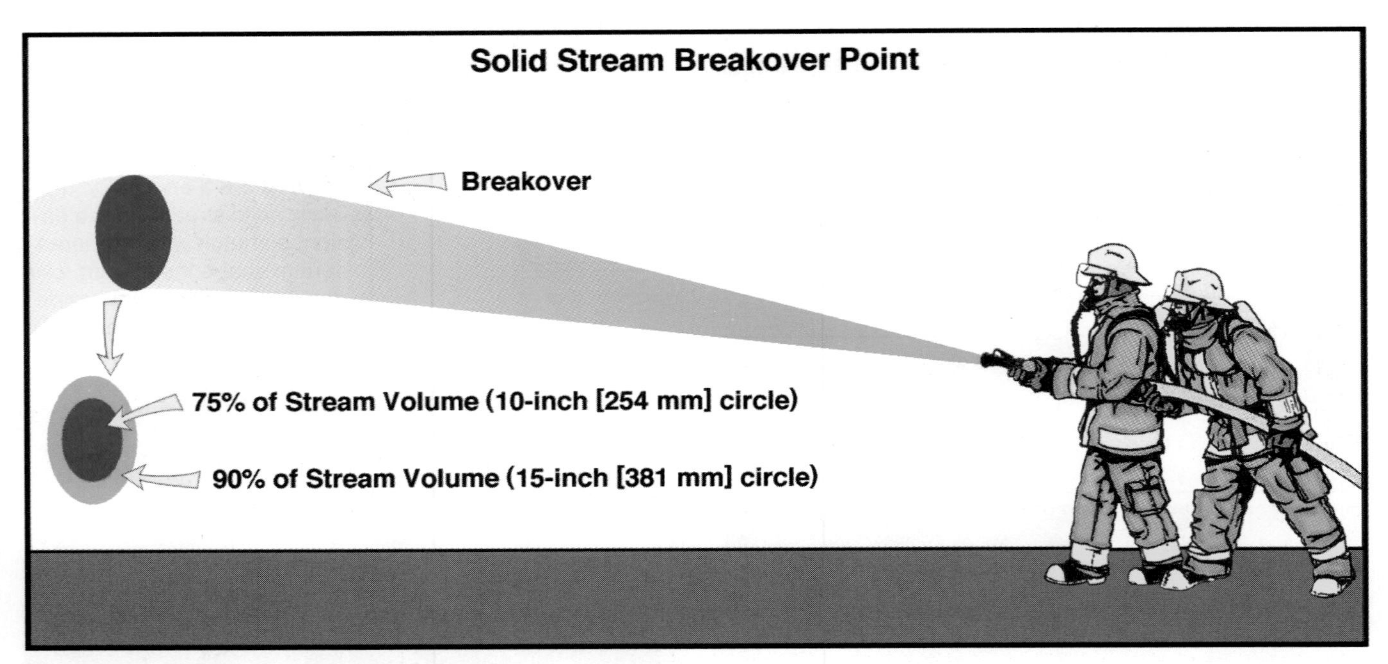

Figure 3.8 The breakover point is that point where a solid stream begins to lose its forward velocity and starts to break up. The stream contains 90 percent of its volume inside a 15-inch (381 mm) circle and 75 percent inside a 10-inch (254) circle.

Figure 3.9 An experienced nozzle person can determine the breakover point by closely observing the fire stream.

types. Because the stream retains its integrity as a column of water, wind and air resistance have a reduced effect on it over longer distances. A 15- to 20-percent increase in horizontal reach can be expected over other stream types. The reach of a fire stream is of great value when interior fire attack is unfeasible because a fire is exceptionally hot or the involved materials cannot be approached safely.

Although these nozzles are relatively simple in design, many factors can adversely affect their performance. Gravity, air resistance, and the velocity of the wind affect the range or reach of the stream. Roughness or deep scratches on the interior surface of the nozzle or dents deforming the orifice also decrease its performance.

Because of the compactness of the solid stream and velocity of the water as it leaves the nozzle orifice, fire and emergency responders directing the stream must exercise great care because damage can result should the stream be misapplied. Uninvolved areas adjacent to the location of a fire can be severely damaged by the force of the stream and large volume of water that can be delivered. Caution must also be taken when solid streams are deployed near electrical equipment. The electrical current can conduct along the stream, causing injury or death to the hoseline crew.

WARNING
Solid streams can conduct electrical current causing injury or death to the hoseline crew.

The smoothbore tip can be used on both handline nozzles and master stream devices such as monitors and ladder pipes. Although small-diameter smoothbore tips were available before the advent of the fog nozzle, they are not commonly used today. The smallest tip that is commercially available is the ½-inch (13 mm) tip; it is found in some standpipe applications and as a breakdown/overhaul tip in combination with the ⅞- or 15⁄16-inch (22 mm or 23 mm) tip. This tip is associated with the Fire Department of New York (FDNY) **(Figure 3.10).** It is a versatile, high-flow, low-maintenance nozzle that provides a volume of flow coupled with reduced nozzle reaction when it is compared to equivalent fog-stream nozzles. When selecting a smoothbore tip size, a quick guideline is that the tip size should not exceed one-half the diameter of the hose. Some tip sizes "cheat" on this guideline and exceed this limit by a 16th of an inch (1.58 mm), allowing an approximate 10 percent increase in flow from the hoseline.

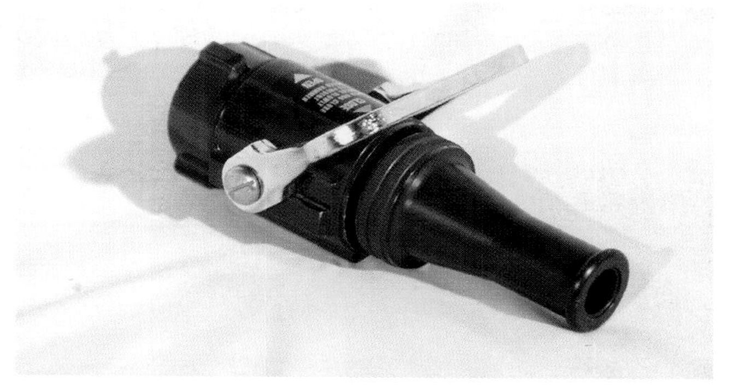

Figure 3.10 The FDNY successfully uses a 15⁄16-inch (23 mm) tip with a 1½-inch (38 mm) ball shutoff. This tip's advantage is that it gives high water flow with the equivalent nozzle reaction of fog nozzles.

Knowledge of the various uses of solid stream nozzles and their special attributes allow them to be used efficiently **(Figure 3.11).** Solid stream nozzles are used in low-volume, handline, and master stream operations, and these nozzles are described in the sections that follow. Information on nozzle reactions when using solid stream nozzles is also given.

Figure 3.11 Examples of solid stream nozzles. *Courtesy of Elkhart Brass Manufacturing Company, Inc., Task Force Tips, and Akron Brass Company.*

Low-Volume Solid Stream Nozzle

As mentioned earlier, low-volume solid stream nozzles are not commonly used for service today. However, in combination with a larger, handline-sized tip (one that delivers over 100 gpm [379 L/min]), smaller, low-volume tips are found. A ½-inch (13 mm) tip is "stacked" with a ⅞-, 15⁄16-, or 1-inch (22 mm, 23 mm, or 25 mm) smoothbore tip. These small tips are used as overhaul or cleanup tips but ***not*** in fire-attack or suppression capacities. See **Table 3.1** for expected flows from various tip sizes.

Tip Size (fraction inches)	Tip Size (decimal inches)	Tip Size (millimeters)	gpm (at 50 psi)	L/min (at 345 kPa) {3.45 bar}
3/8	0.375	9.52	29.54	113.60
1/2	0.50	12.70	52.52	202.17
5/8	0.625	15.88	82.06	316.09

Table 3.1
Low-Volume Solid Stream Tip Sizes and Flows

Handline Solid Stream Nozzle

The flow produced from a solid stream nozzle is dependant upon the diameter (size) of the tip or orifice. Handline tip sizes range from 3/8 inch (10 mm) to 1¼ inches (32 mm), increasing in 1/8-inch (3 mm) increments **(Figure 3.12).** The pumping apparatus driver/operator controls the flow from the fire pump to the nozzle and calculates the pump pressures for the flows produced by each tip size. See **Table 3.2** for expected flows from various tip sizes. These calculations ensure that adequate and correct pressure is continuously delivered to the nozzle. This pressure varies from tip size to tip size, by the lengths of hose deployed in the attack line, and by the elevation of the nozzle relative to the fire pump. The hoseline crew must also carefully tend the attack hoseline. Kinks in the line and numerous turns in the hose affect the flow to the nozzle. The low operating pressure of these attack hoselines often contributes to a greater tendency for kinks in the hose at corners or other sharp turns.

Master Stream Solid Stream Nozzle

Master stream solid stream nozzles are found on monitor or deluge sets **(Figure 3.13).** Monitors or deluge sets can be mounted on many types of fire apparatus including pumping apparatus, aerial devices, mobile water supply vehicles (tankers or tenders), and at other land-based locations. They are also mounted on marine fire-fighting vessels (fireboats) and marine tank vessels or are permanently mounted as stationery monitors (manually or remotely controlled units) on docks and at industrial complexes. They can also be designed to be portable monitor sets that are placed in the street and supported with supply hoselines from a pumping apparatus. Master stream device discharges range from 350 gpm (1 325 L/min) to massive amounts beyond 2,500 gpm (9 464 L/min)

Table 3.2
Handline Solid Stream Tip Sizes and Flows

Tip Size (fraction inches)	Tip Size (decimal inches)	Tip Size (millimeters)	gpm (at 50 psi)	L/min (at 345 kPa) {3.45 bar}
3/4	0.750	19.05	118.17	454.88
7/8	0.875	22.225	160.84	619.14
15/16	0.9375	23.8125	184.64	710.75
1	1.00	25.40	210.08	808.68
1 1/8	1.125	28.575	265.88	1 023.48
1 1/4	1.25	31.750	328.25	1 263.56

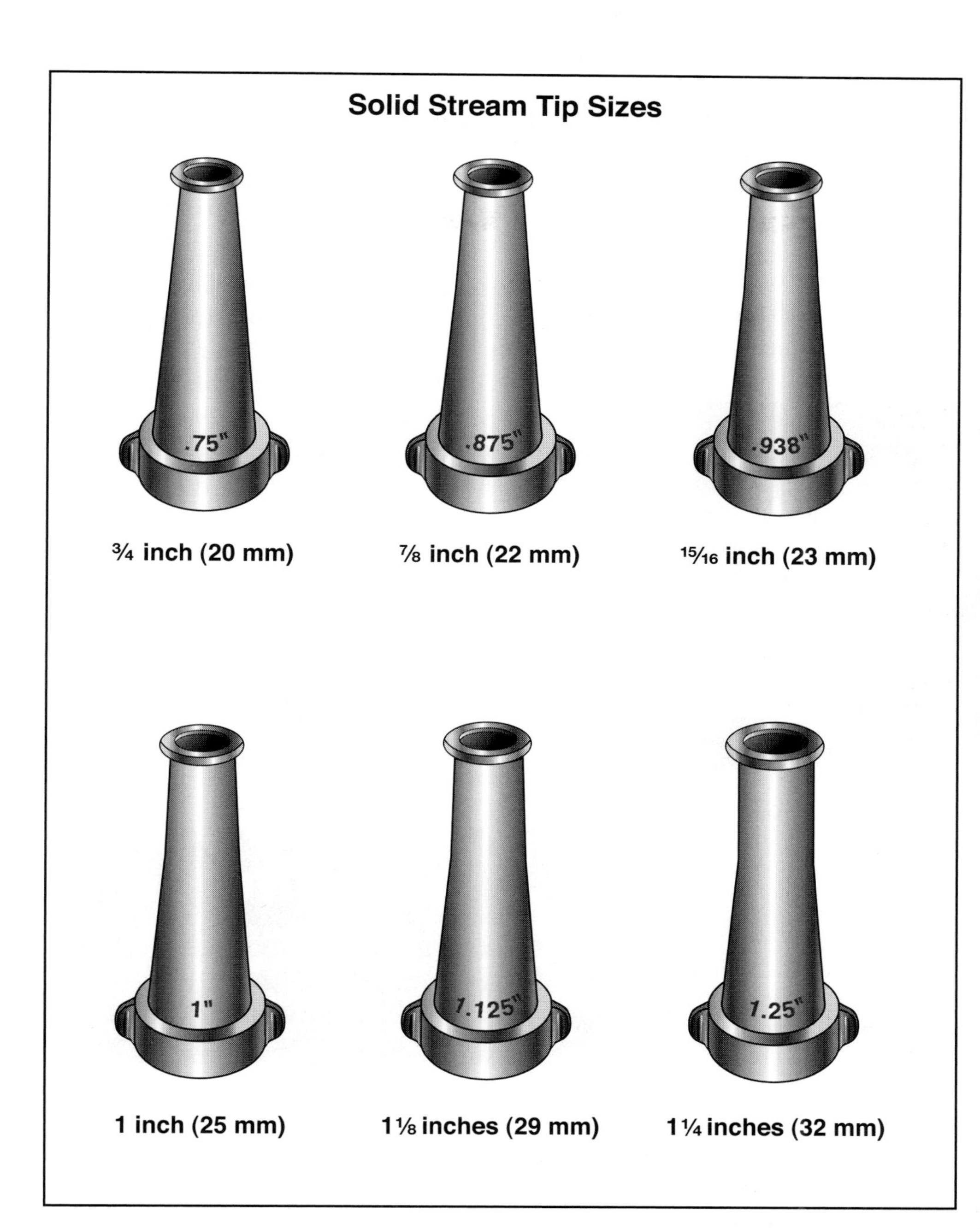

Figure 3.12 Handline solid stream tip sizes range from ⅜ inch (10 mm) to 1¼ inches (32 mm). The fire flow from these handlines is adequate for interior structural fire-suppression efforts.

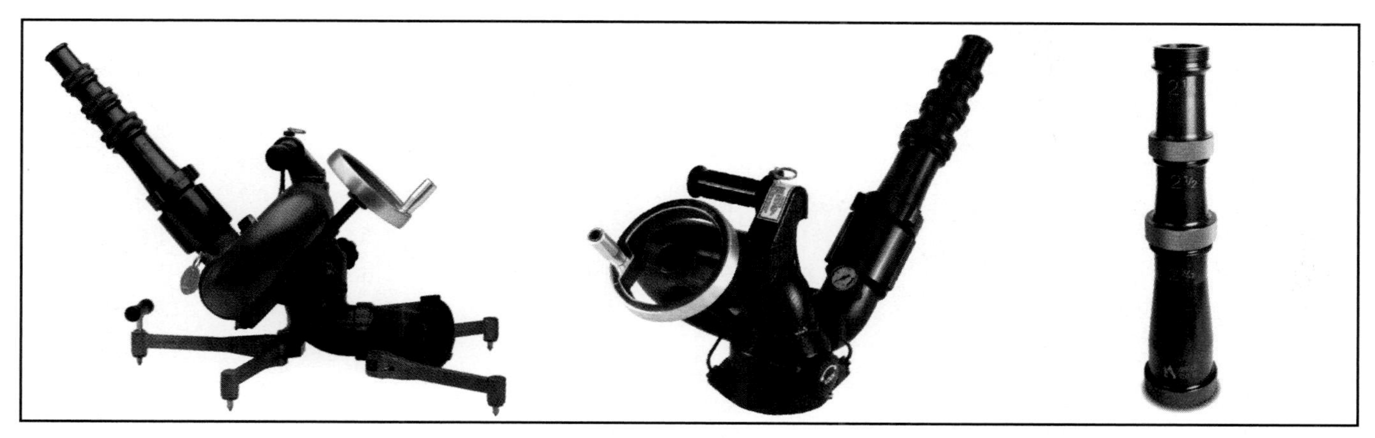

Figure 3.13 Examples of master stream nozzles. *Courtesy of Elkhart Brass Manufacturing Company, Inc. and Akron Brass Company.*

on marine fire-fighting vessels or from specially designed monitors usually associated with industrial applications **(Figure 3.14)**. **Table 3.3** describes the expected flows available from various tip sizes.

Solid Stream Nozzle Reactions

Solid stream low-volume and handline nozzles produce reaction forces that can be maneuvered safely by fire-suppression crews. An operating nozzle pressure of 50 psi (345 kPa) {3.45 bar} produces a reaction force directly proportional to the diameter of the tip size being used. The low-volume and handline operating pressures are the same, so **Table 3.4** defines the reactions for each flow classification.

Figure 3.14 Large master stream devices, capable of tremendous fire flows are found on marine fire-fighting vessels.

Solid stream master stream nozzles — nozzles over 1¼ inches (32 mm) in orifice diameter operating at 80 psi (552 kPa) {5.5 bar} — generate tremendous reaction forces (see **Table 3.5**). These devices, once deployed, are typically stationary and are not repositioned once their initial placements are complete. Although many of these devices can be rotated and redirected when they are part of elevated platforms, great care must be taken to guard against damage or injury from their reaction forces. Deluge sets improperly deployed or placed on uneven surfaces or inclines may suddenly propel in the direction opposite to the nozzle flow. Aerial ladders and platforms may be severely damaged if pushed into objects when master streams are redirected.

WARNING

Nozzle reaction forces can suddenly propel master stream devices opposite to the direction of flow when they are charged with water or during stream redirection.

Table 3.3
Master Stream Tip Sizes and Flows

Tip Size (fraction inches)	Tip Size (decimal inches)	Tip Size (millimeters)	gpm (at 70 psi*)	gpm (at 80 psi*)	L/min (at 552 kPa) {5.5 bar}
1⅜	1.375	34.925	469.96	502.40	1 809.03
1½	1.50	38.100	559.29	597.90	2 152.89
1¾	1.750	44.450	761.25	818.81	2 930.33
2	2.00	50.800	994.29	1,062.94	3 827.37
2¼	2.250	57.150	1,258.39	1,345.88	4 844.01
2½	2.50	63.500	1,553.57	1,660.84	5 980.26
2¾	2.75	69.850	1,879.82	2,009.62	7 236.12
3	3.00	76.200	2,237.15	2,391.61	8 611.58

* 70 psi = 483 kPa {4.8 bar}
80 psi = 552 kPa {5.5 bar}

Table 3.4
Solid Stream Low-Volume and Handline Nozzle Reactions at 50 psi* Operating Pressure

Tip Size (fraction inches)	Tip Size (decimal inches)	Tip Size (millimeters)	Gallons/Minute (L/min) at 50 psi*	Nozzle Reaction (pounds)	Nozzle Reaction (Newtons)
⅜	0.375	9.52	29.54 (111.82)	11.04	47.58
½	0.50	12.70	52.52 (198.81)	19.66	84.68
⅝	0.625	15.88	82.06 (310.63)	30.66	132.39
¾	0.750	19.05	118.17 (447.32)	44.16	190.52
⅞	0.875	22.225	160.84 (608.84)	60.10	259.32
15⁄16	0.9375	23.8125	184.64 (698.93)	69.99	297.69
1	1.0	25.40	210.08 (795.24)	78.50	338.7
1⅛	1.125	28.575	265.88 (1 006.46)	99.35	428.68
1¼	1.25	31.75	328.25 (1 242.56)	122.60	529.23

* 50 psi = 345 kPa {3.45 bar}
Blue: Low-volume solid stream nozzle
Yellow: Handline solid stream nozzle

Table 3.5
Master Stream Nozzle Reactions at 80 psi* Operating Pressure

Tip Size (fraction inches)	Tip Size (decimal inches)	Tip Size (millimeters)	Gallons/Minute (L/min) at 80 psi*	Nozzle Reaction (pounds)	Nozzle Reaction (Newtons)
1⅜	1.375	34.925	502.40 (1 901.78)	237.46	896.52
1½	1.50	38.100	597.90 (2 263.29)	282.60	1 066.93
1¾	1.750	44.450	818.81 (3 099.52)	384.65	1 452.21
2	2.00	50.800	1,062.94 (4 023.65)	502.40	1 896.77
2¼	2.250	57.150	1,345.88 (5 094.69)	635.85	2 400.60
2½	2.50	63.500	1,660.84 (6 286.94)	785.00	2 963.70
2¾	2.75	69.850	2,009.62 (7 607.22)	949.85	3 586.08
3	3.00	76.200	2,391.61 (9 053.20)	1,130.40	4 267.73

* 80 psi = 552 kPa {5.5 bar}

Fog Nozzles

A fog nozzle produces a fire stream made of small droplets of water that leave the tip in a spray or "fog" pattern. Fog nozzles are most often designed to allow fire-suppression crews to adjust the fog tips to form different patterns that allow them to apply fire streams in the most appropriate and efficient manner. The streams formed by a fog nozzle range from a straight stream pattern to a narrow fog (sometimes referred to as a *power cone*) and finally to a wide fog pattern. Most fog nozzles can be adjusted from a wide-angle stream down to a narrow-angle pattern that resembles a solid stream but is called a *straight stream;* however, the narrow fog stream is actually hollow **(Figure 3.15).**

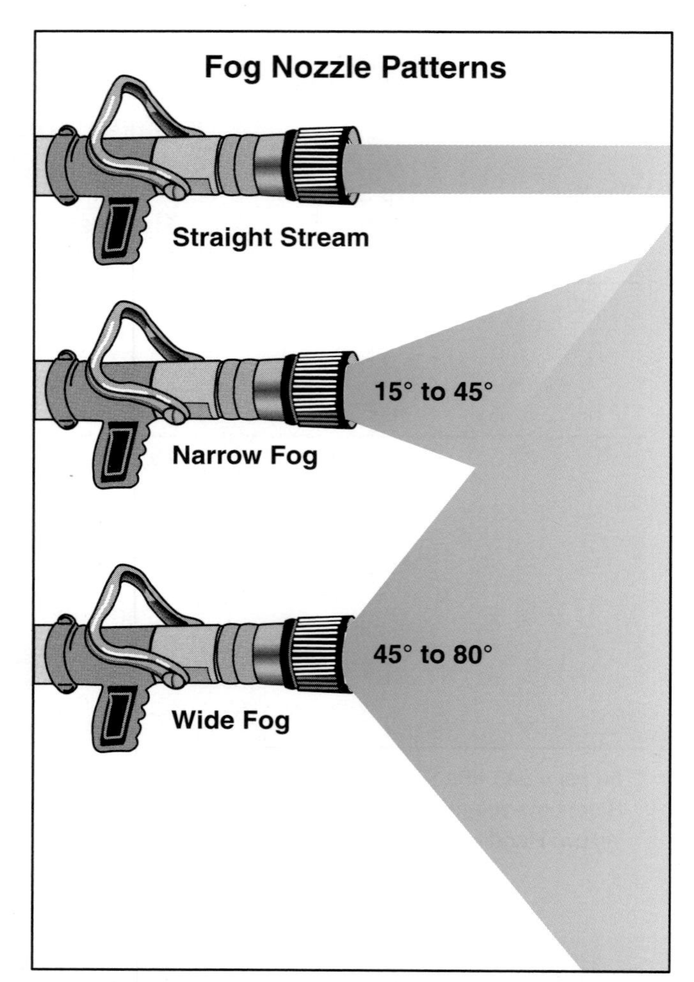

Figure 3.15 Fog nozzles are most commonly adjusted to one of three stream patterns: straight stream, narrow fog, or wide fog.

A fog nozzle produces water droplets that have a total surface area far greater than that produced by solid stream nozzles. When water is delivered to a high-temperature atmosphere, this additional surface area allows water to absorb the heat generated by the fire very quickly. The desired performance of the nozzle is judged by the amount of heat that the stream absorbs and the rate at which the water converts to steam or vapor. A fog stream of water with a certain velocity remains in a solid mass, not losing continuity until it strikes an object, is overcome by gravity, or is slowed and changed by air resistance. The fog stream is produced when it is broken into finely divided particles as it is driven against an obstruction with sufficient force to shatter the mass of water. The angle at which the stream of water is deflected from an obstruction determines the reduction in forward velocity of the stream as well as the pattern or shape that the stream assumes. A wide-angle deflection produces a wide-angle fog pattern, and a narrow-angle deflection produces a narrow-angle fog pattern.

The following terms are important to understand when discussing the principles of fog streams:

- ***Periphery*** — Line bounding a rounded surface; outward boundary of an object as distinguished from its internal regions
- ***Deflection*** — Turning or state of being turned; turning from a straight line or given course; bending; deviation
- ***Impinge*** — To strike or dash about or against; clashing with a sharp collision; to come together with force

Fog streams are formed by deflecting water at the periphery of the nozzle, impinging jets of water, or combining these methods. Periphery-deflected streams are produced by deflecting water from the periphery of an inside circular stem in a periphery-deflected fog nozzle **(Figures 3.16 a and b).** The exterior barrel again deflects the water as it exits the nozzle. As the exterior nozzle barrel rotates, it extends or retracts along the deflecting stem, creating the pattern of the stream **(Figures 3.17 a and b).**

An impinging-stream nozzle directs several jets of water together at an angle and with sufficient pressure to break the water into finely divided droplets **(Figures 3.18 a and b, p. 114).** Nozzles of this type usually produce a wide-angle fog pattern; however, narrower stream patterns are possible (see Multipurpose Nozzles section).

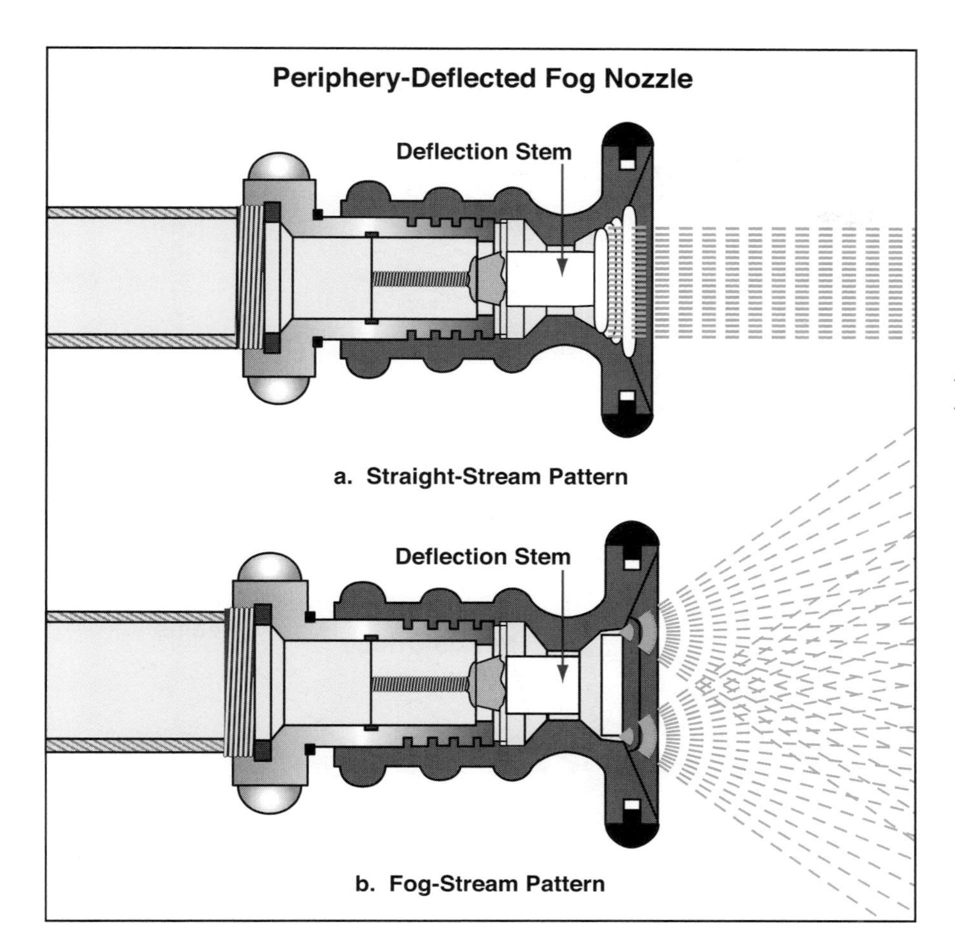

Figures 3.16 a and b The relative positions of the deflection stem and the exterior barrel of a periphery-deflected fog nozzle determine the stream pattern. Examples: (a) Straight-stream pattern (b) Fog-stream pattern.

Figures 3.17 a and b Straight through wide-angle stream patterns produced by periphery-deflected nozzles. *Left* (a) Straight pattern. *Right* (b) Wide pattern.

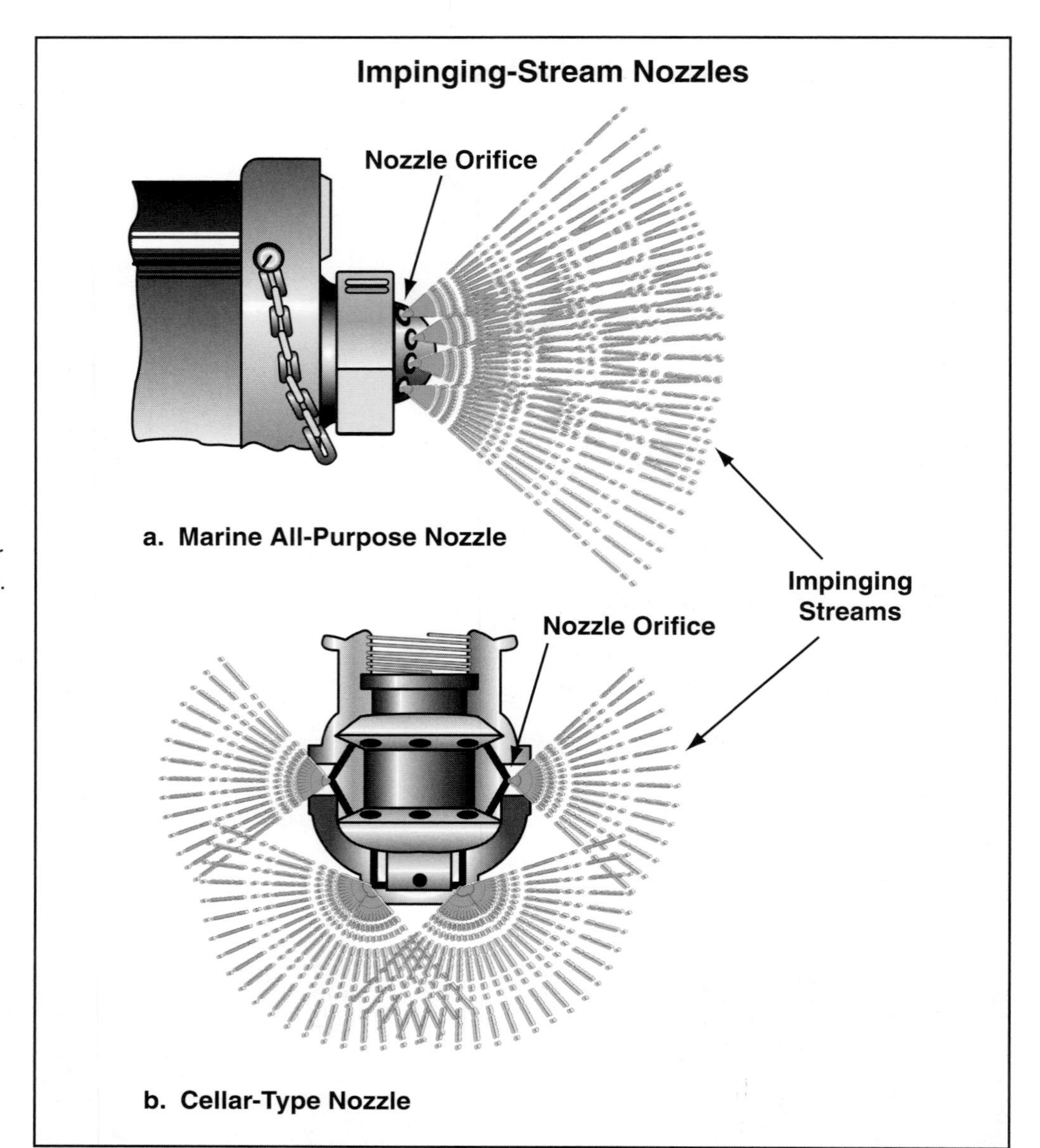

Figures 3.18 a and b An impinging-stream nozzle directs individual jets of water together to form a fog pattern. Examples: (a) Marine all-purpose nozzle (b) Cellar-type nozzle.

The range of a fog fire stream is directly dependent upon the following variables:

- Width (diameter) of the stream
- Size of the water droplets
- Quantity (mass) of the water being flowed
- Elevation to which the stream is directed
- Velocity and direction of the wind **(Figure 3.19)**

As with solid stream nozzles, fog nozzles are used on low-volume hoselines, handlines, and master stream devices. Most fog nozzles are designed to operate at a pressure of 100 psi (689 kPa) {6.89 bar}, which means that the nozzle produces a stream at the rated volume with water droplets of the optimum size for vaporization by the heat of a fire. Some new models (automatic constant-pressure fog nozzles) automatically adjust to variations in nozzle pressure to produce a stream that contains droplets of optimum size. At a nozzle pressure lower than 100 psi (689 kPa) {6.89 bar}, the volume decreases but the nozzle maintains a stream of the same reach and quality as a stream produced at a higher pressure.

Early models of fog nozzles were often designed to produce larger volumes of water when adjusted to a wide-angle fog stream. The narrow-angle stream developed by this nozzle had a reduced quantity of flow when compared to the wide pattern. Because the actual quantity of water flowing

Figure 3.19 The fine water particles of a fog stream are easily affected by the wind.

at any given pattern was not known, the nozzle acquired the name *mystery nozzle*.

The various fog nozzle types are constant-flow, manually adjustable flow, automatic constant-pressure, and master stream automatic; these types are described in the sections that follow. Information on nozzle reactions when using fog nozzles is also given.

Constant-Flow Fog Nozzle

Constant-flow fog nozzles are constructed to flow a specific quantity of water at the operating pressure for which the nozzle was designed. This flow rate is delivered on all stream pattern settings. Most constant-flow nozzles use a periphery-deflected nozzle design to form the stream **(Figure 3.20)**.

The nozzle operator manually selects the pattern of the nozzle flow and rotates the exterior barrel of the nozzle to the desired pattern. As the barrel

Figure 3.20 Examples of constant-flow fog nozzles. *Courtesy of Elkhart Brass Manufacturing Company, Inc. and Akron Brass Company.*

rotates, the space between the deflecting stem and the internal throat of the nozzle remains constant, while the angle of deflection is modified. The volume of water is therefore constant for every stream pattern.

When the nozzle barrel is fully extended, the water is channeled to form a straight pattern. A wide fog is produced when it is fully depressed. Standard pattern settings include straight stream, narrow angle (30 degrees), and wide angle (60 degrees) **(Figure 3.21)**.

Most constant-flow nozzles are designed to operate at a nozzle pressure of 100 psi (689 kPa) {6.89 bar}, although low-pressure fog nozzles with operating pressures of 50 psi (345 kPa) {3.45 bar} or 75 psi (517 kPa) {5.17 bar} are available and becoming more prevalent **(Figure 3.22)**. The low-pressure nozzles offer the advantage of producing lower nozzle reaction forces, making them advantageous for fire-suppression teams who have few members and in areas with high-rise structures where it is difficult to maintain high pressures.

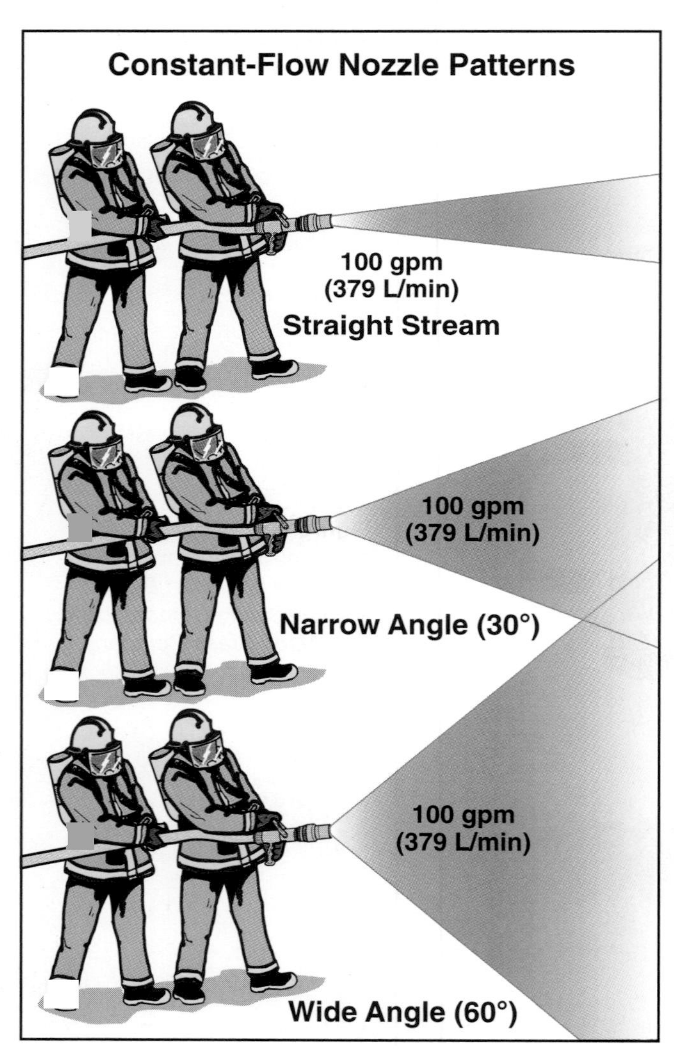

Figure 3.21 A constant-flow nozzle delivers the same flow, regardless of the stream pattern setting.

Manually Adjustable Flow Fog Nozzle

An improvement over the constant-flow fog nozzle is the manually adjustable flow fog nozzle. Several constant-flow settings are available for selection by fire-suppression crews and driver/operators **(Figure 3.23)**. Stream patterns are changed by rotating the exterior barrel of the nozzle like the constant-flow fog nozzles. As the barrel rotates, the pattern is produced as the fire stream passes through the nozzle barrel and is deflected. The design nozzle pressure must be maintained to ensure that it performs as desired. A reduction in nozzle pressure (due to low pump pressure, unforeseen friction loss, head pressure loss, or attack hoseline restrictions) adversely affects the fire-suppression capabilities of the nozzle. Manually adjustable flow fog nozzles designed for use with various attack hoselines have the following flow settings:

Figure 3.22 Constant-flow nozzles are available in low-pressure designs. *Courtesy of Elkhart Brass Manufacturing Company, Inc.*

- ***1½-inch (38 mm) handline*** — 30 or 40, 60, 95, and 125 gpm (114 L/min or 151 L/min, 227 L/min, 360 L/min, and 473 L/min)

  NOTE: Nozzle settings below 95 to 100 gpm (approximately 400 L/min) are not considered adequate for interior structural fire fighting.

Figure 3.23 Manually adjustable flow fog nozzles come in a variety of sizes and flow settings. *Courtesy of Elkhart Brass Manufacturing Company, Inc. and Akron Brass Company.*

Many fire and emergency service organizations also do not consider this flow rate adequate for any fire-suppression activities beyond overhaul or mop-up.

- ***1¾-inch (45 mm) handline*** — 95, 125, 150, 175, and 200 gpm (360 L/min, 473 L/min, 568 L/min, 662 L/min, and 757 L/min)
- ***2½-inch (65 mm) handline*** — 120 or 125, 150, 200, 250, and 300 gpm (454 or 473 L/min, 568 L/min, 757 L/min, 946 L/min, and 1 136 L/min)
- ***Master stream*** — 500, 750, and 1,000 gpm (1 893 L/min, 2 839 L/min, and 3 785 L/min)

Automatic Constant-Pressure Fog Nozzle

Constant-pressure fog nozzles automatically monitor the pressure and rate of water flow through the nozzle while maintaining an adequate nozzle pressure to produce an effective fire stream. As the flow of water changes, the nozzle adjusts automatically to maintain the same pressure and pattern. This feature is the result of the movement of a baffle that moves automatically inside the barrel of the nozzle. This action varies the spacing between the baffle and the nozzle throat **(Figures 3.24 a and b, p. 118)**.

Although a stream from an automatic constant-pressure nozzle may look like a good stream by the pattern appearance, the flow from the nozzle may be inadequate for fire-suppression purposes. The fire-suppression crew, the company officer, and the driver/operator must coordinate their actions to ensure that sufficient water and pressure are maintained in the fire attack hoseline. Most automatic constant-pressure nozzles are designed to operate at 100 psi (689 kPa) {6.89 bar}. However, some nozzles may be designed to operate efficiently at 50 or 75 psi (345 kPa or 517 kPa) {3.45 bar or 5.17 bar}. As with constant-flow nozzles, automatic constant-pressure nozzles are finding increased use in high-rise applications and where reduced nozzle reaction is necessary.

An automatic constant-pressure nozzle serves as a pressure regulator — within its flow limits — for the pumping apparatus as hoselines are added or turned off. In this way, all available water directed to the nozzle is used. If the water supply is inadequate, the maximum volume is that which can be achieved without the pump cavitating. The driver/

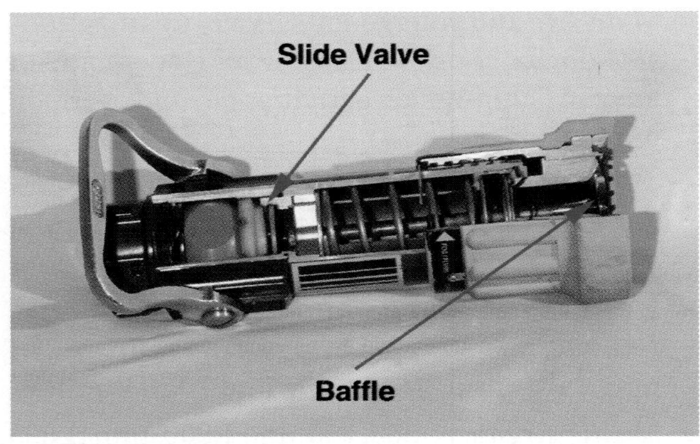

Figure 3.24a Cut-away of an automatic constant-pressure fog nozzle showing slide valve and baffle.

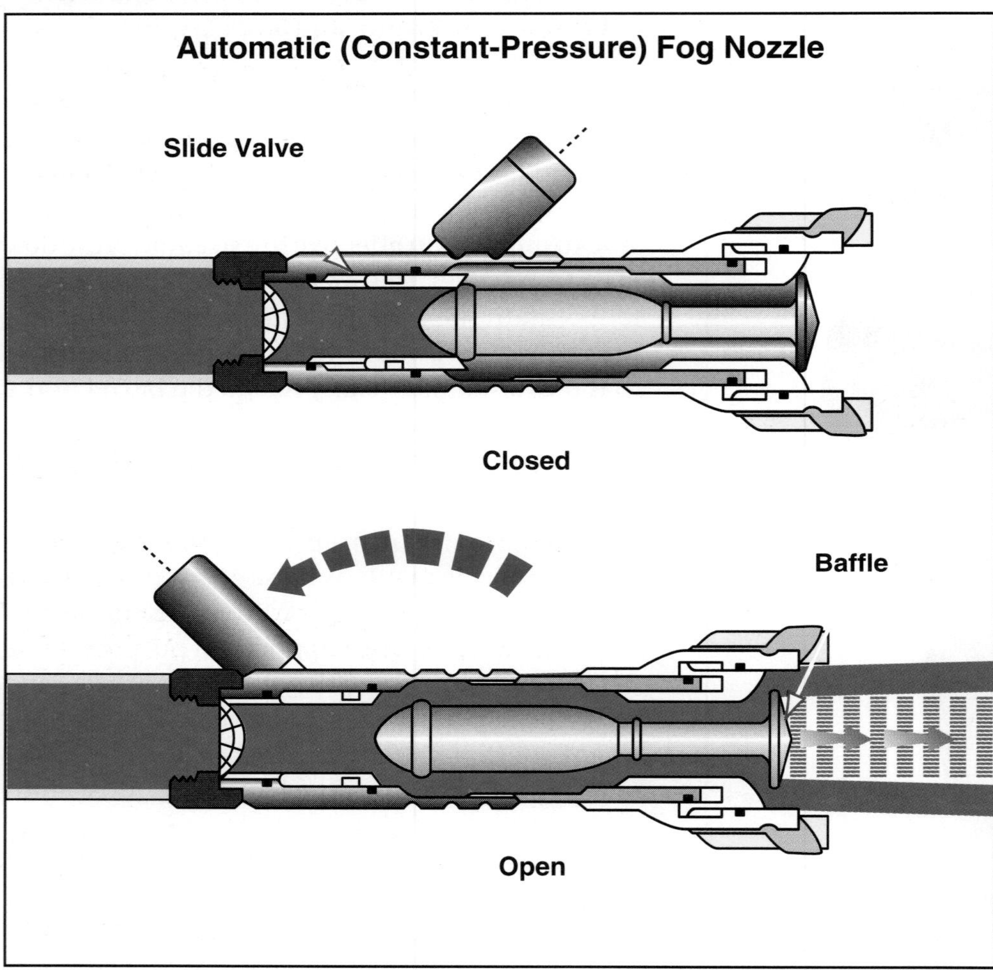

Figure 3.24b One common design for an automatic constant-pressure fog nozzle uses a slide barrel control inside the nozzle. Nozzle is shown in both closed *(top)* and open *(bottom)* positions.

operator can deliver the maximum water flow that is available to the nozzle. Should a situation occur where the quantity of available flow drops, the nozzle adjusts itself and continues to produce an acceptable fire stream pattern.

Master Stream Automatic Fog Nozzle

Master stream automatic fog nozzles supply from 300 to 2,000 gpm (1 136 L/min to 7 571 L/min) and are available in a wide variety of models **(Figure 3.25).** These nozzles operate in a similar manner to handline automatic constant-pressure fog nozzles. Designed to operate at 100 psi (689 kPa) {6.89 bar}, the nozzle "slides" through various flow rates as the fire pump delivers water to it. Should the flow decrease, the nozzle maintains a satisfactory pattern, always providing a good fire stream; however, as flow is decreased, the stream loses its effectiveness

Figure 3.25 Many styles of master stream automatic fog nozzles are available. *Courtesy of Elkhart Brass Manufacturing Company, Inc., Akron Brass Company, and Task Force Tips.*

to absorb the heat generated by a fire. A stream with a good pattern may not have the water flow capable of controlling a fire incident.

Fog Nozzle Reactions

Fog nozzles are subject to the same reaction forces that affect solid stream applications. Because nozzle reaction is a function of the mass of water being discharged and the pressure at which the discharge occurs, the tables for solid stream and fog nozzle reactions of relatively the same flow differ (see **Table 3.6**). This difference is due to the higher pressure at the time of discharge from the fog nozzle. The fog nozzle has a higher reaction force per gallon (liter) than a similar flow from a solid stream nozzle. As fog nozzles with lower operating pressures are developed and put into service, these reaction forces will be reduced and approach the same values found with solid stream nozzles.

Foam Nozzles

With the increasing production and transportation of flammable liquids (Class B fuels), the use of foam has taken on great importance. It is essential that the appropriate nozzle be selected for use with the appropriate Class B foam concentrate. Class A foam concentrates have been used since the 1940s to increase the penetrating capabilities of water; however, only recently has the technology advanced to a point where they can be efficiently deployed for fire suppression in ordinary combustible fuels.

Table 3.6
Fog Nozzle Reactions at 100 psi* Operating Pressure

Gallons/Minute (gpm)	Nozzle Reaction (pounds)	Liters/Minute (L/min)	Nozzle Reaction (Newtons)
15	7.575	60	24.77
30	15.15	120	49.53
45	22.725	180	74.30
60	30.30	240	99.06
95	47.975	380	156.85
125	63.125	500	206.39
150	75.75	600	247.67
175	88.375	700	288.94
200	101.00	800	330.22
250	126.25	1 000	412.78
300	151.50	1 200	495.33
500	225.84	2 000	690.64
750	338.76	3 000	1 035.96
1,000	451.68	4 000	1 381.28
2,000	903.37	8 000	2 762.57

* 100 psi = 689 kPa {6.89 bar}
Blue: Low-volume flow
Yellow: Handline flow
Green: Master stream flow

Class A foam concentrates are effective for fires in structures, wildland, wood chips/sawdust, tire storage, and other incidents involving similar deep-seated fires. Because these concentrates are formulated of special hydrocarbon surfactants that significantly reduce the surface tension (force minimizing a liquid surface's area) of water in the finished foam, better penetration by the water into a burning material can be expected. Coupled with its outstanding insulating properties when employed in a compressed-air foam system (CAFS), Class A finished foam displays tremendous insulating properties that prove invaluable as a fire-spread inhibitor (see section that follows on solid stream foam nozzles).

Fires involving flammable and combustible liquid fuels require the use of Class B foam concentrates to not only extinguish these fires but also to suppress the vapors generated during the combustion process, limiting the possibility of the fuel reigniting. Class B foam can also prevent ignition by sealing the surface and limiting the release of flammable vapors. Many different types of Class B foam concentrates are available, each possessing different performance qualities to address the various characteristics of the materials involved in an incident.

Foam agent application depends upon the proper selection of a foam nozzle that not only satisfies the requirements for proportioning a foam concentrate (mixing with water) and aerating the foam concentrate/water solution (mixing with air) but also on its ability to deliver finished foam (completed product) to the surface of the burning fuel or Class B liquid spill. Low-energy foam generating systems introduce air into the foam solution when it reaches the nozzle just before discharge or at the moment of discharge from the nozzle. They are found primarily in portable or on pumping apparatus-mounted proportioning devices. The fire pump generates all of the pressure on the foam generating and delivery system.

High-energy foam generating systems inject compressed air into the foam solution before it is discharged from the foam hoseline. The turbulence of the compressed air and foam solution while traveling through the hoseline creates finished foam as it discharges from the nozzle.

Finished foam is an effective extinguishing agent because it can separate the fuel from burning vapors and air. This barrier allows the fuel to cool and reduces the quantity of fuel vapors being generated. The flames are smothered, and fuel vapors are restricted from combining with air to form a combustible mixture, thereby eliminating the possibility of reignition. The foam application also protects fuel spills from igniting by limiting the generation of flammable vapors.

Many different types of foam concentrate are available to choose from, depending upon the nature of the fire-suppression task or Class B liquid spill situation. Care must be taken to ensure that the proper foam concentrate and foam nozzle are selected for the particular situation. Some foam concentrates may not be effective on the type of fuel involved. The finished foam may break down or dissolve too quickly to be effective due to the fuel type or conditions encountered. Additionally, the selection of a foam nozzle must be consistent with the type of finished foam being created and its performance expectations (see **Appendix C,** Foam Properties).

A number of foam nozzles are available, depending upon the type of foam concentrate being used and the desired method of application. Solid stream and fog nozzles are effective for several types of foam concentrates and their applications. Additionally, special foam nozzles have also been developed to produce and deliver finished foam in a precise manner for specialized applications. Various foam nozzle types are discussed in the sections that follow. See IFSTA's **Principles of Foam Fire Fighting,** 2nd edition, for further information on the various types of foam concentrates and their applications.

Solid Stream Nozzle

Applying finished foam with a solid stream nozzle can only be done with a Class A compressed-air foam system — a system that entrains large amounts of compressed air and small amounts of water into a foam solution to make finished foam. These systems provide effective fire streams as well as maximum range capabilities. With a solid stream nozzle, a CAFS produces a stream with a range that more than doubles that of any low-

Figure 3.26 A typical solid stream nozzle may be used with a compressed air foam system (CAFS) stream.

Figure 3.27 Compressed air foam system streams have the reach and penetration to be very effective for fire-suppression operations.

energy foam generating system **(Figure 3.26).** It is recommended that the solid stream smoothbore nozzle diameter be increased beyond the usual limitation of using a tip no greater than one-half the diameter of the attack line. Tips ¼-inch (6 mm) smaller than the attack hose diameter produce effective, long-range foam streams **(Figure 3.27).**

Fog Nozzle

Constant-flow, manually adjustable, and automatic constant-pressure fog nozzles can each be used with foam solutions to produce low-expansion (an air/solution ratio up to 20 parts finished foam for every part of foam solution), short-lasting finished foam **(Figure 3.28).** As the foam/water solution passes through the nozzle, a foaming action (agitation) occurs. As the solution leaves the nozzle, it mixes with air and creates finished foam. Fog nozzles are most effective when applying Class A foam or aqueous film forming foam (AFFF) concentrates. Fog nozzles are not effective when used with protein or fluoroprotein foam concentrates, nor should they be used with AFFF concentrates when applied to polar solvent fuel (flammable liquid that is soluble in water) fires because the nozzles do not properly aspirate the foam solution for these applications. Some manufacturers have foam aeration attachments that increase the aspiration of a foam solution from a fog nozzle, producing an improved finished foam that also allows moderate success when it is applied to polar solvent fuels **(Figures 3.29 a and b).**

Figure 3.28 Fog nozzles can produce low-expansion finished foam without special attachments.

Air-Aspirating Nozzle

The air-aspirating foam nozzle inducts air into the foam solution by Venturi action (creates a suction) **(Figure 3.30).** This nozzle produces fully aerated finished foam of excellent quality. A wide variety of foam concentrates may be dispensed through these nozzles, including protein, fluoroprotein, and Class A. The maximum expansion of finished foam is attained using these nozzles, but the velocity of the stream and ultimately its range is reduced substantially from other types of foam nozzles.

Multiagent Nozzle

Multiagent foam nozzles are designed to dispense a variety of foam concentrates and other extinguishing agents in sequence or simultaneously. These applications are complex and best suited when an industrial process requires two or more extinguishing agents to effect fire suppression. Often fire extinguishment is best accomplished by a series application of these agents, each with a specific role in the fire-suppression process: one agent for a rapid fire knockdown, another for vapor-sealing capabilities, one for oxygen dilution or exclusion, and possibly one for fuel cooling.

Twin-agent nozzles provide a simultaneous discharge of a foam concentrate and a dry chemical agent **(Figure 3.31).** The dry chemical agent quickly suppresses the flames while the finished foam provides fuel cooling and seals the vapors, preventing reignition of the product. Most twin-agent nozzles are mounted side-by-side or

Figures 3.29 a and b *Left* (a) Foam aeration attachments on fog nozzles improve the volume of finished foam by increasing aspiration. *Right* (b) Close-up view of foam aeration attachment (showing the air inlet) where it connects to the nozzle.

Figure 3.30 An air-aspirating foam nozzle has a permanent aspirator built into the nozzle so that air is drawn in and mixed with foam solution.

Figure 3.31 Two extinguishing agents (such as dry chemical and halon substitute with foam) can be simultaneously discharged through a twin-agent nozzle.

in an over-under configuration. Each agent is delivered to the nozzle assembly through separate supply (attack line approved) hoses. A control lever allows the dispensing of the extinguishing agent in tandem or separately as needed.

Quad-agent nozzles are designed to discharge four extinguishing agents (water, foam solution, dry chemical, and a halon substitute agent) to interrupt the combustion/reaction process. These nozzles operate on the same principal as the twin-agent nozzles. Each agent is supplied to the nozzle through a separate supply hose. The operator controls each agent with a single pistol-grip operating lever that allows the extinguishing agents to dispense individually or in combination.

Self-educting Handline Nozzle

A self-educting handline foam nozzle operates in a similar manner to an in-line foam eductor in that the foam eductor is built into the nozzle rather than being a stand-alone unit as part of the hoseline or installed on the pumping apparatus **(Figure 3.32).** The foam concentrate must be brought to the nozzle position where it is introduced into the water stream. This process creates a cumbersome and potentially dangerous operational situation. If the nozzle must be repositioned, the foam concentrate must be moved as well. The fire and emergency service responders operating the system cannot move quickly in an emergency and easily maintain the supply of foam concentrate to the nozzle.

The self-educting handline nozzle does offer ease of use to the operators and is less expensive than other eductor systems. Additionally, because the eductor is located at the nozzle, low operating pressures can be used. These nozzles also are available at numerous flow rates.

Self-educting Master Stream Nozzles

When it is necessary to exceed flows of 350 gpm (1 325 L/min), self-educting master stream foam nozzles are required **(Figure 3.33).** These nozzles are available with flow capabilities up to 12,000 gpm (45 425 L/min). A "rich" foam solution (one that has a higher ratio of foam concentrate in the solution than normal) is produced. Deflector plates in the nozzle then dilute this richer-than-normal foam solution as it discharges. As the rich foam solution passes through the hose, a much lower foam stream pressure drop (10 percent or less) can be expected than that obtained when using standard foam eductors.

Master Stream Nozzles/Devices

Large-scale flammable and combustible liquid emergencies can easily overwhelm the capabilities of fire-suppression crews operating handline foam nozzles. Master stream nozzles can deliver adequate quantities of foam during these emergencies. The type of nozzle used can be a standard fixed-flow or automatic fog nozzle. Large-scale industrial foam apparatus and aircraft rescue and fire fighting (ARFF) vehicles may be equipped with special

Figure 3.32 A self-educting handline foam nozzle mixes the foam concentrate at the nozzle instead of a remote location. The portable unit shown is used in wildland firefighting operations.

Figure 3.33 A self-educting master stream nozzle is used when high flow rates are needed. Foam concentrate is brought to the nozzle where emergency responders insert the eduction tube into the foam solution.

aerating master stream foam nozzles/devices. Several types of master stream foam nozzles or devices are available for deployment including manual, automatic, and remote-controlled foam nozzles or devices.

Manual

Manual foam nozzle monitors are the least complex and most easily maintained and operated master stream devices **(Figure 3.34).** These monitors are usually mounted in fixed locations to protect target hazards and can be found in large industrial complexes where large quantities of combustibles are stored. They can also be mounted on pumping apparatus and apply plain water if necessary. The flow capabilities of these master stream nozzles range from 200 to 2,000 gpm (757 L/min to 7 571 L/min). When used in specialized environments such as marine vessels and chemical manufacturing facilities, they can be designed to deliver even higher flow rates.

Figure 3.34 Manual foam nozzle monitors are permanently affixed to water supply pipes. simple to use, and easy to maintain. *Courtesy of Conoco, Inc.*

Automatic

Automatic oscillating foam monitors are typically used in fixed locations where wide-area coverage is desired. The nozzle uses the energy of the foam solution flowing through it to rotate back and forth in a manner similar to the common lawn sprinkler **(Figure 3.35).** Depending upon the manufacturer of the nozzle, an adjustable range from 160 to 200 degrees of operation arc is possible. A flow of up to 2,000 gpm (7 571 L/min) can be provided, and plain water can be used if necessary.

Remote-Controlled

Remote-controlled master stream devices are most often found mounted on fire apparatus of various types including aircraft rescue and fire fighting, fire-fighting watercraft, and specialized industrial vehicles and trailers **(Figure 3.36).** Remote-controlled master stream foam nozzles

Figure 3.35 Automatic oscillating foam monitors operate from the energy of the foam solution flowing through it to rotate back and forth. *Courtesy of Conoco, Inc.*

Figure 3.36 Apparatus-mounted remote-controlled master stream devices allow the operator to direct the flow, regulate the flow rate, and adjust the stream pattern from the apparatus cab or pump panel.

are operated with a joystick-type control, toggle switch, or a combination of both. Remote controls allow the operator to change the direction, flow rate, and type of stream pattern produced by the nozzle. The motors that drive the device are electric, hydraulic, or a combination of both. The devices can flow foam solutions or water up to 3,000 gpm (11 356 L/min).

Water-Aspirating Nozzle

Water-aspirating foam nozzles are used to dispense medium-expansion (water-to-foam concentrate ratios of 20:1 to 200:1) finished foam. These nozzles are similar to other foam delivery devices except that they are much larger and longer. The back of the nozzle is open to allow the introduction of air. The foam solution is pumped through the nozzle in a fine spray that mixes with air to form finished foam. The end of the nozzle has a screen or a series of screens that further divides the foam solution and mixes it with air **(Figure 3.37).** The water-aspirating nozzle is cumbersome but highly effective when used to control a confined-space fire such as one in a basement.

Figure 3.37 A water-aspirating nozzle delivers medium-expansion foam by using a fine spray of foam mixture through the barrel of the nozzle. The fabric screen over the end of the dispenser breaks up the foam solution and completes the mixing process.

Mechanical Blower Generator

Mechanical blower generators (nozzles) look like smoke ejectors at first glance and work in a manner similar to water-aspirating nozzles except that air is forced through the foam spray by a powered fan instead of being pulled through by water movement **(Figure 3.38).** This device produces finished foam with high air content and is used when total flooding of a space is warranted. Its use is limited to high-expansion (water-to-foam concentrate ratios of 200:1 to 1,000:1) finished foam.

Multipurpose Nozzles

While most of the nozzles previously covered have numerous effective uses during fire-suppression operations, nozzles that have limited applications for specific stream operations are classified as *multipurpose nozzles.* For example, exposure nozzles are designed to stop the spread of flame by providing wide screens of water between the fire and unburned exposed areas. Several of these nozzle types provide an extended water delivery. Rockwood nozzles (or marine all-purpose nozzles) with applicators attached distribute water ahead of the hose stream over the fire. When attached to a piercing device, a piercing nozzle delivers a water spray through a wall into an adjacent room or space. Another special group of nozzles (cellar or distributor nozzles) has been designed to drop through holes cut in floors into confined basement spaces or cellar areas below and deliver vast quantities of extinguishing agent into those spaces. A chimney nozzle deploys into a narrow chimney space where it slowly distributes a fine mist of water into the shaft to smother flames without causing additional damage to the lining of the flue. A new combination nozzle is the Akron® Saberjet™ nozzle (shown

Figure 3.38 The mechanical blower is used to produce high-expansion foam when total flooding of a large space is needed.

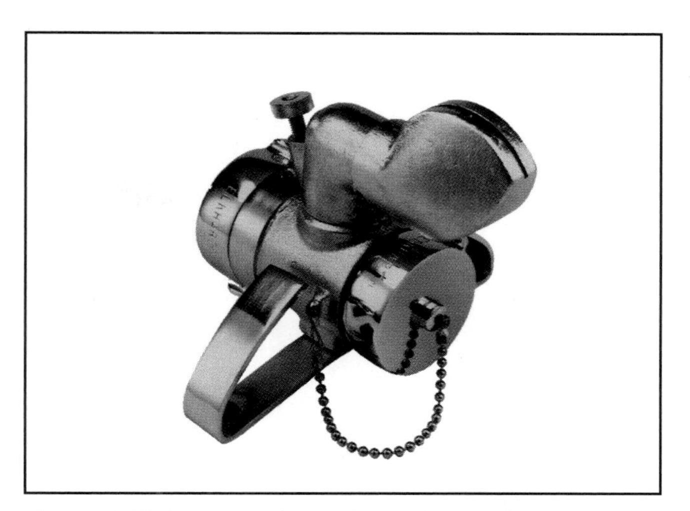

Figure 3.39 An exposure nozzle produces a fan-shaped stream pattern that is directed onto surfaces exposed to radiant heat to provide cooling. This stream does not remove radiant heat on the exposed structure so the constant monitoring of surface temperature is required. *Courtesy of Elkhart Brass Manufacturing Company, Inc.*

earlier) that can deliver a true smoothbore stream, fog stream with all stream patterns, or combination of each of these streams.

Exposure Nozzle

An *exposure nozzle* is one that is specially designed to place between a fire and a nearby exposed surface. The nozzle and hoseline are not normally staffed with fire and emergency services personnel but are deployed and left in place while operating. The nozzle produces a fan-shaped stream called a *water curtain* approximately 35 feet (11 m) wide and 20 to 30 feet (6 m to 9 m or about two stories) high **(Figure 3.39)**. It is designed to protect a building or object from the heat generated by a nearby burning structure. A water curtain does not prevent radiated heat from reaching an exposure. It cools the exposure in two ways: (1) by pulling cool air into the space between the burning structure and the exposure (thus forcing heated air out of the area) and (2) by directly cooling the exposure's surface where the stream is applied. The volume of water that contacts the surface must be of sufficient quantity to absorb the radiated heat striking it. Often these nozzles are overwhelmed because the volume of water discharged is far too low to be effective. When these nozzles are deployed, constant monitoring of the surface temperature of the exposure is essential. Additionally, if the exposure has windows, inspect the interior also because radiated heat pulses can pass through the water stream onto glass and heat nearby interior surfaces to their ignition temperatures.

Marine All-Purpose Nozzle and Applicator

A *marine all-purpose nozzle* (also know as the *Naval All-Purpose* or the *Rockwood nozzle*) is a hybrid nozzle that allows the discharge of several different fire streams from one nozzle. It provides a versatile fire-suppression system very much suited to marine fire situations. Simply moving the shutoff valve to a midrange or a full-open position can produce either a solid or fog stream. Additionally, an applicator tube or pipe can be attached to the fog orifice of the nozzle, allowing for a long overhead extension **(Figures 3.40 a and b, p. 128)**. This long applicator is especially useful when suppressing liquid fuel fires. Finished foam can be applied through the applicator over the burning material. The applicator pipe comes in various head angles (60 to 90 degrees) and lengths ranging from 4 to 12 feet (1.2 m to 3.7 m) for different uses.

Piercing Nozzle

A *piercing nozzle* is a nozzle that can be driven through a surface to deliver water spray into an adjacent space. It has an angled, case-hardened steel tip and a striking surface located behind the tip. It can be driven through a wall, roof, ceiling,

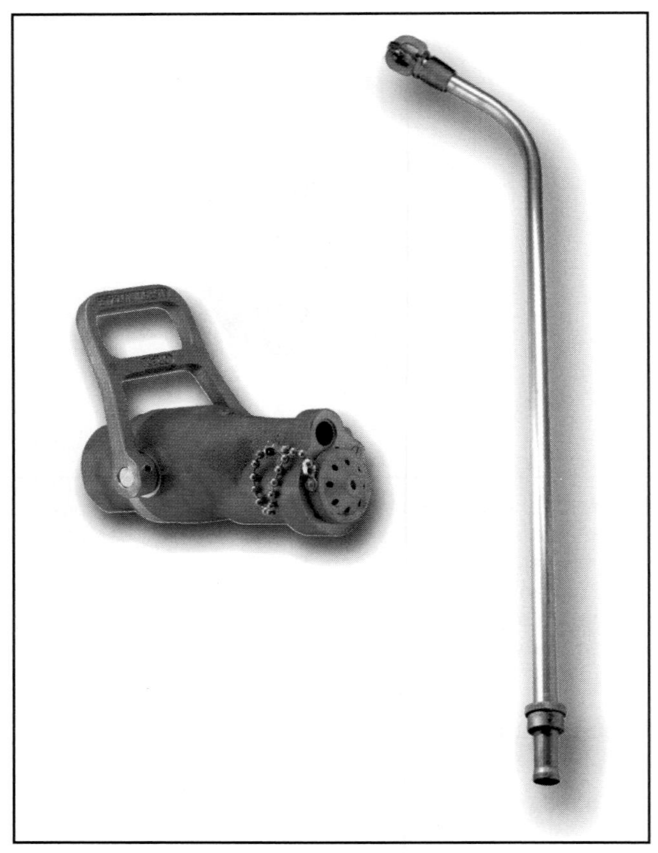

Figures 3.40 a and b *Left* (a) marine all-purpose nozzle. *Right* (b) applicator pipe. *Courtesy of Akron Brass Company.*

bulkhead, or other confined-space surface with a sledgehammer or flat-head axe **(Figures 3.41 a and b).** Small holes bored in the tip permit impinging jets of water to spray outward into the fire area. This nozzle is extremely effective for quick placement of water on a fast-traveling fire above a ceiling, within a wall, or into a confined space. This nozzle is also very effective on automobile or mobile home fires because it can be easily driven through the outside surface into the fire area. Once the confined space has been penetrated, the tip can be removed, creating a solid stream with greater reach and flame penetration.

Distributor Nozzle

A *distributor nozzle* (also referred to as a *Bresnan distributor*) is a blunt-ended nozzle with numerous orifices that are set at various angles. When discharging water, these jets spin the end of the nozzle rapidly, sending jets of water in all directions. This nozzle is lowered through holes in the floor into basements or below deck confined-space areas on vessels. Acting like a giant sprinkler, deep-seated fires can be attacked without exposing fire and emergency service responders to the danger of close approach until the fire is knocked down **(Figure 3.42).** Nozzles of this type are available in 1½-inch (38 mm) and 2½-inch (65 mm) attack line sizes. Total discharge flows at 50 psi (345 kPa) {3.45 bar} are 99 gpm (375 L/min) for the 1½-inch (38 mm) model and 340 gpm (1 287 L/min) for the 2½-inch (65 mm) hoseline.

Cellar Nozzle

A *cellar nozzle* is very similar to a distributor nozzle except that it is comprised of four small spray nozzles attached at an angle to a rotating nozzle base. When discharging, the spray nozzles rapidly rotate

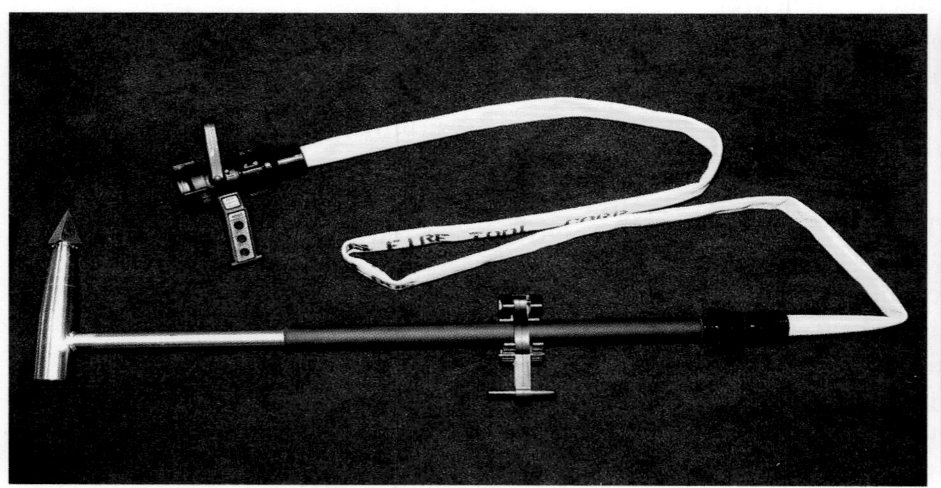

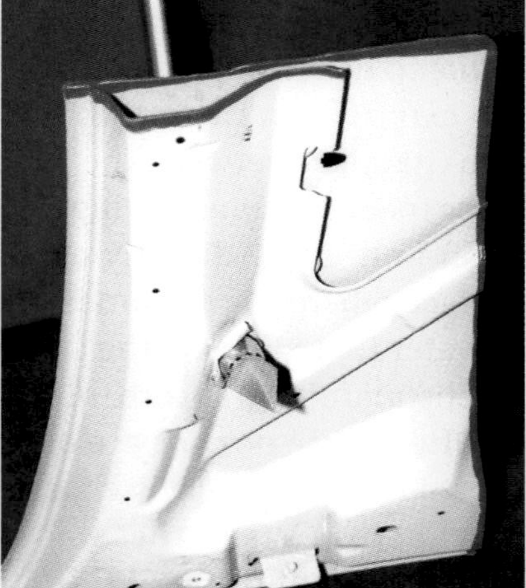

Figures 3.41 a and b *Left* (a) A piercing nozzle penetrates a wall, roof, or other divider and delivers a water spray into the opposite space. *Right* (b) A piercing nozzle penetrating the outside panel of a car door. *Courtesy of Augustus Fire Tool.*

Figure 3.42 A Bresnan distributor (1½-inch [38 mm] shown) can apply large quantities of water to a deep-seated fire in a basement or below deck on a vessel. *Courtesy of Elkhart Brass Manufacturing Company,* Inc.

the end of the cellar nozzle, producing a dense fog cloud in the area of discharge. Cellar nozzles are available in 1½- and 2½-inch (38 mm and 65 mm) attack hoseline sizes and flow 100 and 260 gpm (379 L/min and 984 L/min) respectively. An applicator tube with a shutoff can be attached to the nozzle, allowing for deployment through holes into confined spaces **(Figure 3.43).**

Figure 3.43 The cellar-nozzle, a modification of the Bresnan distributor, has four small spray nozzles mounted at angles on a rotating base. *Courtesy of Elkhart Brass Manufacturing Company, Inc.*

Chimney Nozzle

A *chimney nozzle* is a device developed for the extinguishment of fires that occur inside the flue of a chimney. It is constructed of solid brass or steel with numerous small impinging stream holes and attached to the end of a booster hoseline. The water flow produced at the nozzle's normal operating pressure is very limited, only 1.5 to 3 gpm (5.7 L/min to 11 L/min), which results in a very fine, misty fog cone. This limited flow is necessary to avoid cracking the flue. The nozzle and hose are lowered down into a chimney and then rapidly removed. Care must be taken to protect the booster hose from snagging or contacting the sides of the flue as much as possible as it is lowered into the chimney.

Combination Nozzle

A new innovation in nozzle technology is the combination nozzle, known commercially as the Akron® SaberJet™, which can be adjusted to produce a fog stream, a solid stream, or a combination of each **(Figures 3.44 a–c, p. 130).** The design of the nozzle allows the water stream to be directed around a central bore to a peripheral passage that leads to the fog orifice. The solid stream is formed by passing the water flow directly through a smoothbore stream shaper in the center of the nozzle. The shaper conforms to standard smoothbore stream diameters including ¾-, ⅞- and 15/16-inch (20 mm, 22 mm, 23 mm) diameters **(Figures 3.45 a–c, p. 130).**

Another option available with this nozzle is a combination of fog and solid stream configurations. It provides thermal protection with a fog pattern plus the longer penetrating reach of a solid stream applied simultaneously, which makes this nozzle a versatile tool for the fire-suppression crew. Stream pattern selection is accomplished in two ways: (1) the shutoff handle is used to select between solid stream and fog and (2) the rubber bumper at the front of the nozzle is adjusted to control other fog patterns and combination flows.

Summary

A nozzle completes a hoseline and in many ways defines its use and capabilities. The proper selection of a fire nozzle is essential to the fire-suppression effort. Each nozzle category and fire stream type offers qualities that must be evaluated to ensure that the mission of the using agency is met. Because numerous nozzles are available (providing a wide range of water flows, operating pressures, nozzle reaction forces, and steam ranges from which to select), a variety of fire-suppression problems can be resolved. This chapter provides the vital physical and performance characteristics for each nozzle category and indicates the proper type of hose that is best suited to meet the incident needs.

Figures 3.44 a–c The Akron® SaberJet™ nozzle in action: *Top* (a) Solid stream. *Middle* (b) Both fog and solid stream. *Bottom* (c) Fog stream. *Courtesy of Akron Brass Company.*

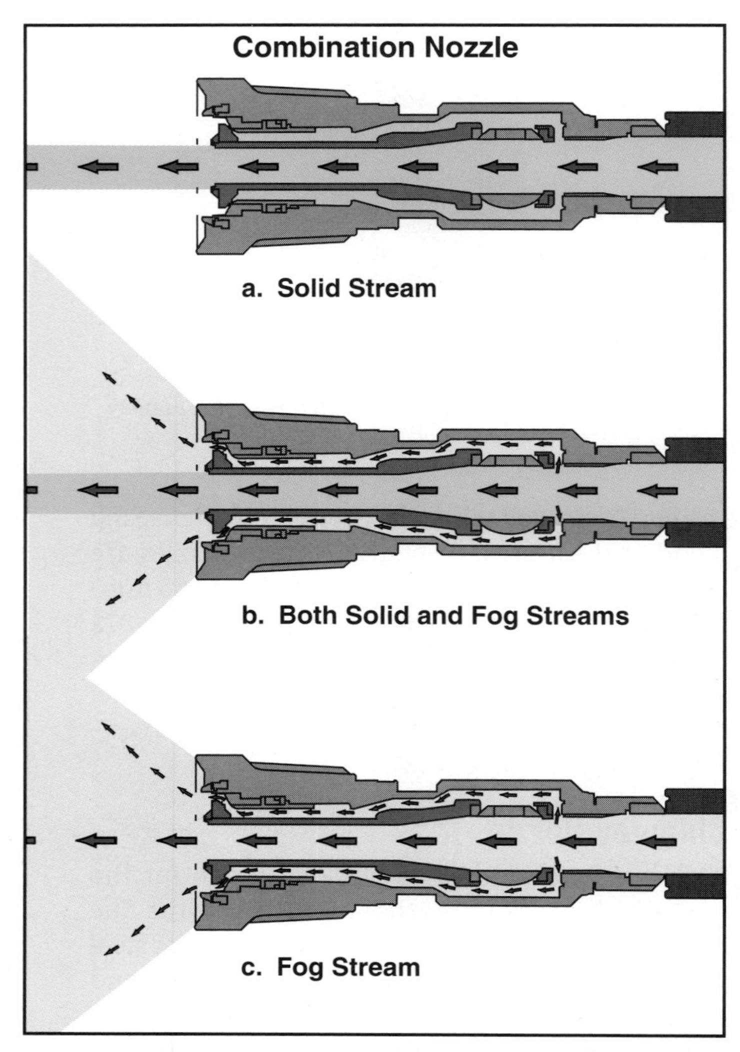

Figures 3.45 a–c Operating principles of the 3 versions of the Akron® SaberJet™ nozzle: (a) Solid stream (b) Both fog and solid streams. (c) Fog stream. *Courtesy of Akron Brass Company.*

Chapter 4
Fire Hose Appliances and Hose Tools

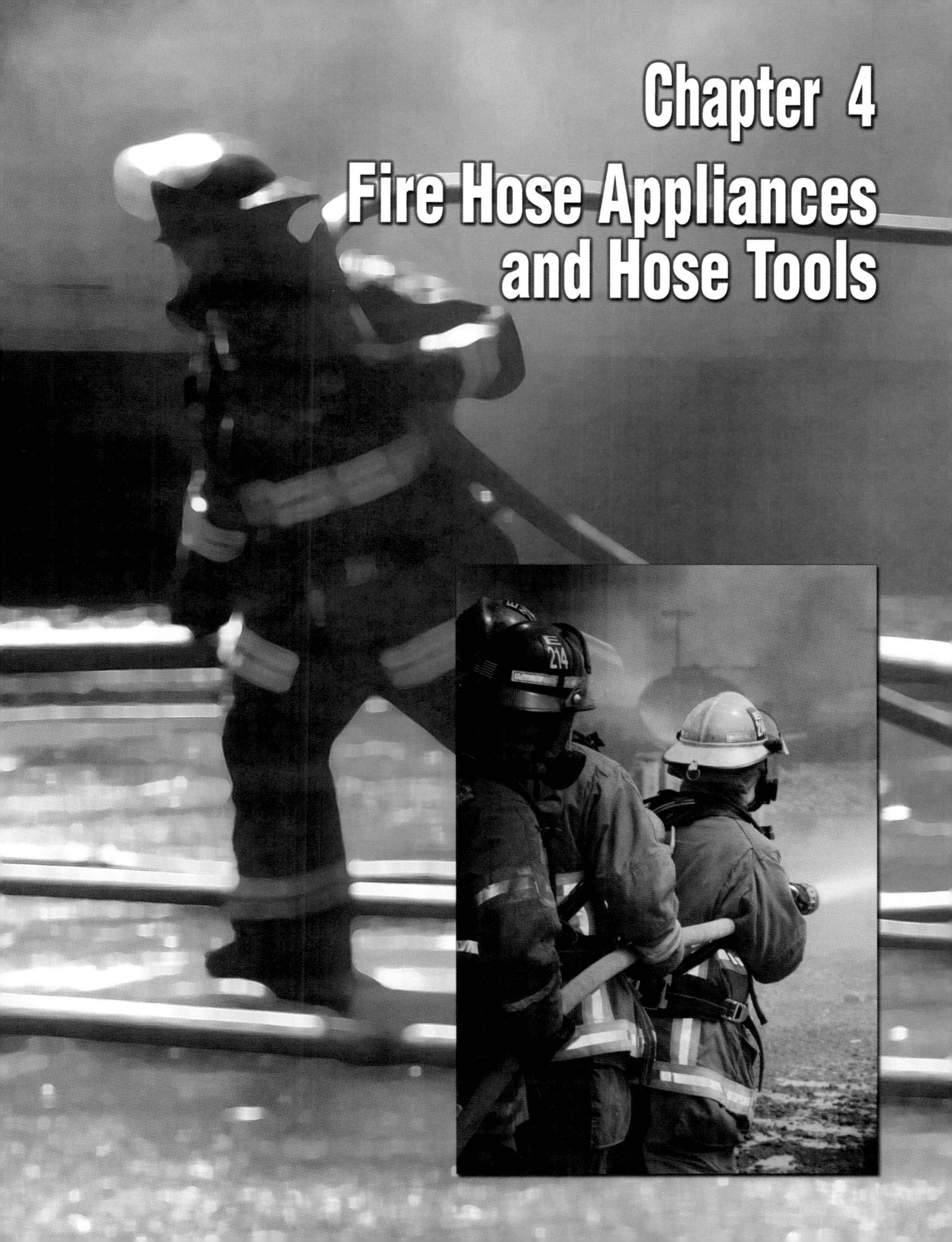

Chapter 4
Fire Hose Appliances and Hose Tools

To control water as it moves from its source to a fire, fire hose appliances such as nozzles and shutoff valves are used. *Fire hose appliances,* therefore, are devices (valves, fittings, and nozzles) other than couplings that are used with fire hose and through which water passes. Chapter 3, Fire Hose Nozzles, discussed the various types of nozzles that are available for use during emergency incidents. This chapter completes the discussion on fire hose appliances by providing a complete overview of valves, valve devices, hose fittings, and portable foam eductors/proportioners. Additionally, the general care and maintenance guidelines for these appliances and devices are also presented.

The last portion of this chapter is devoted to the accessories (fire hose tools and other devices) that ease the job of handling and maneuvering fire hose. This section also includes devices that protect the hose, fire pump, and other appliances from damage, mechanically control the flow of water, or assist fire-suppression crews while they are handling or working with fire hose.

Fire Hose Appliances

Fire hose valves and fittings are placed in the hoseline to direct water or other extinguishing agent to a specific location. Valves control the water or foam solution while fittings configure the hose to direct the flow from its source to the nozzle or other stream application device. Valve devices allow the operator to supplement the water flow and pressure in the hose to compensate for pressure loss caused by friction. Proportioners inject a liquid foam concentrate into the water stream at a rate that is required for the foam being applied and selected by the pumping apparatus driver/operator. The general maintenance guidelines give procedures for proper care of appliances.

Valves

When fire hose is connected to a fire hydrant or pump, a means of directing and controlling the water flow into the hose must be available. This process is accomplished by using an appliance called a *valve:* a device that contains an internal component that can move within the water passage to regulate the flow through the device. Valves control most intake and discharge hoselines to and from the fire pump. Six basic types of valves are available, each named for its internal component: gate, ball, butterfly, floating, piston, and clapper.

Gate

A gate valve has a "gate" that moves into the water passageway when a crank handle is turned **(Figure 4.1, p. 134).** Two types of gate valves exist: *rising stem* and *nonrising stem.* The rising stem gate valve is similar in concept to the outside screw and yoke (OS&Y) valve found in sprinkler fire protection systems. The OS&Y valve is an indicating valve (visually shows open/closed status) that has a yoke and threaded stem that controls the opening and closing of the valve. If the treaded stem is out of the yoke, the valve is open; if the threaded stem is inside the yoke, the valve is closed.

As the gate valve's handle is turned, the gate rises inside the waterway, and the gate stem rises out of the valve assembly, indicating that the valve is open. When nonrising-stem gate valve handles are turned, no stem raises or lowers. The stem remains inside the valve **(Figure 4.2, p. 134).** These valves should be marked with the number of turns required to open or close them fully. Care must be taken should the valve "hang up" or resist opening or closing — an indication that debris is blocking the valve's action or that it is already fully opened

Gate Valve: Cutaway View

Figure 4.1 A gate valve opens and closes the waterway with a gate that moves up and down inside the valve.

Figure 4.2 When operating a nonrising stem valve, the stem stays inside the valve housing in the open or closed position.

or closed. Forcing the valve in either direction may cause serious damage to the entire assembly. Gate valves are relatively heavy and bulky but are well suited for use on a hydrant or pump discharge orifice **(Figure 4.3)**.

Ball

A ball valve has a ball-shaped internal component **(Figure 4.4)**. The ball has a hole through its center, which permits water to flow through with a minimum of friction loss when it is aligned with the waterway. A quarter-turn handle or a push-pull handle (commonly called a *T-handle*) rotates

Figure 4.3 Intake gate valves may be equipped with a handwheel.

Figure 4.4 The internal ball shutoff component has a hole through its center, which provides a waterway designed to minimize friction loss when it is opened fully.

the ball to close the waterway, thus stopping the flow of water. One ball-valve type operates by a single handle attached directly to the valve device, and another type is operated by a set of rod linkages or is electrically operated by remote controlled servos.

The ball valve is the most commonly used valve in the fire and emergency services. Ball valves are found in many applications, including portable units that attach to hoselines and permanent mounts usually found on pumping or aerial apparatus. Hydraulically, pneumatically, or electrically controlled ball valve assemblies, actuated by toggle switches located on the pump panel, are becoming common on new pumping apparatus. Ball valves are used in equipment such as pumping apparatus discharge orifices and "gated" intakes, nozzles, gated and ungated wyes, and valved siamese devices (see Valve Devices section).

Butterfly

The butterfly valve has a disk (baffle) that pivots within the waterway **(Figure 4.5)**. The valve is open when the disk is aligned parallel to the waterway (the handle indicates this position by moving in the same plane). The valve is closed when the disk is turned a quarter-turn across the waterway and is perpendicular to the flow. When open, the baffle is in the center of the waterway turned parallel to the flow. The butterfly valve is most frequently used on 4½-inch (115 mm) and larger pumping apparatus intake orifices and in private water supply systems.

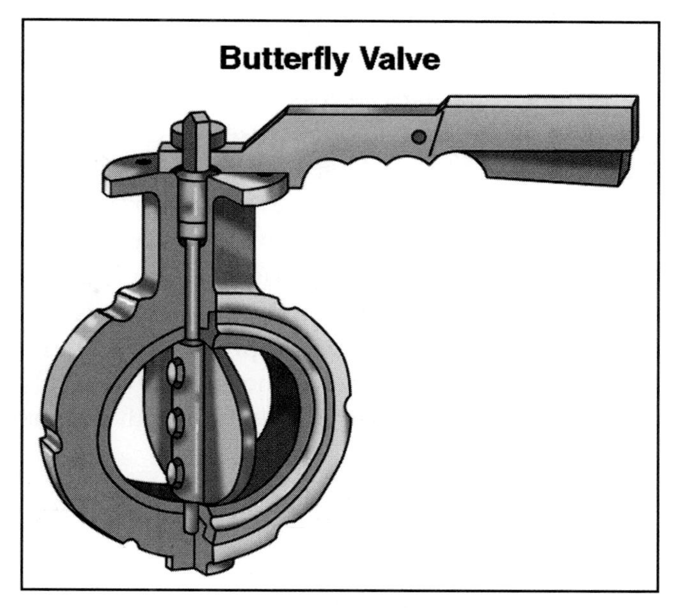

Figure 4.5 The butterfly valve has a disk (baffle) that pivots within the waterway, interrupting the flow.

It is not recommended for use on pump discharges because if the valve is not locked closed, the force of the discharged water can suddenly and violently throw open the valve and its handle.

Floating

A floating valve is a new valve design that allows the connection of a water supply to the pumping apparatus while the pump is operating and under pressure. It has a spring-loaded, dome-shaped disk within the waterway that is held in the closed position by both spring tension and internal water pressure **(Figures 4.6 a–c)**. It is typically used on a

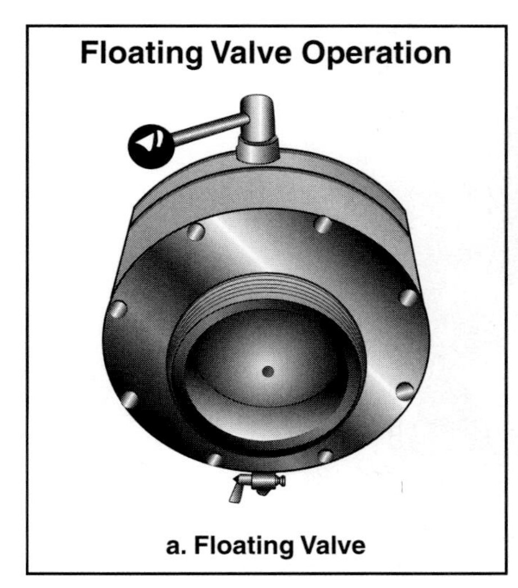

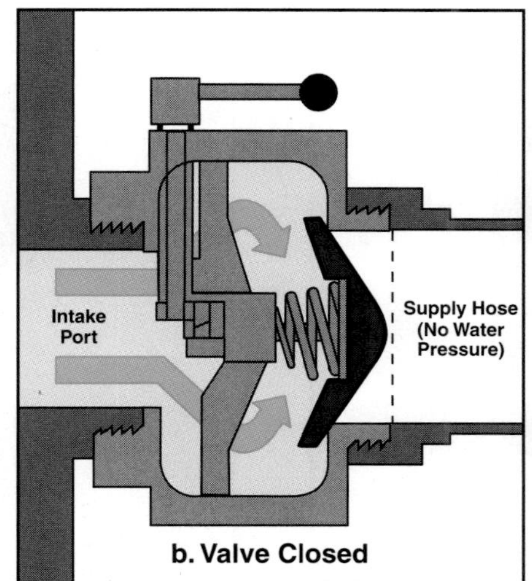

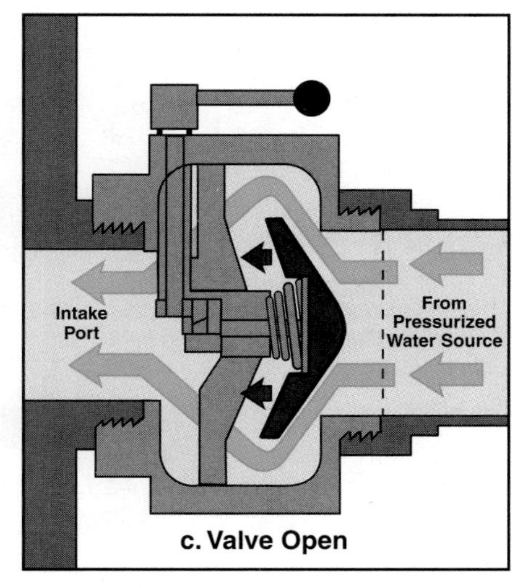

Figures 4.6 a–c (a) A floating valve is typically used on a pump intake port to promote a quick intake hose connection. (b) Valve with spring tension and no water pressure remains closed. (c) Pressurized water flow opens the valve.

main pump intake port to promote a quick intake hose connection. When incoming pressurized water flows against the disk from the outside, it opens to permit water to flow through the valve. A backflow-prevention device operates in a similar manner, allowing water flow in only one direction. It is extremely important to maintain this valve to ensure that it is debris free and that the spring is adjusted to be compatible with water pressures available for fire-suppression use. If the valve spring tension is set higher than the supply pressure, no water passes through the valve to the pump.

Clapper

A clapper valve has a hinged disk that acts as a check valve (automatic valve that permits liquid flow in only one direction). Thus, a *clapper valve* is a hinged valve that allows water to flow in one direction only **(Figure 4.7).** The valve operates automatically, swinging away from the valve seat when a pressurized water supply is provided. A clapper valve opens when water flows in one direction and closes automatically if the water begins to flow in the opposite direction. These valves can be configured with one clapper valve for a single inlet or siamese device located on a sprinkler or standpipe system/fire department connection **(Figure 4.8).** Dual clapper valves are available when it is desirable to have a siamese device located at the intake side of a supply line that may occasionally not have pressure on either inlet during fire-suppression operations. (See Siamese Device section.)

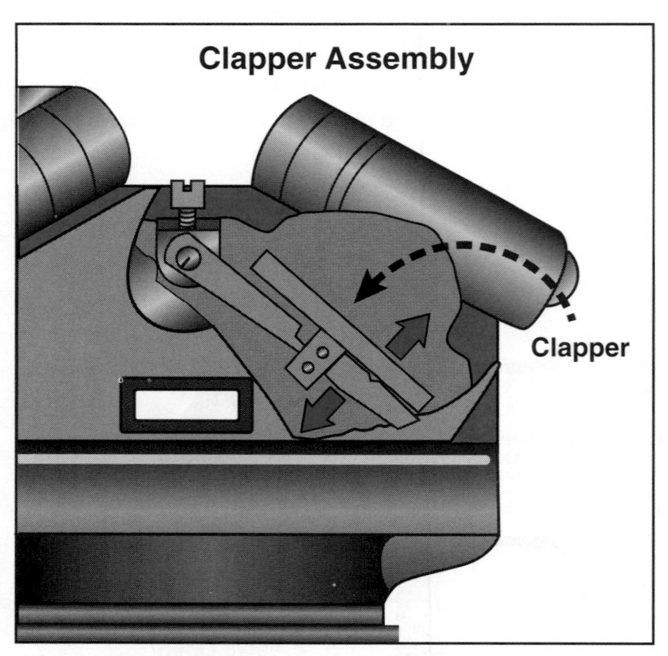

Figure 4.7 A clapper valve is a hinged disk that acts as a check valve that will allow water to flow in only one direction.

Piston

A piston valve has an internal piston that moves within a cylinder to control the flow of water through the valve. It is designed for use with large diameter hose (LDH) **(Figure 4.9).** The valve is

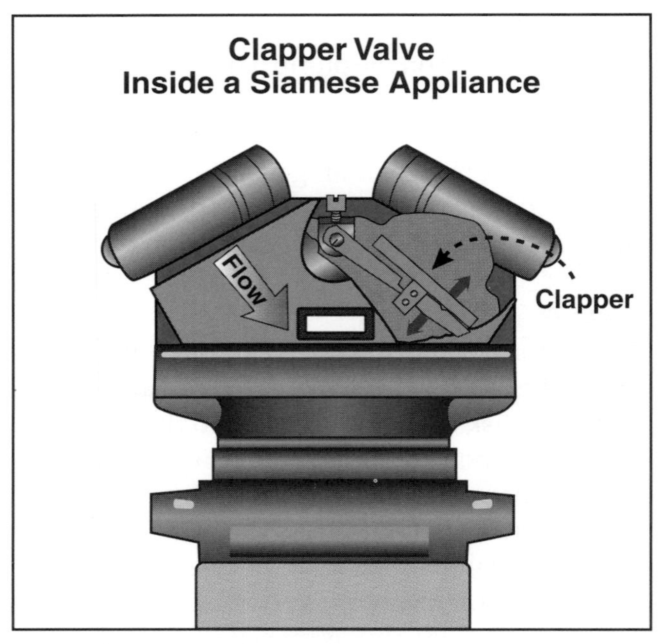

Figure 4.8 Cutaway view of a clapper valve inside a siamese appliance. The clapper valve closes and seals the open side of the siamese when the opposite intake is pressurized with water.

Figure 4.9 A piston moves inside a cylinder to regulate the flow and pressure of water through the piston valve into the pump. *Courtesy of Akron Brass Company.*

opened and closed by rotating a wheeled handle. Rotating the wheel clockwise moves the piston downward into the waterway to block the opening. Rotating the wheel counterclockwise retracts the piston from the waterway.

Valve Devices

The movement of water from its source (hydrant, static reservoir, or mobile water transport) to a pumping apparatus and then from the pumping apparatus or stationary fire pump to an emergency scene is often a complex operation. This operation is especially complex if the distance that the water must travel exceeds 500 feet (152 m). For these long fire hose deployments, the pressure in the hose must be supplemented incrementally at intervals along the hose lay to compensate for the pressure loss caused by water friction and increases in elevation. In most cases, once a water flow is established, it needs to remain uninterrupted by the pressure-boosting effort. A number of valve devices make this situation possible. Valve devices include four-way hydrant valves, automatic hydrant valves, manifolds, water thief devices, wye appliances, siamese devices, inline relay valves, and intake relief valves.

Four-Way Hydrant Valve

A four-way hydrant valve (also known as a *hydrant booster valve* or *Hydrassist™ valve*) allows the first-arriving attack pumping apparatus to connect the valve to the hydrant, connect a supply line, proceed to the emergency scene, and charge the supply line from the hydrant. Upon arrival at the hydrant, a second pumping apparatus connects a supply hose from the hydrant valve's large discharge port to its pump intake. The pump intake shutoff valve is then opened, allowing water to supply the pump. A hoseline from a discharge port on the pump is then connected to the second intake on the hydrant valve **(Figures 4.10 a and b).** The driver/operator on the second pumping apparatus then charges the supply line. As the pressure increases in the supply line, it overcomes the hydrant pressure and closes a clapper valve inside the four-way hydrant valve, directing the pressurized flow through the valve and into the original supply line. All of the available water is then directed to the fire pump **(Figures 4.11 a–c, p. 138).**

Figure 4.10a A four-way hydrant valve allows the water supply from a hydrant to be boosted by a second arriving pumping apparatus without shutting down the hydrant.

Figure 4.10b A hand wheel gear-operated four-way valve incorporating two intakes and two discharges. *Courtesy of Akron Brass Company.*

This procedure allows the attack pumping apparatus to position itself in close proximity to the emergency scene and be available to directly support fire-suppression efforts or other emergency scene operations by dispensing equipment, deploying attack hoselines, and possibly engaging apparatus-mounted master stream devices. Upon arrival and under the direction of the incident commander, the second pumping apparatus may deploy a second supply line to further support the water needs demanded during fire-suppression activities or other emergency scene operations. This second supply line is a charged, pumped hoseline deployed parallel to the first supply line **(Figures 4.12 a and b, p. 139).**

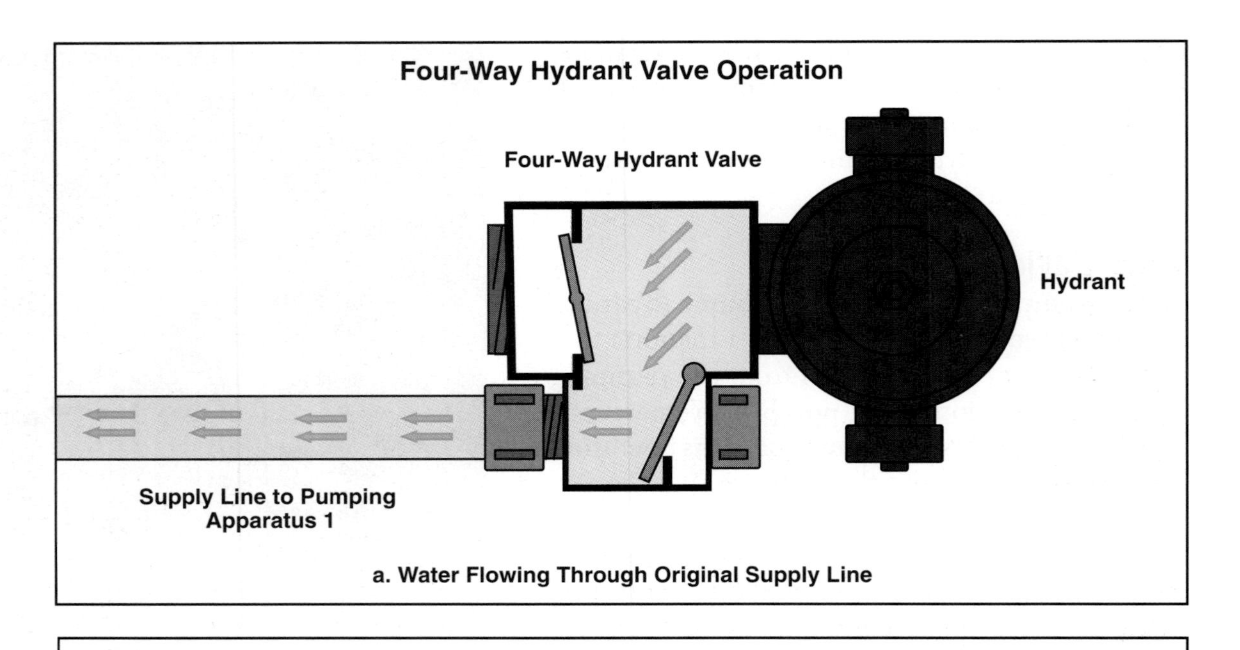

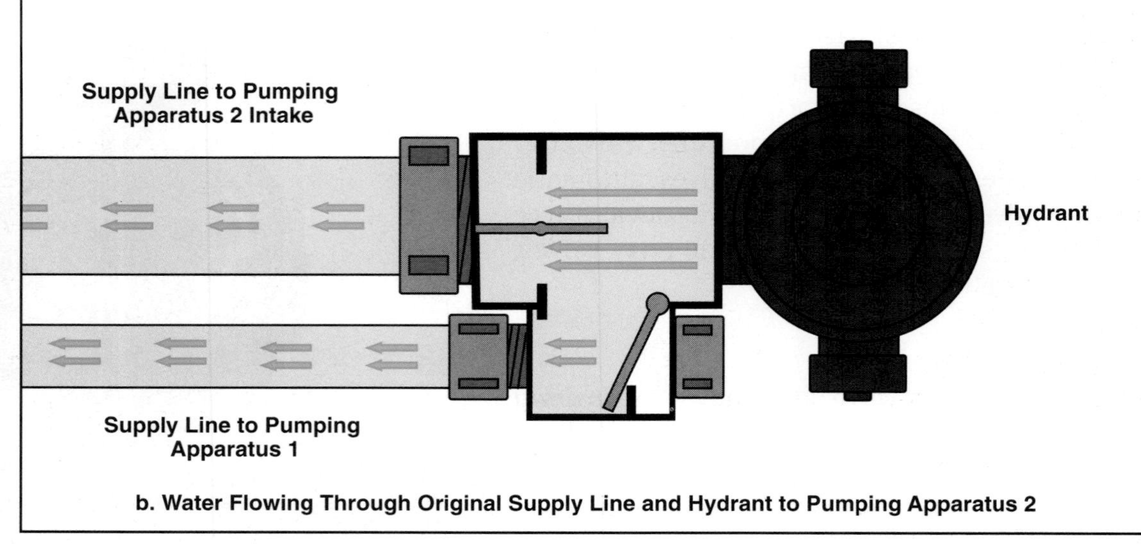

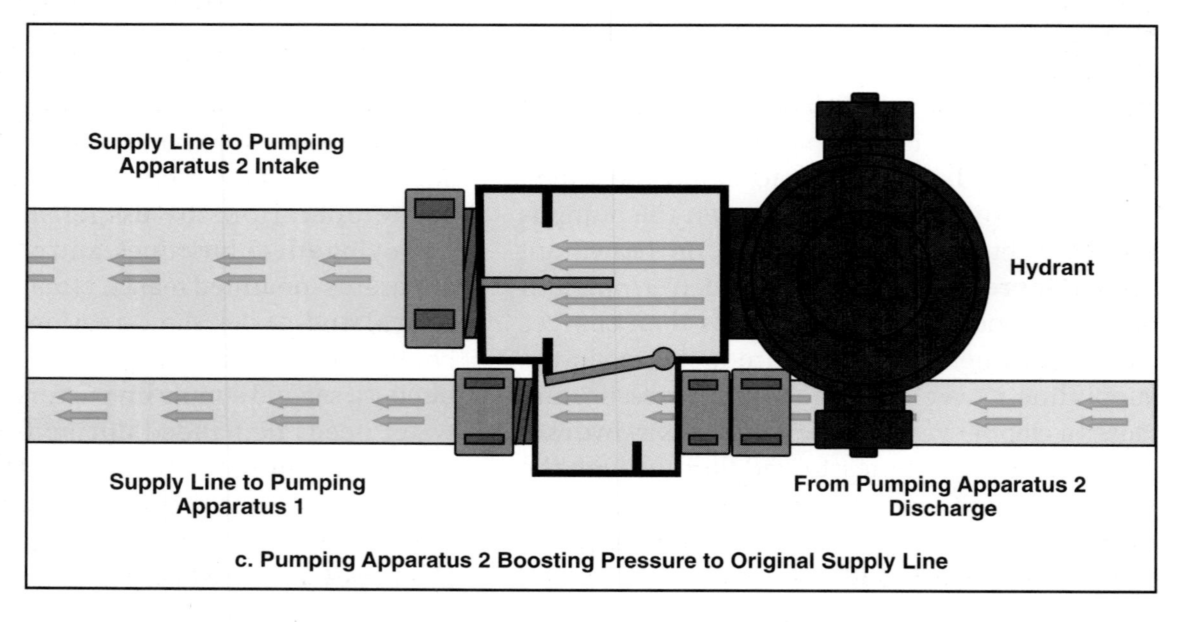

Figures 4.11 a–c This illustration series shows the flow of water through a four-way hydrant valve.

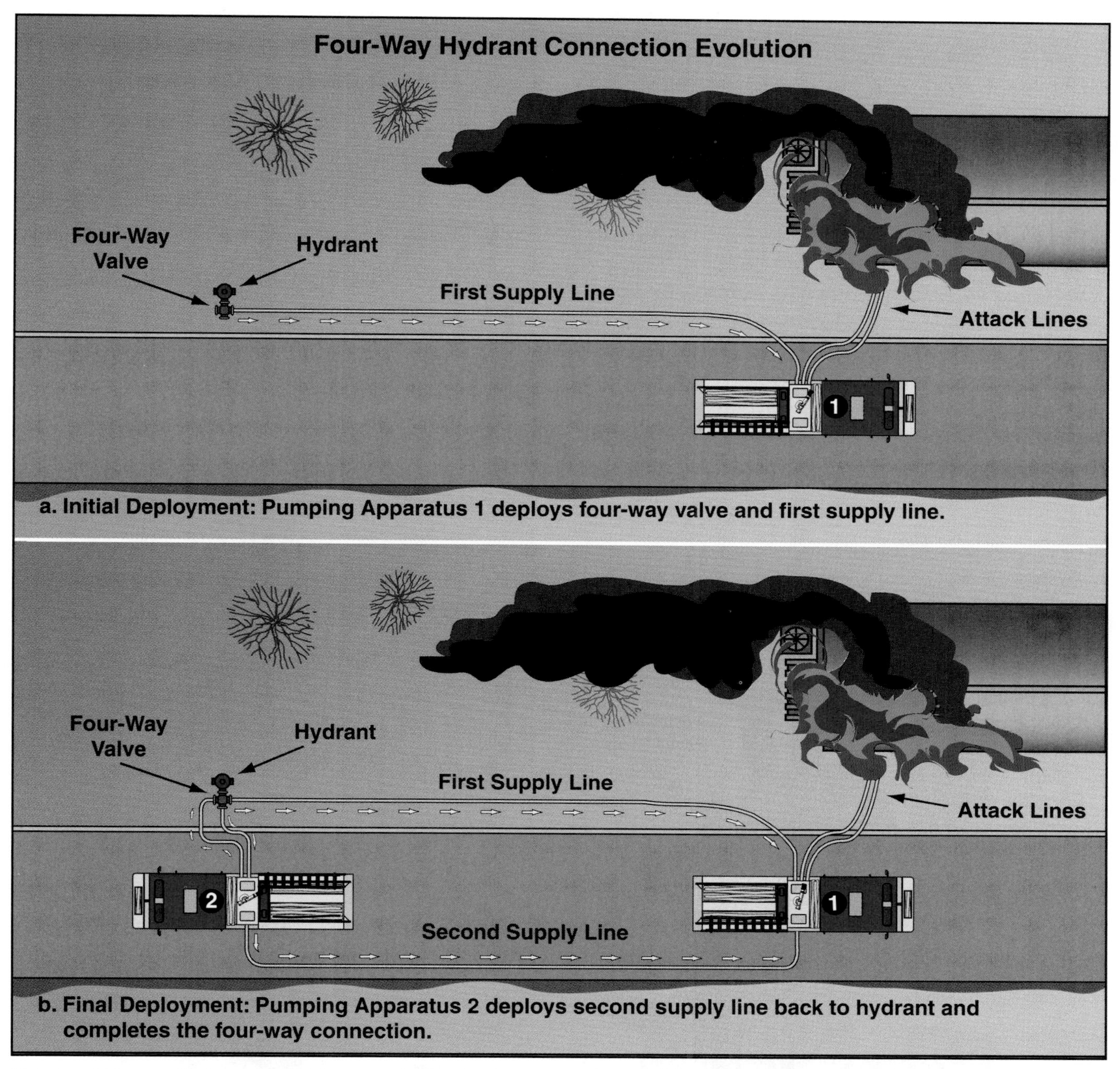

Figures 4.12 a and b A completed four-way hydrant valve connection provides a flexible water supply that can be expanded to meet the fire flow needs of an incident.

Automatic Hydrant Valve

An automatic hydrant valve eliminates the need to leave a person at the hydrant when making the hydrant connection. Otherwise, when a pumping apparatus lays its own supply line from a hydrant to the fire or emergency location, someone must stay at the hydrant until the hose from the hose bed is disconnected and reconnected to an intake. When the connection is made, the hydrant is opened to charge the supply hose. This procedure, in effect, takes one person away from the emergency scene for several minutes during the first critical minutes of an initial attack on a fire or other emergency situation. The procedure for using an automatic hydrant valve is the same as for any hydrant valve, except that the hydrant can be opened immediately, permitting the hydrant operator to rejoin the attack crew immediately.

A *mechanically delayed* automatic hydrant valve automatically opens after a preset period of time **(Figure 4.13).** Normally the hydrant cannot be opened until someone either clamps the hose or connects it to the pump intake. Using the mechanically delayed automatic hydrant valve allows the hydrant to be turned on before the connection of the supply line to a pump intake **(Figures 4.14 a–c).**

Manifold Valve

A *manifold valve* (also known as *portable hydrant, phantom pumper,* or *large diameter distributor*) is a device that receives a supply of water and distributes it through valves to a number of hoses **(Figure 4.15).** Most often this device is associated with the deployment of large diameter supply hoselines. In some situations, a pumping apparatus supplying

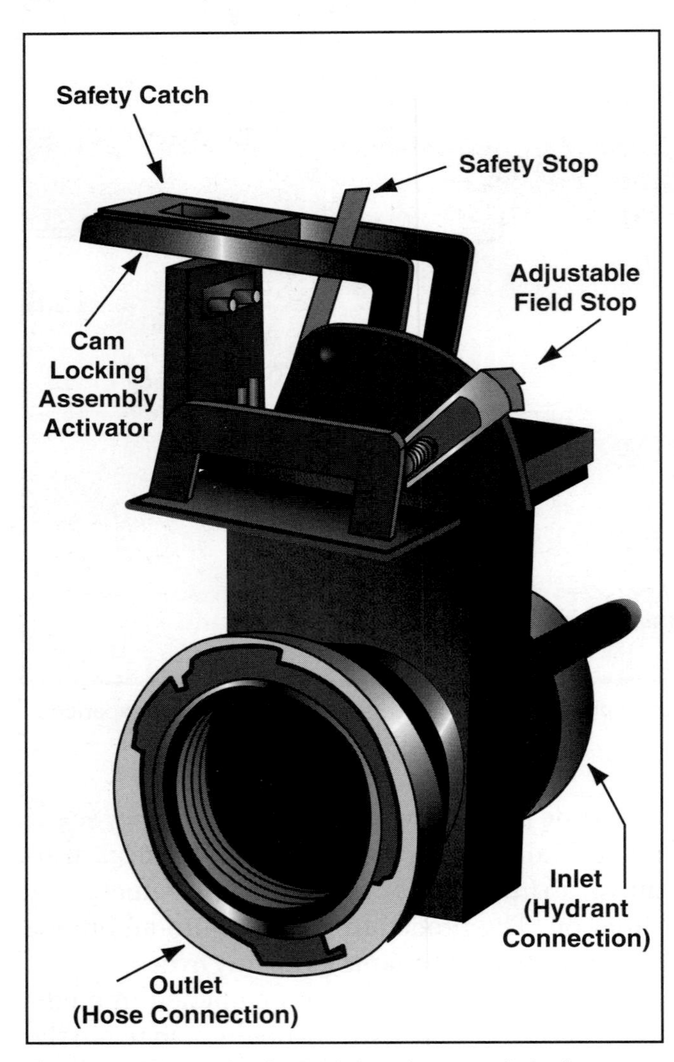

Figure 4.13 A mechanically delayed automatic hydrant valve (also known as a Carlin valve manufactured by Hydra-Shield Manufacturing, Inc.) opens automatically after a preset period of time.

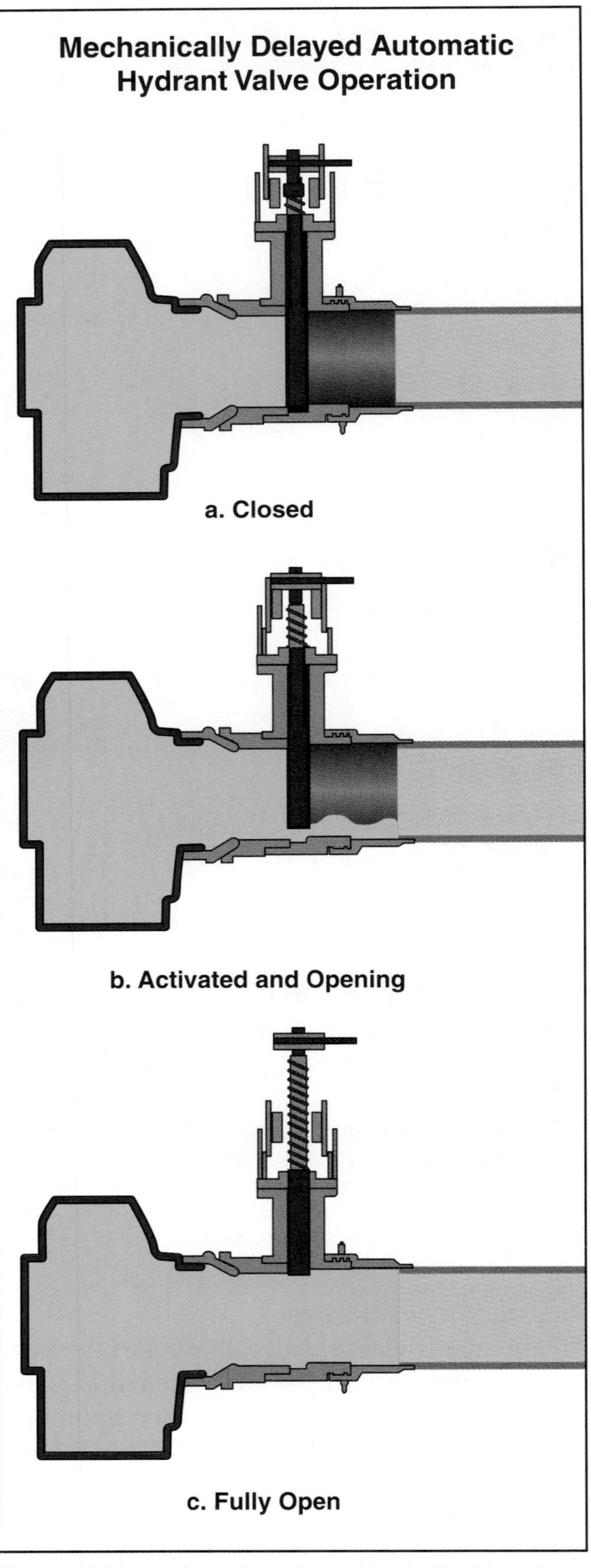

Figures 4.14 a–c Operation of a mechanically delayed automatic hydrant valve, showing (a) the valve closed, (b) activated and opening, and (c) fully open.

Figure 4.15 A large diameter hose manifold distributes water to a number of hoselines.

several attack hoselines must be positioned some distance from the fire or emergency scene (perhaps pumping at a hydrant). Two basic methods provide water to each nozzle: (1) Pump directly to each attack hoseline or (2) pump into a single hoseline that supplies a manifold.

A manifold valve is particularly useful in situations where pumping apparatus have limited access to buildings such as in shopping malls, in large apartment complexes, and on university campuses **(Figure 4.16).** Compared to laying a number of small diameter hoselines, use of a manifold with LDH minimizes the amount of time required to lay hose and usually requires a lower pump pressure. An LDH-supplied manifold can support three or more 2½-inch (65 mm) attack hoselines. A disadvantage, however, is that if the supply line fails, all of the attack hoselines connected to the manifold will be without water. Deploy a new supply line to reestablish the water supply as quickly as possible should that situation occur.

Wye Appliance

A wye appliance has a single female fitting on the intake side and two male fittings on the discharge side. A wye is one of the best appliances to use when it is necessary to supply water directly to two attack hoselines with a single supply hoseline. The wye may or may not be gated, although most modern wyes are. Usually this device is configured with manually operated ball or gate valves to permit separate control of water to each discharge hoseline. A reducing wye provides water for two attack lines smaller in diameter than the supply hose. The most common gated wyes are 1½- by 1½- by 2½-inch (38 mm by 38 mm by 65 mm) models. A second type is a 2½- by 2½- by 2½-inch (65 mm by 65 mm by 65 mm) design **(Figures 4.17 a–d, p. 142)**.

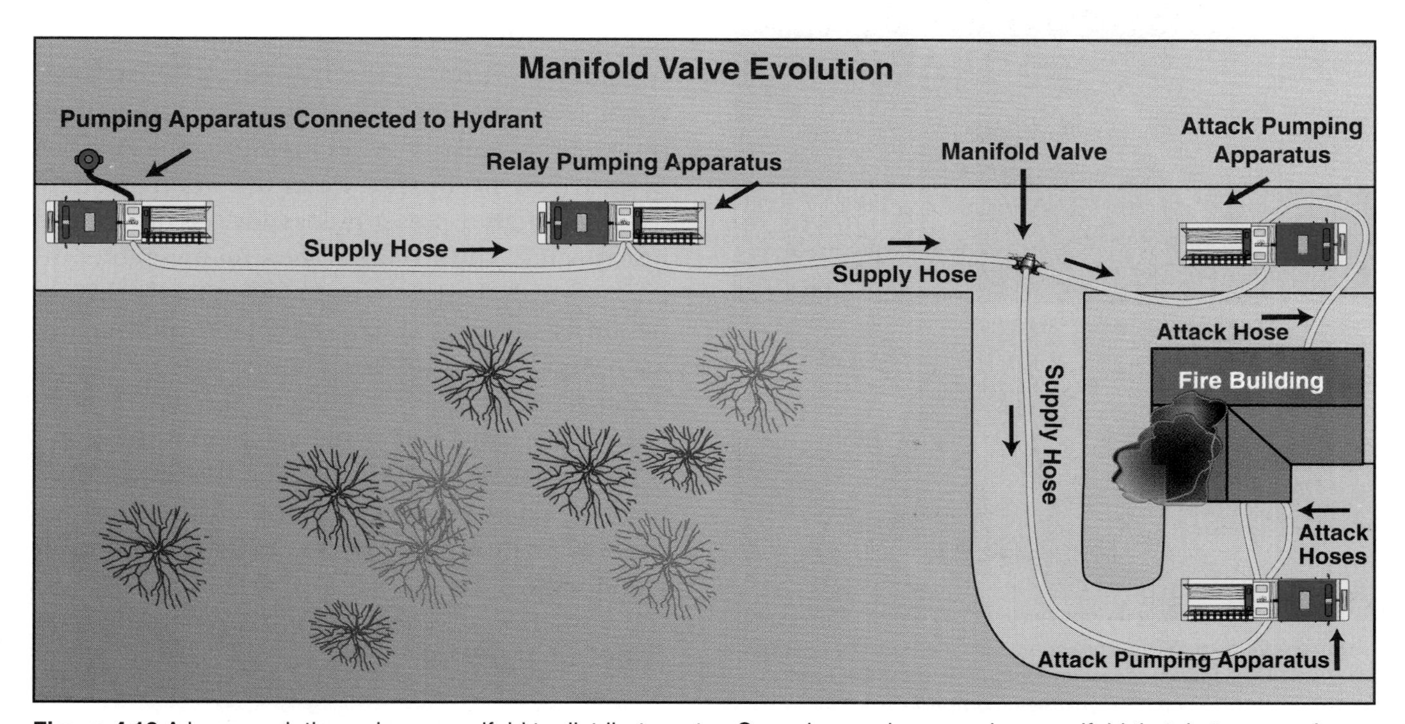

Figure 4.16 A hose evolution using a manifold to distribute water: Some hose relays supply a manifold that, in turn, sends smaller supply hoses to multiple attack pumping apparatus.

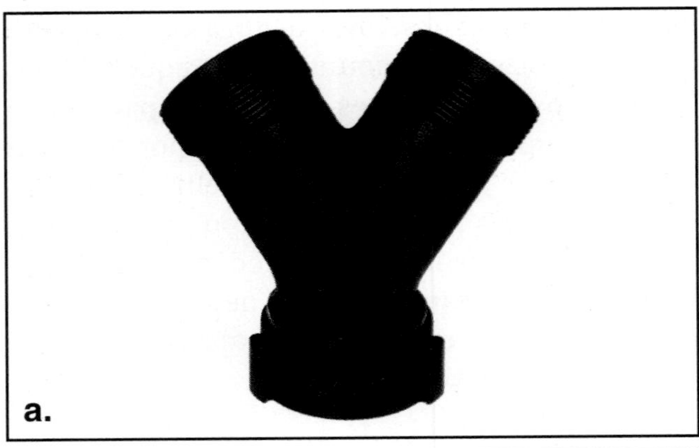

a.

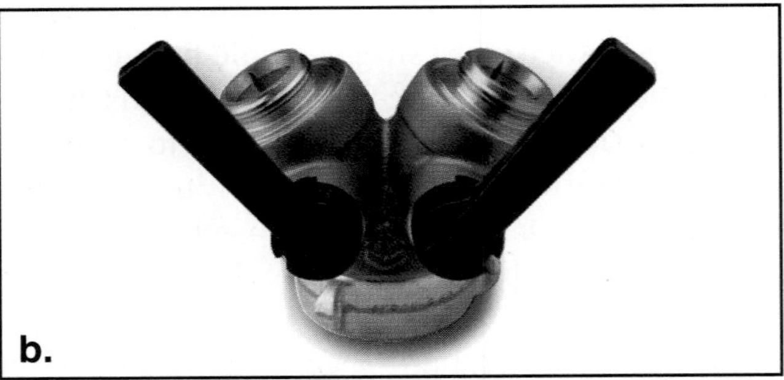

b.

c.

d.

Figure 4.17 a–d Wye appliances: (a) Simple wye. *Courtesy of Elkart Brass Manufacturing Company.* (b) Gated reducing wye. *Courtesy of Akron Brass Company.* (c) 2½- by 2½- by 2½-inch (65 mm by 65 mm by 65 mm) gated wye, and (d) Storz wye. *Courtesy of Sam Goldwater.*

With the growing use of large diameter supply hoselines, it is becoming common to see large (some as large as 4 inches [102 mm]) gated wyes. The fire-suppression crew and the driver/operator must exercise care when deploying these devices. Improper fire-pump pressures may not supply the required volume of water to the nozzle. Also, nozzles must be opened and closed slowly to prevent causing either a loss of water or an overpressure situation with the other nozzle. Operating nozzles and valves too quickly can also cause water hammer that can damage equipment and cause serious injuries to fire and emergency services responders.

Siamese Device

A siamese device has two female couplings on the intake side and one male coupling on the discharge side **(Figures 4.18 a–c).** A siamese allows the laying of two parallel hoselines for a portion of the hose lay. This deployment allows the use of 2½-, 3-, and 3½-inch (65 mm, 77 mm, and 90 mm) supply hoselines to deliver high flows at lower total friction loss than using a single hoseline. A siamese device brings the hoselines together.

One situation where a siamese is used is when friction loss in a supply hose needs to be reduced. Most siamese appliances have check valves (clapper valves) so that if unequal pressure is transmitted through the two parallel hoselines, water from the hose of greater pressure cannot return to the hose of lesser pressure. Because two pressurized streams of water are being brought together, higher pressure loss for the device can be expected. The mixing of the streams creates increased turbulent flow and subsequently greater fiction loss within the siamese device. Additionally, as the total volume of the streams is restricted, the water must increase its velocity to pass through the device. The greater the velocity of the water through the siamese device, the higher the friction loss experienced. For additional information, see IFSTA's **Pumping Apparatus Driver/Operator Handbook.**

Water Thief Device

A water thief (a variation of the gated wye) has three gated outlets and quarter-turn gate valves on its 1½-inch (38 mm) threaded outlets and usu-

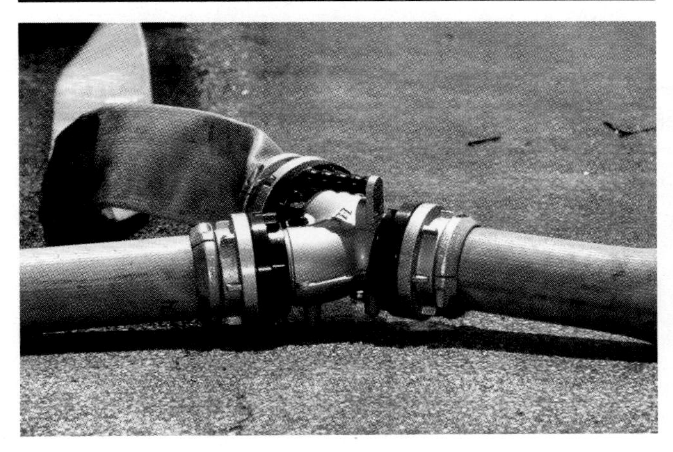

Figure 4.18 a–c *Top* (a) A siamese device allows two lines of the same size to be brought together to form one stream. *Courtesy of Task Force Tips. Middle* (b) A triple siamese can be attached to the main intake on a pumping apparatus with two or more lines going into the pump. *Courtesy of Elkhart Brass Manufacturing Company. Bottom* (c) A clapper valve siamese allows the connection of one or two supply hoses while preventing water from entering the uncharged line. *Courtesy of Task Force Tips.*

ally a quarter-turn valve on its 2½-inch (65 mm) threaded outlet **(Figure 4.19).** A water thief is used on a 2½-inch (65 mm) or larger hoseline, most often near the nozzle. This placement allows fire hose with 1½- and 2½-inch (38 mm and 65 mm) threads to be used as needed for fire attack or other emer-

Figure 4.19 A water thief device is used to extend 1½- or 1¾-inch (38 mm or 45 mm) hoselines while retaining a 2½-inch (65 mm) or larger attack hoseline capability.

gency scene use. The water thief can also be found with only 1½-inch (38 mm) threads.

When using a water thief that has a 2½-inch (65 mm) inlet and two 1½-inch and one 2½-inch (38 mm and 65 mm) discharges, avoid the use of all discharges at one time unless the device is within one section of hose to the pumping apparatus. Large diameter water thief devices have been introduced with the increasing use of large diameter supply hose (3½ inches to 6 inches [90 mm to 150 mm]). If used in conjunction with attack hoselines, the supply hoselines must meet the minimum standard for attack hoseline operating pressures as outlined in NFPA 1961, *Standard on Fire Hose* (2002).

Inline Relay Valve

An inline relay valve placed along the length of a supply hose permits a pumping apparatus to connect to the valve to boost pressure in the hoseline **(Figures 4.20 a and b, p. 144).** When a supply line is exceptionally long (over 500 feet [150 m]) or deployed up a steep grade, friction and elevation head pressure loss can significantly deplete water pressure, which reduces the amount of water delivered to the terminal end of the hoseline. One of the best ways to boost pressure in the hose is to position one or more pumping apparatus along the line to "relay" the water. When a supply hoseline is laid before the arrival of the relay pumping apparatus, inline relay valves are placed at regular intervals within the hose lay. This procedure allows the pumping apparatus to connect to the supply hoseline and boost its pressure without interrupting the water flow **(Figure 4.21, p. 144).**

Figure 4.20a An inline relay valve is placed along the length of a supply hose, which permits a pumping apparatus to connect to the valve and boost pressure in the hoseline. *Courtesy of Jaffrey Fire Protection.*

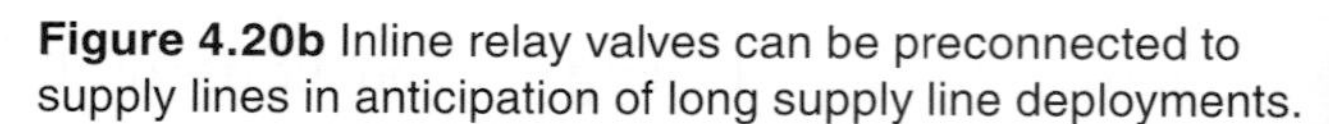

Figure 4.20b Inline relay valves can be preconnected to supply lines in anticipation of long supply line deployments.

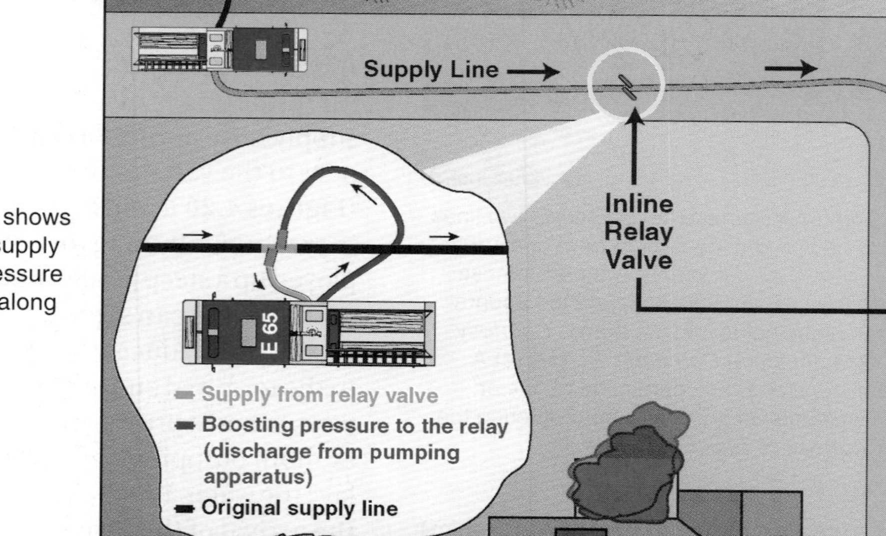

Figure 4.21 This diagram shows how an inline relay valve supply line evolution. It allows pressure to be boosted at intervals along the supply line.

Intake Relief Valve

An external intake relief valve has a spring-loaded internal component that diverts any sudden pressure surge away from the fire pump **(Figure 4.22).** The force of water hammer in a large diameter hose is significantly greater than the force in smaller hoselines. For this reason, the potential for damage to a fire pump is also greater when large diameter hose is used as a supply hoseline. One of the best ways to prevent this damage, of course, is to open and close valves slowly. Another way to prevent damage is to place a pressure relief device such as an intake relief valve on the pump intake.

Fittings

As fire hoselines are deployed during fire-suppression or other emergency operations, special situations often arise that require the use of a category of hose equipment classified as fittings. *Fittings* are devices that facilitate the connection of hoselines of different sizes to provide a continuous, uninterrupted supply of water from a source to the nozzle. Occasionally, because of differing fire hose loads, the diameter of hose, threads per inch on the coupling, or the need to attach two female or male couplings together, the use of special adapter fittings is needed to successfully complete hoseline deployment.

Fittings also include elbows that are used on fire pump intakes and discharge valves to reduce kinking of hoselines and to allow preconnection of suction sleeves. Caps and plugs are fittings used to seal the ends of hoselines and protect the threads of discharge and intake orifices on pumping apparatus. Caps are for male intakes/discharges/couplings, and plugs are for female intakes/discharges/couplings. Blind caps seal the ends of hoses equipped with sexless couplings. Reducers are adapters used to connect a small hose to a larger one. Increasers are adapters used to attach a large hoseline to a smaller one.

Figure 4.22 An external intake relief valve operates when excess pressure is received at the intake.

Adapter

An *adapter* is a fitting for connecting hose couplings with dissimilar threads or coupling types but with the same inside diameters. The double male and double female adapters that connect two threaded couplings of the same thread type, size, and sex are possibly used more often than any other hose fitting **(Figures 4.23 a and b).** As the use of Storz or sexless couplings becomes more popular, adapters to make connections with fire hose equipped with threaded couplings are needed **(Figure 4.24, p. 146).**

Figure 4.23 a and b Adapters connect two threaded couplings of the same thread type, size, and sex. *Top* (a) Double male. *Bottom* (b) Double female.

Figure 4.24 Large diameter hose equipped with Storz couplings requires an adapter to connect to a hydrant. *Courtesy of Sam Goldwater.*

Additionally, in locales where different coupling threads are used by fire and emergency service organizations that mutually respond to emergency incidents together, adapters that have male threads of one type thread and the female threads of the other are necessary. Often, this need is the result of a community's fire and emergency services organization having a nonstandard thread size, while buildings were equipped with National Standard Threads on standpipe and sprinkler connections during their construction. This situation often occurs when a government building is erected using federal or state building standards that are not subject to local review. It is still the responsibility of the local fire and emergency services organization to make hoseline connections to these buildings. Structures constructed outside the political jurisdiction of a fire and emergency services organization may not be equipped with the same thread size; therefore, adapters are required to either complete hoseline connections from the pumping apparatus to the buildings' systems or to extend standpipe packs from internal standpipes.

Elbow

An *elbow* is a hose fitting that provides support for an intake or discharge hose at the pumping apparatus **(Figure 4.25)**. The elbow eliminates the possibility of the hose kinking at the intake or discharge point of the pumping apparatus and relieves stress on the fire hose coupling. When large

Figure 4.25 Elbow fittings reduce stress on hose and allow the flow of water to be redirected without the hose kinking.

diameter supply hose is employed, the use of an elbow at the point of connection to the fire pump is vital in limiting the potential of coupling failure.

Reducer/Increaser

A common fire hose fitting is the reducer. When large diameter hoselines are deployed, a *reducer* can be attached and a smaller hoseline extended. Reducers are also found on pumping apparatus discharge outlets, allowing a small handline to connect directly to the fire pump. The large diameter side of the fitting is on the intake end, which is the female coupling. The extension of a hoseline using a reducer limits the extended line to the size of the reducer. In this situation, a gated wye (if available) is a better selection because it gives the option of extending an additional hoseline from the open gate **(Figure 4.26)**.

An *increaser* performs the same function as the reducer, except in reverse. The large diameter coupling is on the male side of the fitting. When using an increaser, friction loss is decreased, but it is not usually an efficient way to accomplish this

Figure 4.26 A reducer allows a large diameter hoseline to be reduced in size and smaller hoselines to extend toward the fire.

decrease. Increasers are used most often when a supply hose is brought to a large pumping apparatus pump connection that is equipped with a larger diameter intake.

Cap and Plug

A *hose cap* closes male couplings at the end of a fire hoseline and is also commonly found on pump discharges **(Figure 4.27).** Caps prevent the unwanted flow of water while protecting the exposed male threads of the coupling or discharge.

A *hose plug* has male threads and is attached to the female coupling. It performs essentially the same function as a cap by preventing water from escaping from the open "butt" or coupling. Plugs also prevent dirt and other debris from entering the coupling and hoseline **(Figure 4.28).**

Portable Foam Eductors/Proportioners

Portable foam proportioners mix water and a predetermined quantity of foam concentrate by eduction (induction) or injection into a fire stream to create a foam solution. These two methods are defined as follows:

Figure 4.27 Hose caps close and protect the male couplings at the end of a hoseline or pump discharge port.

Figure 4.28 A hose plug has male threads and prevents dirt and other debris from entering the coupling and hoseline.

- ***Eduction (Induction)*** — The pressure of a water stream flowing through an orifice reduces atmospheric pressure, which inducts or draws foam concentrate into the water stream by the Venturi principle.
- ***Injection*** — Pressurized proportioning devices inject foam concentrate into a water stream at a predetermined foam concentrate/water ratio at a higher pressure than that of the water.

As described earlier in Chapter 3, Fire Hose Nozzles, the foam solution is carried to the nozzle where it is mixed with air (known as *aspiration*) and discharged as finished foam. For this process to successfully produce good finished foam, the proportioning equipment must operate within strict design specifications. Improper operation of even the best foam-proportioning equipment results in poor quality finished foam or none at all.

The manual proportioning method of *batch mixing* is the simplest method of proportioning: Pour a premeasured quantity of foam concentrate directly into the water reservoir, and pump the tank in the normal manner. Foam solution then pumps through any open hoseline and discharges through a regular nozzle. Some types of foam concentrates are not designed for batch mixing because of their polymer components. See **Principles of Foam Fire Fighting,** 2nd edition, for information on foam concentrates.

Some foam concentrates mix readily with water and retain their foaming qualities for extended periods, but other foam concentrates lose their foaming properties after 24 hours once they are mixed with water. If agitated or mixed too rapidly, lathering may occur when water refills the tank, resulting in pump cavitation (when air cavities are created in the pump or bubbles pass through it). The bubbles move from a high-volume point in the pump into the pressurized portion of the pump where the bubbles collapse and fill with fluid. The high velocity filling of these bubbles or cavities causes a severe shock to the pump. Damage to the pump can result in extreme cases or with prolonged exposure to these conditions. Priming difficulties may also be experienced because of the entrained air in the foam concentrate and water solution.

Although the batch-mixing method is very simple to deploy, flow duration is limited to the amount of foam concentrate that can be mixed in a tank. On large incidents, continuous foam solution flow cannot be maintained until the foam concentrate supply is replenished. It is almost impossible to maintain the proper water/concentration ratio unless the tank is completely empty.

Premixing is another commonly used method of proportioning: Premeasured portions of water and foam concentrate are mixed in a container. The most common application for the premixing foam concentrates method is in fire-extinguishing equipment that uses stored pressure for discharge energy — for example, handheld wheeled fire extinguishers or skid-mounted twin-agent units and rapid intervention vehicle-mounted systems. With this equipment, the foam solution is stored in a pressure vessel charged with nitrogen. The agent is discharged when the nozzle is opened. The disadvantage encountered is that the supply of agent is limited and the entire unit must be recharged to continue finished foam application.

Proportioning equipment is generally divided into two types: portable foam eductors and apparatus-mounted/fixed-system proportioners. Portable foam eductors are the most simple and common forms of foam proportioning devices in use today. Three types of portable foam eductors in current use are as follows:

- In-line foam eductor
- Self-educting handline foam nozzle eductor
- Self-educting master stream foam nozzle eductor

In-line Foam Eductor

An in-line foam eductor is attached directly to the pumping apparatus pump panel discharge or at some point in the hoseline after it has left the apparatus. The safe operating pressure of the pump, hoseline, and nozzle along with the required eductor operational pressure must always be monitored by the driver/operator **(Figure 4.29).** It is extremely important to follow the manufacturer's operating instructions regarding inlet pressures and maximum hoseline lengths between the eductor and foam nozzle.

The Venturi principle is employed by in-line eductors to draw or draft foam concentrates and induct them into the water streams. When water passes through the eductor, the stream is narrowed, speeding the velocity of the water. This action creates a low-pressure region inside the eductor. A pickup tube connects the foam concentrate supply to this low-pressure region of the

Figure 4.29 The eductor is attached either directly to the pumping apparatus pump panel discharge or at some point in the hoseline after it has left the apparatus. The driver/operators monitors the operation.

eductor and allows concentrate (that is at a higher pressure) to draw up the tube and into the high-speed water stream passing through the eductor body **(Figure 4.30)**.

Self-educting Handline Foam Nozzle Eductor

The self-educting handline foam nozzle eductor operates on the same principle as the in-line system. However, it is built into the nozzle rather than being at a remote location in the hoseline. As a result, its use requires the foam concentrate container to be close to the nozzle (no more than 100 feet [30 m] away). Should it become necessary to move the nozzle, the foam concentrate container also needs to move. The logistical problems of relocating the nozzle are magnified by the quantity of foam concentrate that may need to be moved also. Use of a foam nozzle eductor could compromise fire and emergency service responder safety: Emergency responders cannot always move quickly, and

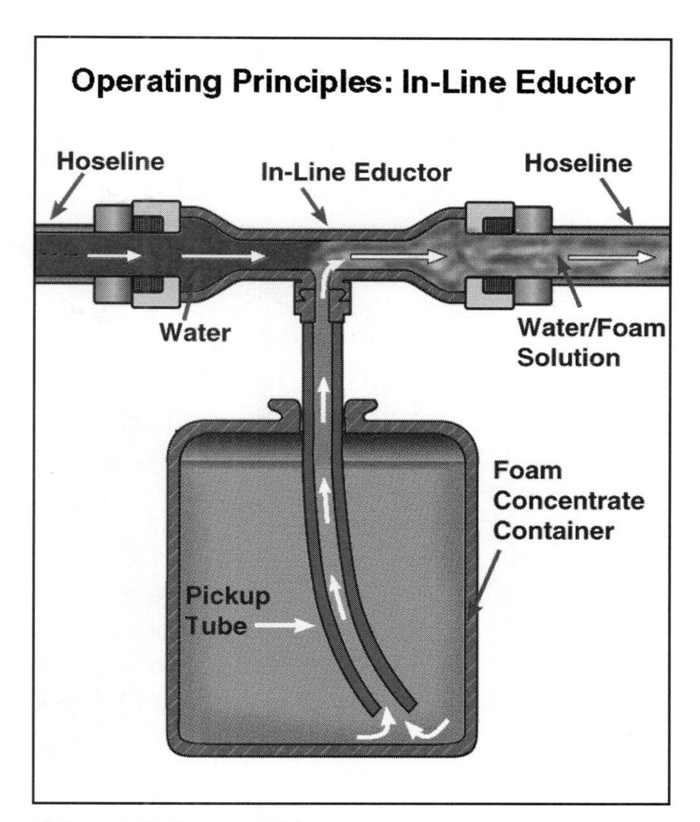

Figure 4.30 Operating principles of an in-line eductor: Foam concentrate is drawn into the eductor using the Venturi principle.

they would have to leave the foam concentrate container should they need to quickly withdraw for any reason.

Self-educting Master Stream Foam Nozzle Eductor

The self-educting master stream foam nozzle eductor is deployed when foam stream flows of 350 gpm (1 325 L/min) or more are necessary **(Figure 4.31, p. 150)**. These nozzles have flow capabilities up to 12,000 gpm (45 425 L/min). As presented in Chapter 3, Fire Hose Nozzles, this nozzle also uses the Venturi principle in its design.

A jet ratio controller supplies foam concentrate to the nozzle. This controller is a type of in-line eductor that allows the foam concentrate supply to be as far as 3,000 feet (914 m) away from the self-educting master stream foam nozzle. Fire and emergency service responders who are involved in operating fire pumps and maintaining the foam concentrate supply can remain at a safe distance from a fire or emergency scene when this device is used.

Figure 4.31 Self-educting master stream foam nozzles with flow capacities up to 12,000 gpm (45 425 L/min) are often mounted on top of apparatus.

General Care and Maintenance

As with any fire-fighting equipment, valves, valve devices, fittings, and foam proportioners need to be well maintained to prevent failure under emergency conditions. The following maintenance guidelines apply to most fire hose appliances:

- Thoroughly inspect appliances on a scheduled basis as well as after each use.
- Never use grease or oil to lubricate moving parts. Use lubricants such as silicone or graphite that do not injure gaskets, O-rings, or other rubber or elastomer parts of the appliance.
- Lubricate all threads to facilitate the process of connecting appliances to hoselines and prevent fittings from freezing together when coupled during cold weather.
- Do not allow foam concentrates or foam solution to freeze either before or during application.
- Replace worn gaskets in female fittings. Carry spare gaskets on the pumping apparatus. Turn gaskets over to ensure that they experience uniform wear.
- Make connections hand tight; overtightening can flatten gaskets and cause leaks. If a leaking connection must be tightened with a tool designed to perform this task (usually a spanner wrench; see next section), exercise care to tighten the appliance only enough to stop leaks.
- Carry appliances with valves open to prevent the accumulation of dirt and debris inside the valve mechanism or waterway.
- Do not use brass polish or any other abrasive on appliances made of hardened aluminum alloy.
- Repaint appliances as needed to prevent rust and corrosion.
- Service test an appliance that was damaged and subsequently repaired according to NFPA 1962, *Standard for the Inspection, Care, and Use of Fire Hose, Couplings, and Nozzles and Service Testing of Fire Hose* (2003), and NFPA 1963, *Standard for Fire Hose Connections* (2003).

Hose Tools and Other Devices

A number of accessories are available that make the handling of fire hose more efficient, easier, and safer for fire and emergency service responders. Not only does the use of these devices (hose straps, ropes, and chains) assist in the deployment of attack and supply hoselines, they are also specifically designed to protect the hose and couplings from damage during fire-fighting operations. Other devices (hose bridges and rollers) are designed to protect the hose from external abuse. Hose clamps and jackets help when problems arise while using fire hose such as small holes appearing in hose, couplings that are out-of round, or unwieldy hoselines. Other devices help protect fire hose against unnecessary wear and damage. Often these devices are very simple in nature, but the jobs they perform are invaluable to fire and emergency service responders in making their jobs easier and safer. Devices of this type include spanner and hydrant wrenches, hose control devices, suction hose strainers, and hose wringers.

Spanner and Hydrant Wrenches

The primary purpose of a *spanner wrench* (also know as simply *spanner*) is to tighten and loosen hose couplings **(Figures 4.32 a and b).** As with many fire-fighting tools, a number of other features have been built into these devices to assist fire and emergency service responders. These features include the following:

- Wedge for prying
- Opening to fit gas valves
- Slot for pulling nails
- Flat surface for hammering

Hydrant wrenches are primarily used to remove caps from fire-hydrant outlets and to open fire-hydrant valves **(Figures 4.33 a and b).** A five-sided (pentagon-shaped) wrench opening is provided. This opening fits over the hydrant operating nut. Many hydrant wrenches are adjustable so that even though differences in the size of the operating nut may exist, hydrants can still be opened by one tool. Some hydrant wrenches are designed with an extension that can also be used to tighten and loosen couplings.

CAUTION

Open a fire hydrant with a hydrant wrench specifically designed for the task. Serious damage can occur to the hydrant valve and operating nut if opening is attempted using other tools or "cheater bars."

Hydrant wrenches should be between 16 and 18 inches (406 mm and 457 mm) long but not so long that opening the operating nut or removing

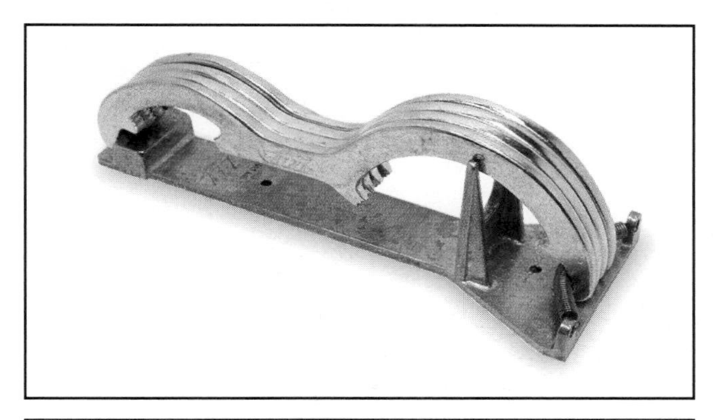

Figures 4.32 a and b Use a spanner wrench to tighten or loosen a coupling connection. *Top* (a) Use on storz connections. *Courtesy of Task Force Tips. Bottom* (b) Use on threaded connections. *Courtesy of Elkhart Brass Manufacturing Company.*

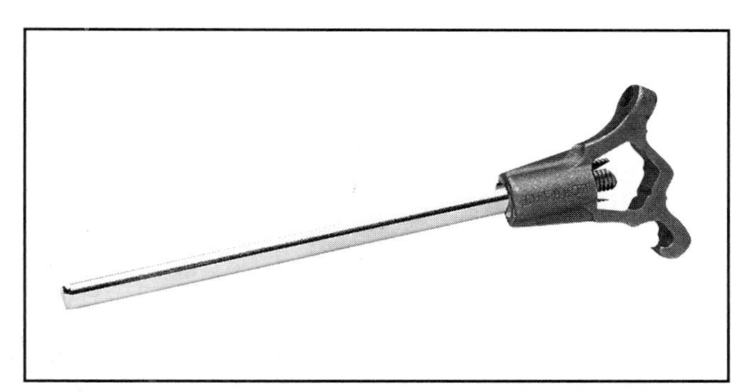

Figure 4.33a The pentagon-shaped hydrant wrench. *Courtesy of Elkhart Brass Manufacturing Company.*

Figure 4.33b A hydrant wrench removes the caps and opens a fire hydrant.

hydrant nozzle caps is impeded due to the length of the wrench. Hydrants should be installed in a manner that allows a minimum of 16 inches (406 mm) from the center of the operating nut or the center of the cap to the ground as the wrench is turned. Additionally, most municipalities require a clear 3-foot (1 m) circle around a hydrant measured from the center of the operating nut **(Figure 4.34).**

Hose Strap, Rope, and Chain

One of the most useful tools to aid in handling a charged hoseline is a hose strap. Similar to the hose strap are the hose rope (rope/hose tool) and hose chain tools **(Figure 4.35).** These devices can carry and pull fire hose, but the singular value of using them is to provide a more secure means to handle pressurized hose when applying water. Another important use of these tools is to secure hose to ladders, in stairwells, and to other fixed objects. Additionally, ropes and chains can be used to secure hose loops and multiversals. Exercise great care when using these devices to secure charged hoselines. Improper securing methods can allow a charged, flowing hoseline to release unexpectedly. The result can be serious injury to fire and emergency service responders and damage to equipment, including the hoseline itself. Regular inspection and maintenance of these devices is essential to ensure that they are ready for service.

Hose Control Device

In situations where a hoseline must be kept in a static position for an extended length of time, a hose control device can hold the nozzle end of a 2½-inch (65 mm) hoseline. The device is stable enough to leave the hoseline and nozzle unattended as an exposure protection line when staffing is short or conditions are too hazardous for emergency personnel to be in the area. These devices are not used as commonly today as they were in the past; however, they are still found in some small municipalities to assist in the deployment of 2½-inch (65 mm) exposure hoselines.

Hose Roller (Hoist)

A hose roller (hoist) consists of a metal frame with two or more rollers. Fire hose can be damaged when it is dragged over sharp surfaces such as those found on roof edges and windowsills, but this device allows the hose to pass over a set of rollers, preventing the hoseline from abrading at the edge of a roof or parapet. The roller can also be used to protect ropes when they are deployed for rescue situations and hoisting equipment to support fire-suppression operations. Attach the hose roller (hoist) to a stabilized object and secure with a rope or locking carbineer. Often a strap is attached to the roller assembly to keep the hose centered on the rollers as it advances or retreats.

Hose Jacket

A *hose jacket* is a hose tool composed of a two-piece metal cylinder with heavy rubber gaskets located on each end. It is often referred to as *Coopers hose jacket* or *Boston hose jacket.* The device hinges open

Figure 4.34 Proper placement and installation of a fire hydrant is essential to its use.

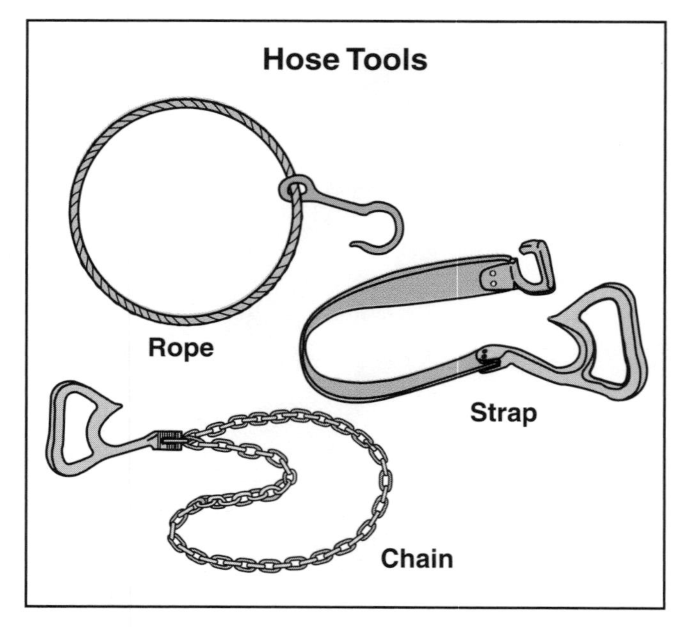

Figure 4.35 Typical hose tools.

Figure 4.36 A hose jacket in use.

and close with a clamp-retaining catch located on the open side **(Figure 4.36).** When a section of hose ruptures or develops a leak, a hose jacket can be installed on the hose at the point of rupture to allow the hoseline to continue transporting water effectively. The rubber gaskets form a watertight seal to prevent leakage. The clamp-retaining catch locks the cylinder closed during use. Other uses of this device include connecting hose with mismatched threads, connecting damaged screw-thread couplings, or connecting male-to-male or female-to-female couplings. Hose jackets are made in two sizes: 2½ inches and 3 inches (65 mm and 77 mm). They are not available for large diameter supply hoselines. The proper sized hose jacket must be used with the appropriate hose diameter.

The most practical way to permanently and safely correct a ruptured hoseline is to shut down the line and replace the damaged section of hose. When fire-fighting conditions or other emergency operations are such that it is not possible to shut down the hoseline and replace the bad section, a hose jacket can enclose the hoseline temporarily so that it can continue to operate at full pressure until a replacement line is deployed.

Hose Clamp

Three types of hose clamps are available based on the method by which they work: *screw-down, press-down,* and *hydraulic-press* **(Figure 4.37).** Each of these designs works with different hose diameter sizes. The hydraulic-press and screw-down types are most often used to clamp large diameter supply hoselines. The press-down type uses a scissorlike action to clamp the hose, and it is limited to smaller hoselines sized 3 inches (77 mm) and less. A hose clamp is used to stop the flow of water in a hoseline in the following situations:

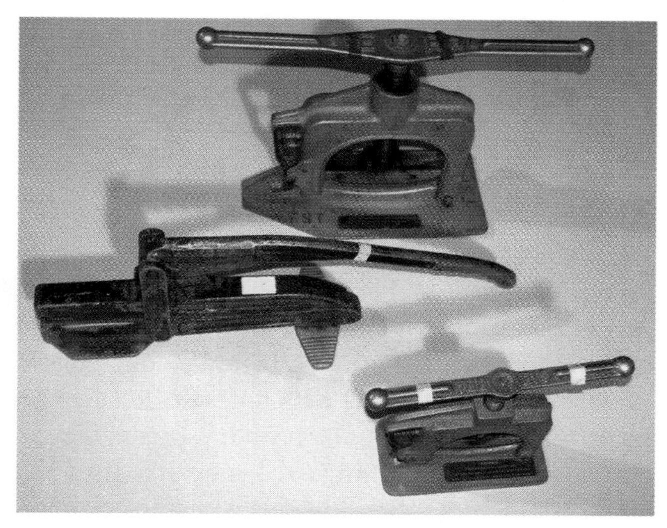

Figure 4.37 Various types of hose clamps.

- Prevents charging the hose bed during hose-deployment operations by stopping the flow of water from a hydrant supply hoseline
- Allows replacement of a burst hose section without turning off the water supply
- Allows extension of a hoseline without turning off the water supply

It is important to remember that a hose clamp can cause injury to fire and emergency service responders or damage fire hose if it is not used correctly. Some general guidelines that apply to all hose clamps are as follows:

- Apply the hose clamp at least 20 feet (6 m) behind the pumping apparatus.
- Apply the hose clamp no farther than 5 feet (1.5 m) from the coupling on the incoming water-side.
- Stand to one side when applying or releasing the press-down type of hose clamp (the operating handle is prone to snapping open suddenly).
- Place the hose evenly in the jaws of the hose clamp to avoid pinching it.
- Open and close the hose clamp slowly to avoid water hammer.

Suction Hose Strainer

A suction hose strainer is attached to the drafting end of a hard suction (sleeve) to keep debris from entering the fire pump. Otherwise, debris would pass into and then through the pump and down the hoseline, damaging the pump and possibly clogging the nozzle. Three types of strainers are available for use in drafting operations: *barrel, floating,* and *low-level.*

Barrel

The barrel strainer is possibly the most recognizable type of strainer because it has been a standard piece of equipment since power fire pumps first came into service **(Figure 4.38).** It is constructed of either a metal cylinder with holes drilled around its full circumference or heavy wire screen. It has a threaded female coupling to connect it to the hard suction drafting tube. The opposite end is a solid plate with an eyelet for attaching a rope or chain to assist in the deployment/recovery of the drafting tube assembly. The base plate also allows the tube to suspend in a manner that keeps the strainer above the bottom of a water source, helping to keep mud and other debris from being pulled into the pump.

These devices are only used when sufficient water is available to maintain a vacuum in the drafting tube. Drafting can cause a whirlpool to form; and if the water level is low (less than 2 feet [6 m] of coverage), air will be pulled into the suction tube. A small quantity of air can pass through the pump with little effect; however, sustained airflow results in the loss of vacuum, causing drafting to cease.

Figure 4.38 Barrel strainer.

Floating

The floating strainer draws water into the suction hose while floating on the surface of a pool or stream **(Figure 4.39).** A floatation device keeps the strainer on the surface, and a swiveled suction intake with holes on only the bottom provides the drafting waterway. In order for a floating strainer to work properly, it must float freely in the water and not be constrained by the rigidity of the intake hose. The major disadvantage of this device is that it is limited to drafting water through only one side of the intake. Because of this limitation, the rated capacity of the pump may not be realized. The pumping apparatus driver/operator must closely monitor the pump to avoid the possibility of pump cavitation.

Low-Level

The low-level strainer operates in a portable water tank or swimming pool and is capable of drafting water down to a depth of 2 inches (51 mm). The strainer sits on the bottom of the tank or pool. The use of this device must be monitored to ensure that debris is kept away from the intake portion of the strainer. It is also essential to deploy enough drafting suction hose to ensure that the device lays flat on the bottom, keeping its sharp square edges away from the material of the portable tank or pool sides. During deployment and extraction, the driver/operator must exercise care so that the portable tank or pool is not damaged, which could result in a catastrophic loss of water **(Figure 4.40).**

Figure 4.39 Floating strainers are used in shallow water.

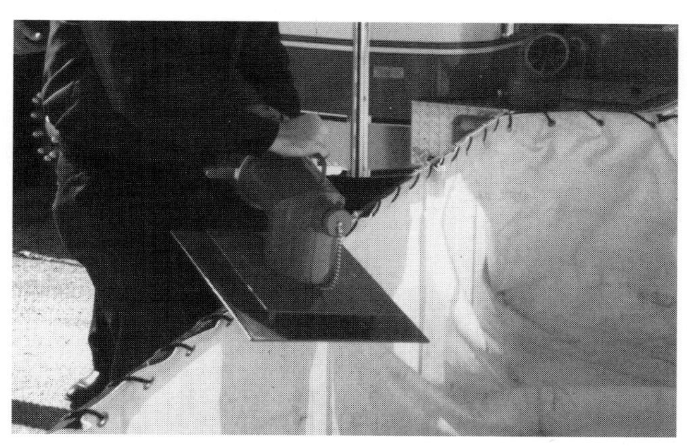

Figure 4.40 Low-level strainers are most commonly used when drafting from portable water tanks.

Hose Bridge

A hose bridge (also called *hose ramp*) helps prevent injury to fire hose when vehicles have to cross it. It is used whenever a hoseline crosses a street or other area where vehicular traffic cannot be diverted. Some nontraditional uses for hose ramps include placement over railroad tracks as crossing ramps and under hose as chafing blocks. Some ramps can also be positioned over small Class B liquid spills to keep hoselines away from potentially damaging chemicals. Large diameter hose ramps are available but extremely cumbersome due to size and weight, often limiting their deployment to long-term campaigns. It is critical that fire and emergency service responders or law enforcement officers direct traffic over hose bridges in an appropriate manner. This fact is especially true with hose bridges over large diameter supply hose because the height that a vehicle may bounce on its springs could damage the hose if the vehicle crosses the bridge at an excessive speed.

Hose Wringer

A hose wringer removes water and air from large diameter hose before it is reloaded. Two fire and emergency service responders place it over the hose and then walk along the hoseline, one on each side with the wringer clamped over the hose by one emergency responder's grip **(Figure 4.41)**.

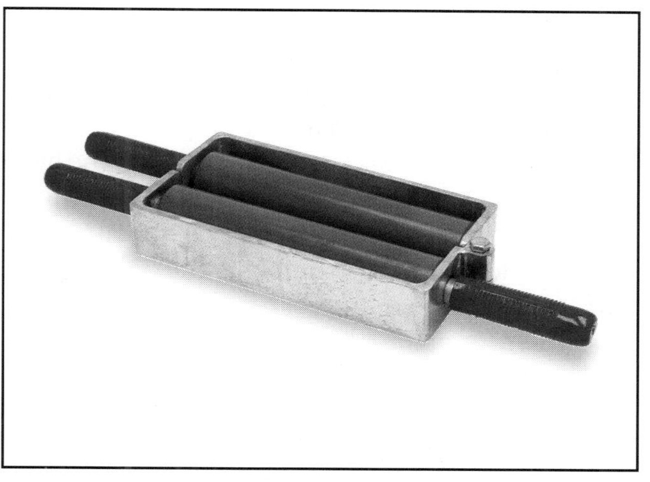

Figure 4.41 A hose wringer drains water and removes air from large diameter hose before reloading it into the hose bed. *Courtesy of Task Force Tips.*

Summary

The process of moving water from its source to a fire or emergency scene requires equipment that assists in directing and controlling water to and through a fire pump and then to attack or exposure hoselines. By using valves (and other appliances), water flows to the pump and then to the various discharge points required to control the emergency situation. Additionally, valves can open, slow, or completely interrupt or stop water flow through the hoselines. A driver/operator can split the supply of water using a wye and direct it to two discharge points or combine hoselines with a siamese to increase the flow to a nozzle or device. Adaptors can adjust the stream to conform to a new condition. Often it is desirable to reduce the size of an attack hoseline to make it more manageable; reducers are employed to accomplish this task. Increasers do the opposite. Large hoselines produce lower friction loss when moving water. Increasers allow small hose streams to attach to larger hoselines to reduce the pressure loss due to friction.

Because fire-suppression efforts or liquid spill control measures may involve flammable liquids or chemicals that do not respond well with water applications, finished foam is used to control these incident types. For foam applications to be effective, an eductor/proportioner is employed to produce finished foam. This device measures and mixes foam concentrate with water so that finished foam can be dispensed through a nozzle. Many types of foam concentrate are available, each possessing different attributes that mitigate different fire-suppression or liquid spill situations. Eductors/proportioners were developed to produce the desired finished foam.

Special tools are also used by fire and emergency service responders to work with fire hose, hose appliances, and foam proportioning devices. These tools, often developed to meet a single need, allow emergency responders to efficiently adjust the hose and attached appliances to ensure a continued water supply to a fire or other emergency situation. An example of this type of tool is the hose clamp. Additionally, other tools such as hose bridges have been developed to protect the hose and appliances from damage during fire-suppression or other emergency operations. Other devices (hose jackets) have been designed to temporarily fix a problem with a hose so that an uninterrupted hoseline operation can continue. Each of these hose valves, adaptors, proportioners, or tools provides an essential component in any emergency operation. Without their availability, the ability of fire and emergency service responders to perform their jobs would be severely hampered, and in many instances, emergency operations could be in jeopardy.

Chapter 5
Basic Methods of Handling Fire Hose

Chapter 5
Basic Methods of Handling Fire Hose

The use of hoselines is central to all fire-suppression and other emergency operations and must be a prime focus of training evolutions. The deployment of hose at the scene of a fire or flammable liquid spill can be complicated by many variables: size of the fire/spill, location of the water source, quantity of extinguishing agent available, type of extinguishing agent being used, number of personnel, and extent of equipment resources. No matter how involved the hose deployment may be, the success of the operation depends upon executing a series of basic tasks: carrying or dragging hose, connecting couplings, and attaching nozzles and other appliances. This chapter is devoted to the discussion and illustration of a number of widely accepted methods for handling and operating fire hose, including the various hose rolls, hoseline deployment and advancement, and hose connections (nozzles, couplings, fixed fittings, and monitors). These fundamental hose-handling operations are essential to the success of any fire-suppression or other emergency effort and must be carried out with efficiency and effectiveness by each fire and emergency service responder.

Hose Connections

The process of connecting and disconnecting a hose connection is for the most part a simple matter of screwing together a threaded male and female hose coupling or joining two sexless (Storz) couplings to make a continuous water conduit. Under fire-fighting or other emergency conditions, however, hose must be connected quickly and efficiently. The need for speed and accuracy under adverse conditions requires training. This training includes practice not only in the specific techniques of connecting/disconnecting hose, but also in connecting hose to discharge gates, fixed fittings, nozzles, master stream devices, and sprinkler and standpipe systems (see From Standpipes section under Hoseline Advances). Procedures for loosening a tight connection are also given.

Connecting/Disconnecting Hose Couplings

A fire and emergency services responder must be able to connect, disconnect, add, or replace fire hose as efficiently as possible during fire-suppression or other emergency operations. Although this procedure seems to be a very simple process, training and practice are vital. Although similar in concept, the process of connecting threaded and nonthreaded couplings are different. Some of the following basic guidelines apply when making most threaded hose connections:

- Make all connections on threaded hose couplings hand tight without the use of spanner wrenches so that the hose coupling can be disconnected later by hand **(Figure 5.1, p. 160)**. The anticipated twist of the hose during the process of charging the line is in the direction that will tighten the coupling connection. Not everyone at an emergency scene carries spanner wrenches, and spanner-tight connections could create a delay when the person taking the assembly apart at the connection must stop to find a spanner wrench.

  NOTE: It may be necessary to use spanner wrenches or other devices to tighten hose connections during drafting and hose-testing operations.

- Check for the presence of a gasket when connecting any type of swiveled coupling. A visual check can be done quickly **(Figures 5.2 a and b).** If sight is obscured by darkness or smoke, feel for the gasket with a finger. Check for worn gaskets on hand-tightened couplings that are leaking and replace gaskets instead of tightening leaking couplings with spanner wrenches.
- Connect sections of fire hose so that the hose edges are in the same plane or crease so that the uncharged hose lays flat from one section to another when it is uncharged **(Figure 5.3).** This practice makes hose easier to handle and load by fire and emergency service responders.

Various one- and two-person methods of connecting/disconnecting hose couplings are available. Making use of the Higbee cut/indicator also helps when making coupling connections.

Higbee Cut/Indicator

Making a hose connection on threaded couplings is often a difficult task to perform given all of the possible conditions that might be present when a connection needs to be made. Although the threads on a coupling, nozzle, or appliance are usually coarse enough to keep them from jamming or "cross-threading," a thread design was developed to further assist fire and emergency service responders in completing the connection. The *Higbee cut* (often referred to as a *blunt start*) is a tapered end of the threaded portion of the male and female coupling threads. A *Higbee indicator* or *Higbee mark* is cast into the male coupling lug and the female swivel coupling lug. These couplings are designed so that when the lugs are brought together (adjacent to each other) the coupling's Higbee cuts are aligned and the couplings can be

Figure 5.1 Make all connections on threaded hose hand tight.

Figure 5.3 Load fire hose so that all sections remain in the same orientation with the creases aligned along the same plane.

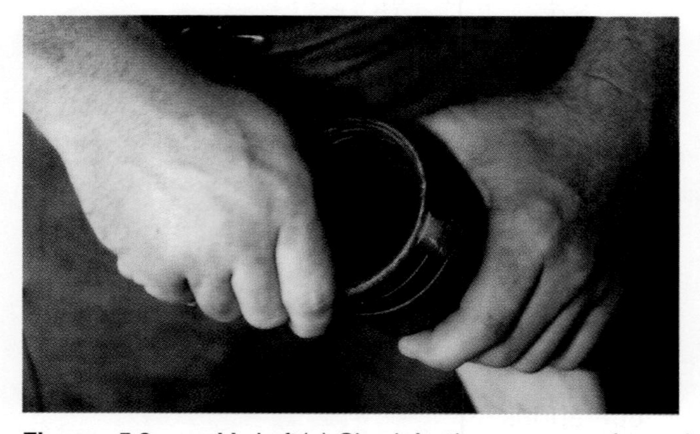

Figures 5.2 a and b *Left* (a) Check for the presence of a gasket when connecting any type of coupling. *Right* (b) Flex the gasket to ensure that it remains pliable with no cracks or splits.

tightened without cross-threading. The Higbee indicator can be located by sight or by feel if the connection is being done at night or in low-light situations.

WARNING
Personal protective equipment must be worn at all times during emergency operations. The Higbee indicator is molded into the rocker lug of the coupling and can (with practice) be found while wearing gloves. Do not remove gloves to locate the Higbee indicator.

One- and Two-Person Methods

Under fire-fighting or other emergency conditions, it is common that a fire and emergency service responder works alone or in pairs to connect hose sections, appliances, or fittings. Each of the methods presented allows a person to connect the hose alone, or connections can be accomplished using the two-person method if staffing permits. Three one-person methods and two two-person methods for connecting sections of hose are given: foot tilt, between the feet, and across the leg, plus the two-person methods for connecting both threaded and sexless (Storz) couplings. The same procedures used for these methods, completed in reverse order, can also be used to disconnect couplings.

Foot tilt. This method is a quick, easily accomplished hose- or appliance-connection technique. The procedure is possibly the simplest of the three described, allowing economy of movement and low effort on the part of a fire and emergency services responder. This method is described in **Skill Sheet 5-1.** It is most often employed when the hose has been deployed on the ground and is being connected for reloading onto the pumping apparatus.

Between the feet. This connecting procedure is ideal for attaching nozzles and other large appliances to a hoseline (see **Skill Sheet 5-2**). Although awkward when first attempted, this method allows a fire and emergency services responder to approach the connection straight down. When attempting to align a coupling, heavy nozzle, or appliance, some find this method superior to the others outlined because the coupling supported by the responder's feet carries the weight of the top device.

Across the leg. This method (see **Skill Sheet 5-3**) is primarily used when the coupling connection must be accomplished when the coupling is not on the ground. This situation occurs when hose is being loaded, and connections must be made as the hose is being placed into the hose beds. This method requires somewhat more upper-body strength than the other methods and may be slightly more difficult for some fire and emergency service responders to accomplish. It is, however, a quick reliable method of connecting couplings together and is quite useful when the coupling end and/or appliance cannot be placed on the ground.

Two-person methods. Two-person connecting methods can be employed when sufficient personnel are available (see **Skill Sheet 5-4** for threaded couplings). Although procedures for connecting sexless (Storz) couplings (commonly used on large diameter supply hose) are somewhat different, they can be connected in an efficient manner when two fire and emergency service responders work together (see **Skill Sheet 5-5** for sexless couplings). The two-person connection processes are much faster and more efficient ways for making hose connections than the single-person methods.

NOTE: Pictures of most hose-handling methods are referenced to right-handed persons. In every case, left-handed persons can reverse the hand positions shown to connect/loosen connections.

Connecting to a Fixed or Stationary Fitting

The procedure for connecting a female coupling to a fixed or stationary male fitting (such as a pump discharge valve) is similar to that used to connect a female hose coupling to a male hose coupling. Because the fitting is fixed and unmoving, it is quite simple for one fire and emergency services responder to make these connections. It is extremely important that the fire and emergency responder performing this procedure do so with the knowledge of the pumping apparatus driver/operator. No valve or pump control device should be operated during this operation unless authorized by the driver/operator. Care must be constantly taken to ensure that the connection is completed correctly and that the hose, appliance, or fitting is in proper working order to avoid a failure at the fire pump control panel. If the connection is made to a large diameter supply hose, connect an elbow fitting to relieve the stress on the coupling created by the bending action of the hose where it is attached to the fire pump. See **Skill Sheet 5-6.**

Connecting a male hose coupling or fitting to a fixed or stationary female fitting (such as that found on the auxiliary intake at the pump panel) requires a slightly different procedure. The same precautions taken when connecting to a fixed male fitting must be taken during these procedures also. This procedure is also appropriate for fire department connections (FDCs), standpipes, and sprinkler connections that have female fittings and (in some instances) are permanently mounted siamese fittings. If these fittings are equipped with plastic breakaway caps, fracture the caps by striking them sharply with a spanner wrench or other striking tool. Clear away the broken cap pieces, and then attach the hose with a counterclockwise rotation of the swivel **(Figures 5.4 a–c). Skill Sheet 5-7** gives the procedures for this evolution.

NOTE: Before attaching hose to a standpipe or sprinkler connection with the caps missing, look and feel inside for debris that could have been pushed inside the waterway. If debris is present, remove it before making the hose connection. Be sure to check for any debris trapped behind the clapper valve of the siamese.

Attaching a Nozzle

Nozzles may be attached to fire hose couplings by methods similar to those used for connecting couplings. Nozzles have female threads and attach to the male end of the hose or device. Because the nozzle is usually quite heavy relative to the weight of other devices, position the body to assist in aligning the threads while supporting the nozzle as the connection is completed. Two methods that fire and emergency service responders may use to accomplish this task are described in **Skill Sheets 5-8** and **5-9.**

Connecting Portable Monitors

Portable monitors are capable of discharging large volumes of water when high-volume nozzles are used. While a portable monitor can be an effective tool when combating an intense fire, it can also become a lethal weapon if it is not properly secured. The flows of large volumes of water at 70 to 100 psi (483 kPa to 689 kPa) {4.8 bar to 6.89

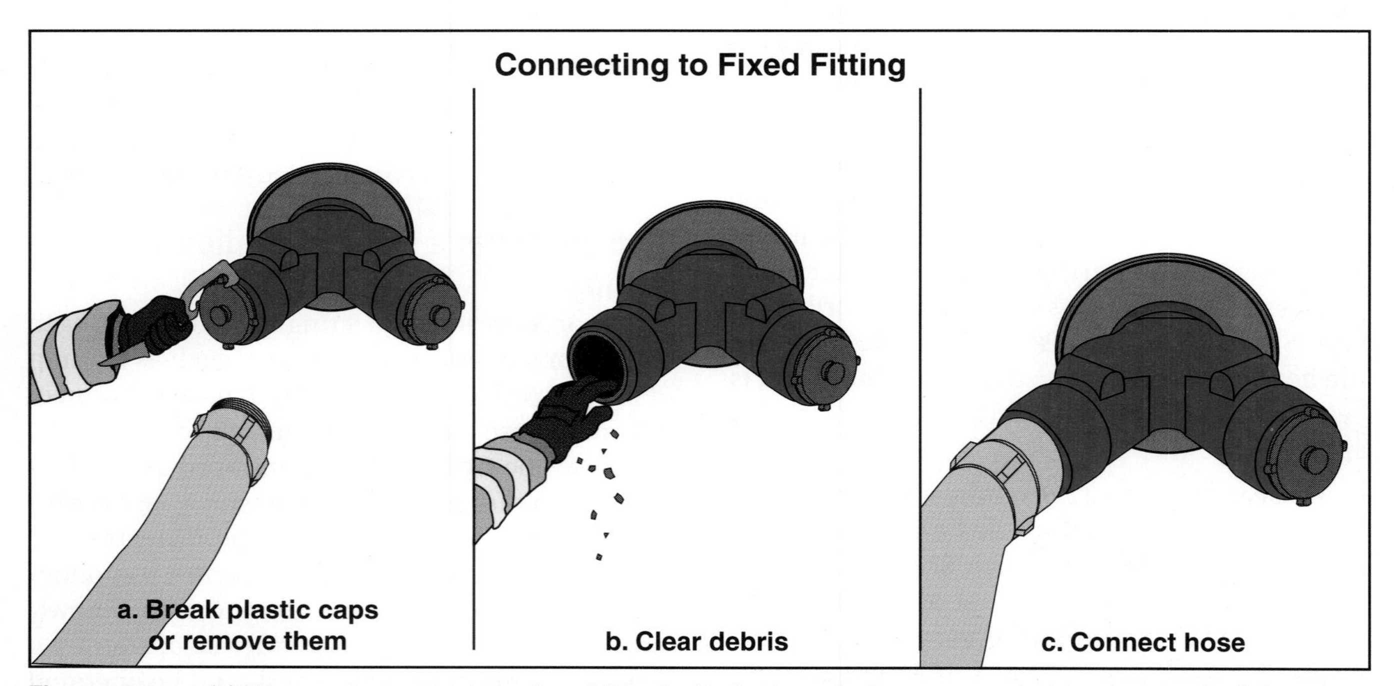

Figures 5.4 a–c (a) Use a spanner wrench or other striking tool to fracture plastic caps on sprinkler or hose-outlet fittings or simply remove them. (b) Once the cap is broken or removed, clear away the broken pieces from the waterway. (c) Make the hose connection.

bar} nozzle pressure produce significant nozzle reaction forces. A poorly secured monitor can be propelled opposite to the direction of flow with catastrophic effects.

Master stream nozzles were originally designed with two or more female threaded swivel intakes. This configuration has been slowly changing as the use of high-volume large diameter hoselines to supply these nozzles has increased. Current designs use single or possibly double Storz couplings, eliminating the need for multiple supply lines.

When connecting hoselines to portable monitors using 2½- or 3-inch hoselines (65 mm or 77 mm) (attended or not), it is prudent to loop the supply hoses around to the front of the devices. Rotate the nozzles, reversing the inlets of the monitors so that they are aligned with the direction of fire stream flow, cross the supply lines (if two or more are used), and make connections to the devices. Tie the supply lines together with a rope or hose strap where they cross to provide additional stability when the hoselines are charged and the monitor is flowing **(Figure 5.5).** Another advantage to arranging the hoselines in this manner is that it is easier to move the monitor backward should fire conditions deteriorate.

Loosening a Tight Hose Connection

It may sometimes become necessary to loosen a tight screw-thread connection. This operation can be accomplished with or without the assistance of spanner wrenches. The following three basic procedures can be employed to loosen tight couplings:

- ***Knee-press method (one person)*** — With the hose on the ground, grasp the hose behind the female coupling, and stand the connection on end with the male coupling on the bottom. Bend the hose close to the male coupling. Place one knee on the hose and shank of the female coupling while keeping the leg above the knee directly above the couplings, and apply weight to the connection. The compression of the hose gasket permits the swivel to turn more easily. See **Skill Sheet 5-10.**
- ***Stiff-arm method (two persons)*** — This method is used when two persons are working together and no spanner wrenches are available. Holding the arms in straight, rigid positions makes them act as levers to move the hands in a more efficient and powerful manner than if the arms are bent. The two persons face each other (feet shoulder-width apart) with the hose coupling held

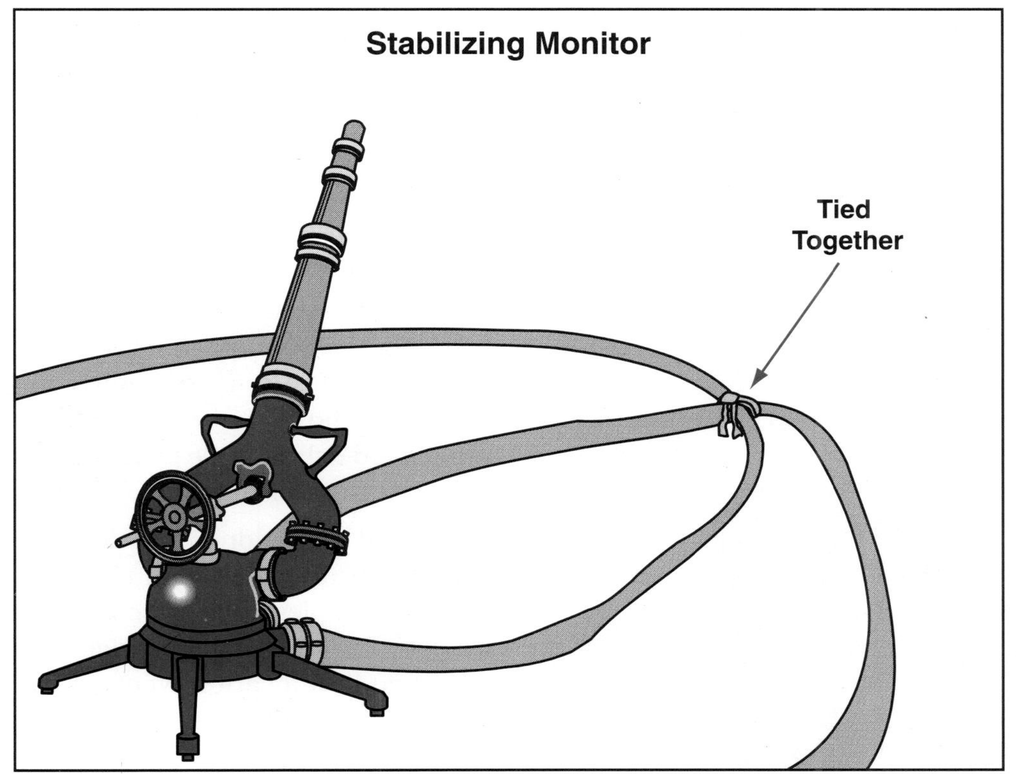

Figure 5.5 Bring supply lines around to the front of the monitor, reverse the nozzle (placing the inlets toward the fire), and tie hoselines together with a rope or hose strap where they cross to prevent the monitor from moving.

between them. With firm, two-handed grips, one person holds the male shank, while the other holds the female coupling (fingers can be locked together to tighten the grip). Hold arms rigidly straight. Each person begins by leaning slightly to the right, and then applies force while leaning to the left to produce a counterclockwise force on the couplings. Body weight supplies most of the force to loosen the connection. See **Skill Sheet 5-11.**

- ***Two-spanner-wrench method*** — Use spanner wrenches to loosen tight connections when other methods fail. Place one spanner across the lugs on the swivel of the female coupling, and place a second spanner across the lugs on the shank of the male coupling. Hold the male coupling spanner wrench firmly in place, and rotate the female coupling spanner wrench counterclockwise to loosen the coupling. See **Skill Sheet 5-12.**

Hose Rolls

A number of different methods for rolling/unrolling hose are available: straight (street) and donut (regular donut, twin donut and self-locking twin donut). Each roll type or variation has advantages over other rolls based on its intended storage or deployment method. In all methods, care must be taken to protect the hose and couplings from damage. During hose-rolling and deployment operations, fire and emergency service responders must resist dragging or dropping the hose. When possible, remove debris that is found in the area where these operations are conducted and do not permit it to come in contact with the hose. Great care must be given to avoid injury to emergency personnel during these operations. Hose deployment from a roll while emergency operations are being conducted can be extremely hazardous because a multitude of separate operations are being performed simultaneously.

WARNING

Exercise extreme caution while deploying hose during emergency operations. Ensure that all emergency personnel are in safe locations away from the area where hose is being deployed.

Straight Roll

The straight roll (also known as the *street roll*) is a one-person hose roll that is the fastest and most easily accomplished roll used by the fire and emergency services **(Figure 5.6).** The straight roll is simple and fast to construct and can have either the male or female coupling exposed. See **Skill Sheet 5-13.** All sizes of fire hose lend themselves to this roll. It is suitable for hose that is going to be handled in one of the following ways:

- Placed in rack storage
- Returned to the station for washing when dirty following deployment
- Reloaded on apparatus after use at the scene (rolling the hose before reloading purges any air and water trapped inside the hose so that it is flat, which allows a more compact load)

Numerous end uses exist for hose rolled in this manner, including storage and preparation for loading on the apparatus. Several related but distinctly different ways to complete the roll are available. When hose with screw-thread couplings is rolled into a straight roll for storage, start rolling the hose at the end with the male coupling so that the coupling is protected within the core of the roll. If the roll is made to facilitate loading directly into the hose bed, make the roll so that the appropriate coupling appears at the outside of the roll (this choice depends on how the hose bed is configured).

NOTE: Before rolling hose in a straight roll after a fire-fighting or emergency operation, it may not be necessary to drain the hose beforehand if it rests on a fairly level surface. When the hose is rolled, its weight compresses the hose and pushes any remaining water out of the hoseline. It is, however, recommended that the draining procedure not be omitted because the hose can be inspected for damaged or embedded debris while performing a manual hose-draining exercise.

If hose is deployed on a slope, roll the hose downhill to further drain water from the hose. Once the hose roll is complete, place it to the side, out of the path of traffic and away from debris or contaminants. Place in a location where it can be easily picked up or reloaded on the pumping apparatus.

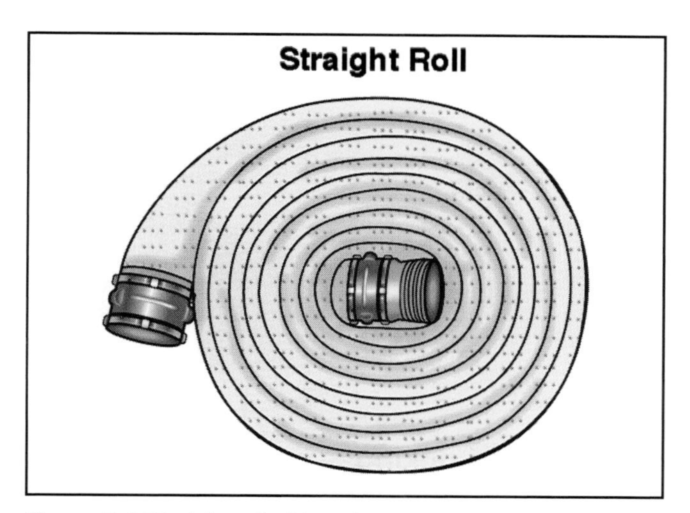

Figure 5.6 Straight roll with male coupling on the inside.

A traditional way of unrolling the straight roll is to push the roll away while grasping the outside coupling. This method can lead to serious damage to the coupling in the center of the roll. The coupling will snap down onto the ground from the momentum of the unrolling action and possibly damage the threads on the male coupling or the swivel on the female coupling. Additionally, there is a danger of injury to the fire and emergency services responder deploying the hose and also to those who may inadvertently be struck as the hose unrolls quickly.

Donut Rolls

The donut roll is different from the straight roll in that both couplings are on the outside of the roll. The donut roll is commonly used in situations where the hose is going to be deployed directly from the roll. This roll makes it easier to unroll the hose for loading into the hose bed or for extending a working hoseline. Also, the hose in a donut roll is less likely to spiral or kink when unrolled. The roll is designed so that the couplings will be within 12 inches (305 mm) of each other upon completion of the roll **(Figure 5.7).** The hose can be deployed by holding both couplings and pitching the roll away so that it unrolls. This method allows for a quick deployment of the couplings (similar to the straight roll), but care must be taken to ensure the safety of other fire and emergency service responders during this procedure. Variations of the donut roll include one- and two-person methods, twin donut method, and self-locking twin donut method.

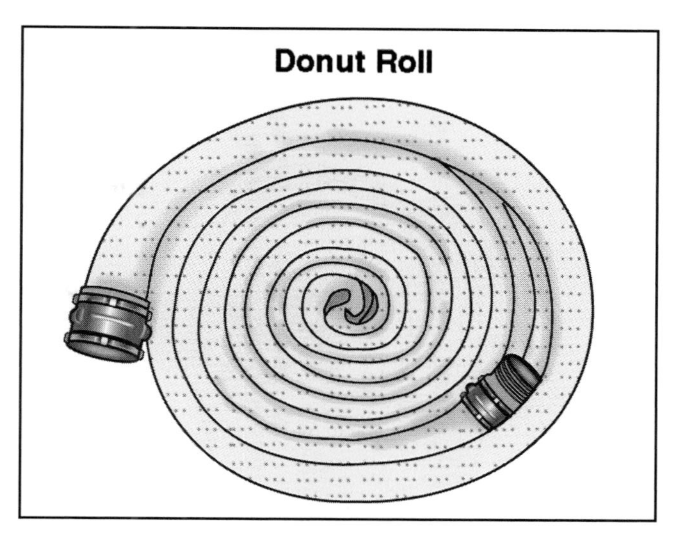

Figure 5.7 When the donut roll is completed, the hose couplings will be on the outside of the roll and within 12 inches (305 mm) of each other.

One-Person Methods

Two methods exist for rolling the donut roll by one person. It is extremely important to prepare the hose for rolling in the proper manner. Although the rolls are not particularly difficult, **Skill Sheets 5-14** and **5-15** outline both methods.

Two-Person Method

The procedure for making the donut roll with two persons is given in **Skill Sheet 5-16.** This method produces a tight roll that allows for easy storage in an apparatus compartment. The roll also is easy to carry long distances by hand without it slipping apart.

Twin Donut

The twin donut roll (sometime known as the *double donut roll*) is smaller in height but wider than a standard donut roll **(Figures 5.8 a and b, p. 166).** Although best used for 1½- and 1¾-inch (38 mm and 45 mm) hose, larger sizes of hose up to 3 inches (77 mm) can also be rolled in this manner. The purpose of a twin donut roll is to provide a compact roll that is easy to carry for special applications such as high-rise operations. The procedure for making the twin donut roll is given in **Skill Sheet 5-17.**

Self-Locking Twin Donut

The self-locking twin donut roll is a twin donut roll that has a built-in carrying strap formed from the hose itself **(Figure 5.9, p. 166).** This strap locks over

Figure 5.8a The twin donut roll allows both the male and female couplings to be available for connection when deployed.

Figure 5.8b By offsetting the couplings by 1 foot (0.3 m), they can be coupled together (protecting the threads) and then tied using a hose strap.

the couplings to keep the roll intact for carrying. The length of the carrying strap may be adjusted to accommodate the height of the person carrying the hose. Refer to **Skill Sheet 5-18** for a complete description of the procedures required to construct this hose roll.

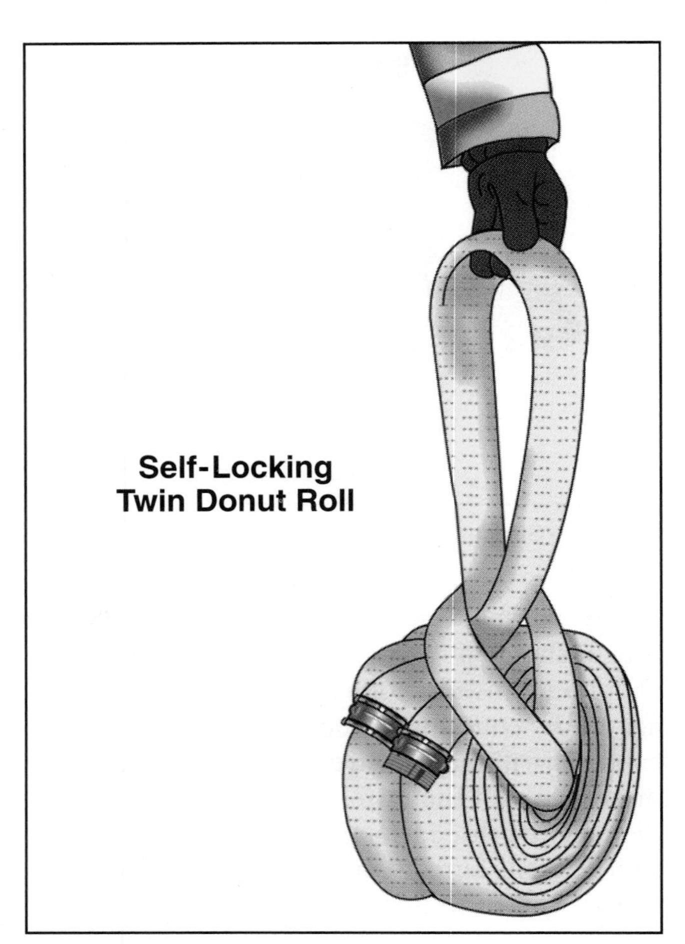

Figure 5.9 The self-locking twin donut roll provides a built-in carrying strap.

Manual Deployment of Apparatus-Carried Hose

One of the first steps for putting hose into service is to move it from the pumping apparatus to the place of use. Although one person can move small diameter (1-, 1½-, 1¾- 2-, 2½-, or 3-inch [25 mm, 38 mm, 45 mm, 50 mm, 65 mm, or 77 mm]) hose easily, larger sizes (3½-, 4-, 5-, or 6-inch [90 mm, 100 mm, 125 mm, or 150 mm] and large supply hose) often require more than one person to safely and efficiently complete the operation. Different types of materials and hose construction (including its weight per section and roughness of its surface) can also affect the ease with which it can be moved. Sections of hose are usually moved in two basic ways: carrying and dragging. Once the hose arrives at the place it is needed on the emergency scene, it must be deployed in an efficient, safe manner using proper lifting techniques.

Hose Carries

Carrying is the preferred method of moving hose because it subjects the hose to less abrasion than when it is dragged. A general rule when making a preconstructed bundle for carrying is that the farther the bundle must be carried, the more it should be designed for comfort. When using the methods shown in the skill sheets, it may be advantageous to tie the hose bundles with rope or straps to stabilize them for carrying. Another general rule is that if the speed of deployment is a factor, design the bundle so that it can be advanced in a progressive manner, limiting the chances of it catching on obstacles or becoming entangled.

Several hose-carry methods exist to accomplish this task. These methods are designed to be flexible so that regardless of hose type or method of loading, the hose can be effectively deployed. The various methods that can be employed are described in the sections that follow and include the shoulder-loop, accordion shoulder (3 variations), modified accordion shoulder, horseshoe shoulder, and horseshoe underarm.

Shoulder-Loop Method

The shoulder-loop carry from the ground method **(Skill Sheet 5-19)** can be used by one person to carry a single section of hose or by several persons to carry a number of interconnected sections **(Figure 5.10).**

Accordion Shoulder Method

The accordion shoulder carry is particularly useful for carrying a large volume of hose because several persons can carry interconnected bundles, and then deploy the hose directly from the bundles. One person can transport a section of hose with this carry. Three methods for the accordion shoulder carry are described in the paragraphs that follow. The first method starts with hose removed from the hose bed and laying on the ground. The other two methods start with the hose still in the hose bed: one loaded with an accordion load and the other loaded with a flat, horseshoe, or hose reel load. Hose bed loads are described in greater detail in Chapter 6, Supply Hose Loads and Deployment Procedures. Each person performing an accordion shoulder carry usually carries one section of hose.

From hose on the ground. This method requires that the hose be laid out in a straight line so that it can be picked up while walking along the line. This method is also used to pick up hose after a fire or emergency incident, and it works well to also drain water from hose as it is being picked up **(Figure 5.11).** See **Skill Sheet 5-20.**

From hose in an accordion-loaded hose bed. When a hose bed is loaded with hose in an accordion load, a number of folds can be moved directly from the hose bed to the shoulder in a ready-made bundle. This procedure will not work, however,

Figure 5.10 Shoulder-loop method: Form successive loops until the desired amount of hose is loaded.

Figure 5.11 Accordion-shoulder method: Guide the hose over the shoulder, alternating the loops in front of and behind the body until the entire section is loaded.

when the pumping apparatus has an extremely long hose bed in which the folds exceed 10 feet (3 m) in length. Folds longer than 10 feet (3 m) may drag on the ground when carried by a person of average height. Because the hose sections are connected in the hose bed, several persons can carry and lay out the hose directly from the shoulder. Installation of a nozzle is optional but is included in the procedure given in **Skill Sheet 5-21.**

From hose in flat, horseshoe, or hose reel loads. A different technique is required when hose is taken from a load other than an accordion load such as a flat load, a horseshoe load, or even from a hose reel system. In these cases, one person stands on the tailboard to pull hose from the hose bed and helps load hose onto the shoulders of each advancing person carrying a fire hose load. The description in **Skill Sheet 5-22** outlines the procedures that the person working from the tailboard employs while assisting others with shoulder loads.

Modified Accordion Shoulder Method

The accordion shoulder carry can be modified so that the couplings are protected inside the folds. This simple, preconstructed bundle is also designed for moving 2½-inch, 3-inch (65 mm, 77 mm), or larger hose long distances such as up the stairs of a high-rise building. This bundle can be prefolded, tied, and stored ready to use.

Horseshoe Shoulder Method

The horseshoe shoulder bundle is useful for carrying a single section of hose on the shoulder or under the arm. An advantage of this load is that both couplings are carried in front of the person, making them available for connection to extend the line or attach a device **(Figure 5.12)**. Always use proper lifting techniques when lifting hose from the ground. The procedure for this carry from the ground is described in **Skill Sheet 5-23.**

Horseshoe Underarm Method

A horseshoe shoulder bundle can be modified for carrying under the arm by adding one more step — folding the bundle once again to make a more compact load **(Figure 5.13).** This hose carry is especially useful when deploying fire hose while wearing full personal protective equipment including self-contained breathing apparatus. This method (described in **Skill Sheet 5-24**) allows the fire and emergency services responder to prepare and carry the hose without interference from the respiratory protection equipment. As with the horseshoe shoulder method, caution must be exercised when picking up the hose from the ground by using the legs to lift the bundle.

Manual Hose Deployment

Once hose has been loaded onto the person's shoulders, a coordinated procedure is followed to ensure that hose is deployed in an efficient manner. Each of the methods described — the shoulder-loop deployment and accordion-shoulder deployment methods — are designed to allow fire and emer-

Figure 5.12 Horseshoe shoulder method: Pick up the bundle at its center and place it on the shoulder.

Figure 5.13 Horseshoe underarm method: Cradle the bundle under one arm.

gency service responders to quickly place the fire hose in service. Each method has an advantage over the other depending upon the intended deployment of the hoseline.

The shoulder-loop method allows a person to more accurately estimate the required amount of hose needed to complete the deployment because it is laid on the ground before being loaded onto the shoulder. Additionally, the bottom loop tends to slide over obstructions rather than becoming entangled and snagging. The accordion shoulder load is often a more rapid hose deployment method because it is loaded directly onto the shoulders of fire and emergency service responders from the pumping apparatus hose bed.

Shoulder-Loop Method

The deployment of hose carried in shoulder loops by several persons is relatively simple but requires coordination and communication. As it deploys, the hose peels off the top of each person's shoulder to the ground behind. To deploy hose carried in shoulder loops, a group walks to the place where the hose is to be connected to the water supply (for instance at a pump discharge valve). One person can attach the last coupling in the load to the water supply or another can hold the coupling to anchor the hose. Members of the group then walk toward the destination point while maintaining firm control of the hose loops on their shoulders. The hose deploys from the last person's shoulder first. To aid in dropping the loops, this person lifts the top loop from the shoulder and drops it when the hose pulls taut against its anchor **(Figure 5.14)**. As the last fold of hose is deployed, the last person signals the next person to start dropping loops. Each subsequent person continues the dropping process in the same manner: top loop, then the next loop, and so on until all hose bundles are deployed.

Figure 5.14 Shoulder-loop method deployment: Lift the top loop from the shoulder and drop it when the hose pulls taut against its anchor.

Accordion Shoulder Method

To deploy fire hose carried in an accordion shoulder load, a group of emergency responders carry the shoulder loaded hose to the place where it is to be connected. This place may be a water supply source, fire pump or hydrant, or possibly a pump intake or sprinkler/standpipe connection. One person either connects the last coupling in the load to the connection or has someone hold the coupling to anchor the hose. Members of the group then walk toward the destination point while maintaining firm holds on the hose bundles. The hose pays off the last person's bundle as the top fold pulls loose **(Figure 5.15)**. As the last fold on the

Figure 5.15 Accordion shoulder method deployment: Walk toward the destination point so that the hose pays off the rearmost person's bundle.

last person deploys, the next person is signaled to release the hose so that it too can deploy. The top fold, the next fold, and so on pull loose from each person's bundle as the group walks on. The group repeats this process until the entire length of hose deploys.

Hose Drags

Dragging is a fast way to move hose from one place to another. While carrying is the preferred method for transporting hose, fire-fighting and other emergency operations sometimes require fire and emergency service responders to move hose from the apparatus to the fire scene with the greatest speed possible. For this reason, dragging is still a necessary method of handling hose. Although it is often necessary to use a hose drag method, it must be emphasized that these procedures pose a severe risk of damage to the hose. When these methods must be employed, exercise great care to avoid contact with debris or other abrasive material that can damage the hose.

The simplest method of dragging hose is to grasp both couplings and walk forward, dragging the doubled hose behind. If the hose is large and heavy, drape the hose ends over the shoulder with the couplings in front so that the upper body, rather than the arms, carries most of the hose's weight. Do not pull the hose from the folded end and drag the couplings.

An important fact to remember when dragging hose is that the less hose that contacts the ground, the easier it is to drag. The methods outlined in single-section and multiple-section hose drags take this fact into account and are designed to ease the dragging task.

Single-Section Methods

This single-section method is frequently known as the *street drag.* It starts with a section of hose laid out straight. The couplings are carried in each hand, and the hose is dragged behind. See **Skill Sheet 5-25.**

A second single-section drag method also starts with a length of hose laid out straight. Draping the body with hose as outlined in **Skill Sheet 5-26** distributes its weight equally over a person's upper body. The procedure is designed as a single-section hose drag method; however, a single person can deploy two hose sections when necessary. Attempting to drag two hose sections increases the risk of hose and coupling damage.

Multiple-Sections Methods

When several people must move several interconnected lengths of hose, one of the fastest ways is to pull the hose directly from the pumping apparatus hose bed and immediately proceed to the location where it is needed. Each person drags one section of hose by grasping the connected couplings. For the process to progress smoothly, one person stands on the tailboard to help deploy the hose from the hose bed. See **Skill Sheet 5-27** for an outline of this procedure.

Several persons can drag a number of interconnected sections of hose by first pulling the hose from the hose bed and laying it on its edge in folds on the ground. Each person then picks up several folds and proceeds to the destination. See **Skill Sheet 5-28.**

Hoseline Advances

Once hoselines have been deployed and connected for fire-fighting or other emergency operations, they must be advanced to a position appropriate for applying water or beginning a fire attack. Advancing the hoseline is best accomplished before it is charged with water because of the additional weight and reduced maneuverability that charged hoselines create. The various advancing hoseline situations include uncharged hoselines, charged hoselines, into a structure, up a stairway, down a stairway, up a ladder, hoseline hoists, booster lines, from hose standpipe connections (hose outlets, convenient outlets, etc.), and hose roll-a-loop.

Uncharged Hoselines

Several ways are available to advance an uncharged hoseline with a nozzle attached. While protecting the nozzle from damage, a single fire and emergency services responder can safely and efficiently move a hoseline and nozzle to a position for use by using one of several methods. The length of the hose as well as type of conditions present where the hoseline is to be deployed determine the best method that an emergency responder should use. Do not advance hose by simply holding onto the nozzle and walking forward because the hose could snag and pull the nozzle from the hands.

A more secure method is to drape the hose over one shoulder with the nozzle in front. Holding the nozzle on the opposite side of the body increases the ability of an emergency responder to control the nozzle while at the same time allowing the use of upper body strength to help pull the hoseline. See **Skill Sheet 5-29.** Additionally, should the hoseline be inadvertently charged with water, the emergency responder would be better prepared to control both it and the nozzle. When it is necessary to have the hands free, such as when climbing a ladder or while carrying equipment, hang the nozzle over the shoulder and rest it on the back. Exercise care to prevent the nozzle from tangling with or damaging components of the emergency responder's self-contained breathing apparatus.

Charged Hoselines

Advancing a charged line is more difficult than advancing an uncharged line because hose becomes stiff, heavy, and unwieldy when pressurized and filled with water. **Table 5.1** indicates the weight of water in a charged hoseline. These calculations are of the water weight only and do not include the weight of the hose or couplings. Nevertheless, moving a charged hoseline is often necessary. For example, when a hoseline is advanced to the door of a structure fire, the nozzle is opened to knock down fire, and the hose is then moved forward to extinguish fire deeper into the building. As with an uncharged hoseline, always protect the nozzle. More importantly, however, is the need to maintain control of the nozzle at all times. Before changing positions, close the nozzle slowly to reduce the possibility of water hammer. Open the nozzle slowly to minimize the effects of nozzle reaction (force that pushes the hose backward) that makes the hose difficult to control.

Never operate a hoseline or nozzle with less than two persons during fire-suppression attack operations. This practice is a matter of safety for emergency responders operating the pressurized hoseline and is also consistent with the "two-in, two-out" rule and accountability systems required today (see information box). No emergency responder performs operations alone in a dangerous environment. When it becomes necessary, however, for one emergency responder who is not participating in direct fire-suppression activities to maneuver or advance a charged hoseline outside a hazardous area, one of the best methods is to place one arm under the hose and grasp the opposite arm to lock the hoseline into place (arm-lock method). The opposite hand is always on the nozzle shutoff to maintain control of the water flow.

Safety Requirements

The Occupational Safety and Health Administration has concluded that for interior structural fire fighting, a buddy system for workers inside the immediately dangerous to life or health (IDLH) atmosphere and standby personnel outside that atmosphere are necessary [see Title 29 (Labor) *CFR* 1910.134 (g) (4) (i)].

Table 5.1
Weight of Water in a Charged Hoseline

Diameter (inches and mm)	Length (feet and meters)	Gallons per 100 feet	Liters per 30 meters	Pounds per 100 feet	Kilograms per 30 meters
1½ (38)	100 (30)	9.2	34.7	76.5	34.7
1¾ (45)	100 (30)	12.5	47.2	103.6	47.2
2 (50)	100 (30)	16.3	61.7	136.0	61.7
2½ (65)	100 (30)	25.5	96.5	211.6	96.5

A second method for advancing hose without help is to attach a hose strap or rope tool to the hose and use it to pull the hose forward with the body (hose-strap/rope method). Attach the strap to a point on the hose so that when the loop is placed over the shoulder, the nozzle is within easy reach.

WARNING

Only use single emergency responder hoseline advances in areas that are not within the danger zone, and never use it within a burning structure.

When two people advance a hoseline, similar methods are used to control it. The nozzle person can use either the arm-lock method or the hose-strap/rope method. A second person on the same side of the hose pulls the hose forward with the hands or with a hose strap **(Figure 5.16).** If other persons are needed to pull the hose (such as for the large sizes of hose), they space themselves on the same side of the hose at 10-foot (3 m) intervals and pull the hose in the same manner.

NOTE: It is not absolutely necessary that the first and second persons stand on the same side of the hose, but this position makes it easier for them to move through a doorway.

Figure 5.16 Advancing charged hoseline: The nozzle person directs the stream; a second person on the same side of the hose pulls the hose forward with the hands or with a hose strap.

Into a Structure

For maximum safety, fire and emergency service responders must be alert to the potential dangers of backdraft, explosion, rollover, flashover, and structural collapse during hoseline advances into structures on fire. The general safety guidelines to observe whenever hoselines are deployed into a building are as follows:

- Place an emergency responder on the nozzle and the backup emergency responders on the same side of the attack hoseline **(Figure 5.17).** Do not allow emergency responders to become "trapped" or "pinned" on the inside of a curve by the hoseline at a doorway or other turn.
- Check doors for heat before entering to begin fire-suppression operations. This check may give an indication of whether extreme heat is on the opposite side of the door. Be alert to the possibility of backdraft and flashover conditions inside the structure if heat is present. Additional ventilation or other safety procedures may be required if extreme heat conditions are found.
- Release (bleed) air from the hoseline by directing the stream towards the floor once it is charged and before entering the building or fire area **(Figure 5.18).**
- Stay low when entering the structure, taking all possible precautions to avoid blocking ventilation openings, especially doorways and windows **(Figure 5.19).** This guideline applies to the entire fire-suppression team.

Up a Stairway

Advancing a hoseline up a stairway can be a difficult task because of the tendency for a charged hose and its couplings to snag on stairway turns. Maneuverability is greatly impaired by the numerous turns and obstacles that are found in stairways. If a stairway is an emergency exit for building occupants during an emergency operation, deploying the hose may be nearly impossible to accomplish against the traffic flow of those leaving the building. Additionally, emergency responders advancing against this exiting traffic flow may expose themselves and civilians to greater risk than if the operation focused first on the evacuation of building occupants and then aggressive

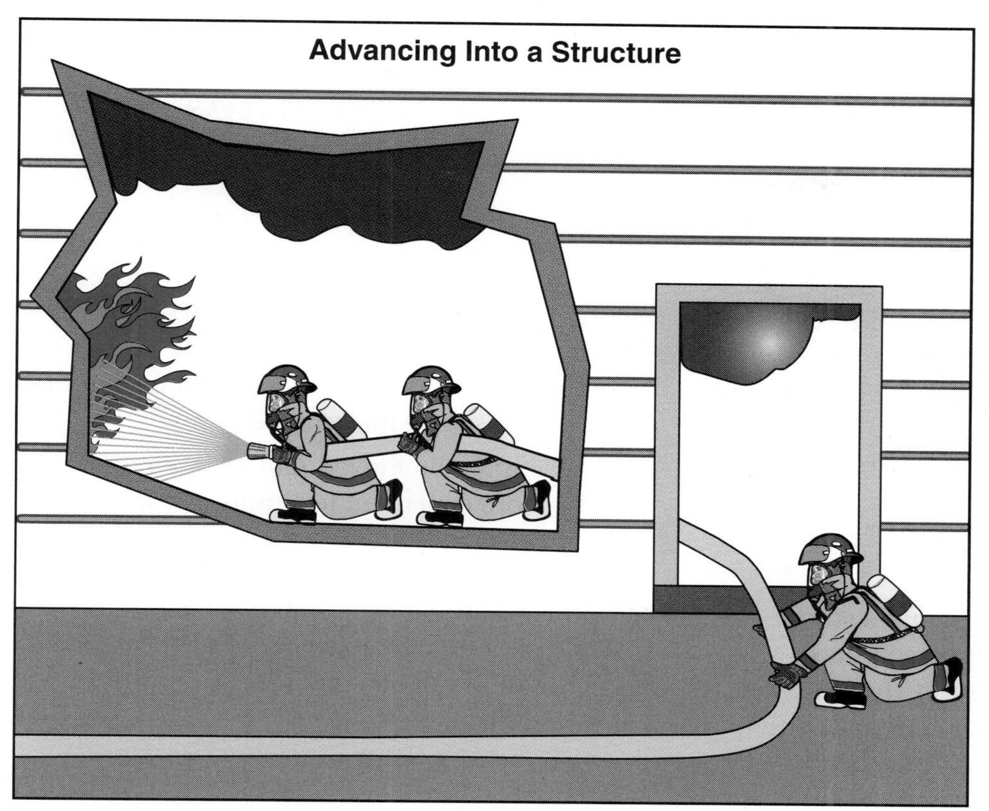

Figure 5.17 Advancing into a structure: All emergency responders stay on the same side of the attack hoseline.

Figure 5.18 Advancing into a structure: Release (bleed) air from the hoseline and check the nozzle by directing the stream towards the floor once it is charged and before entering the structure.

hoseline deployment. For this reason, advance hose up a stairwell before charging the hoseline. The accordion-shoulder carry is particularly well suited for this operation because the hose is laid progressively as the emergency responder climbs the stairs. Lay the hoseline against the outside wall to avoid sharp bends **(Figure 5.20a, p. 174).**

Figure 5.19 Advancing into a structure: Stay low when entering the structure, taking all possible precautions to avoid blocking ventilation openings.

In some buildings the stairwells have no walls to separate the staircases. This arrangement creates an opening in the center that makes it possible to extend hose without laying it on the stairs. In this situation, the easiest way to extend hose from the upper floors to the ground floor may be to carry a bundle of hose to the required floor and then lower the hose down the balustrade opening **(Figure 5.20b).** A standpipe pack described in Chapter 7, Attack Hose Loads, Finishes, Hose Packs, and Deployment Procedures, is particularly well suited for this purpose. Also see From Standpipes section later in this chapter. Another method that can be employed is to use a rope bag to hoist the hose up the stairwell through the balustrade opening.

Advancing a charged hoseline up a stairwell is a difficult task; therefore, only attempt it for short distances of one flight or less. If the hose must be moved more than one floor, assign additional emergency responders to each location in the stairwell or doorway where the hose makes a turn. Additionally, deploy an adequate amount of hose to allow the fire-suppression team to attack the fire once team members are in position next to the fire. Stairwell hoseline deployments use more hose than standard deployments because of the number of turns required to deploy it to upper floors of structures. It is extremely difficult to advance or retreat a hoseline once it is in position, so extending a line by adding more hose is not only time-consuming but dangerous because the waterway has to be interrupted.

If hose must be moved more than one floor, it may be more expedient to simply clamp the hose behind the nozzle, detach the nozzle, and attach additional uncharged hose to the end. The fire-suppression team must ensure that a charged protection hoseline is in place before the waterway

Advancing Up a Stairway (Dragging)
Loop to floor above fire provides excess for advancing.
Hose Against Outside Stair Wall

Figure 5.20a Advancing up a stairway (dragging): Lay the hoseline against the outside wall to avoid sharp bends.

Figure 5.20b Advancing up a stairway (by hand): Carry a bundle of hose to the required floor and then lower the hose down the balustrade opening.

is interrupted. Extend the additional hose up the staircase, attach the nozzle, and recharge the hose as soon as the nozzle is positioned for the attack. If there is an open stairwell, a charged hose may be passed up hand-over-hand by persons standing at each landing.

It is generally not advisable to stretch more than two hoselines up a stairway because multiple hoselines impede foot traffic and tend to become tangled. If more than two hoselines are needed on upper floors, take additional lines up other staircases, up ladders, and through windows or hoist lines by using rope.

Down a Stairway

Advancing an uncharged hoseline down a flight of stairs is considerably easier than advancing a charged line. Because the necessity for advancing a hoseline down a flight of stairs is usually into a belowground location for fire-suppression purposes, the line must be charged and ready to operate not only as a fire-suppression tool but also more importantly as protection for the nozzle crew. Because of this danger, always charge a hoseline and have it ready for use whenever it is advanced down a stairway during fire-suppression activities.

Advancing a charged hoseline down a stairway is difficult because it is extremely awkward, and the fire-suppression crew is usually moving through a high heat environment to a belowground area **(Figure 5.21).** The heat is rising through the stairway from the fire floor, which is similar to moving down a chimney directly through the heat for the fire-suppression team. To ease the deployment operation, position fire and emergency responders at critical points to help feed the hoseline to the team and keep it in a position that will not impede the hose team in case a rapid retreat is needed.

Up a Ladder

Advancing hoselines up ladders can be best achieved before they are charged with water. If a charged hose must be moved up a ladder, it is not only easier but also safer to drain the hose before

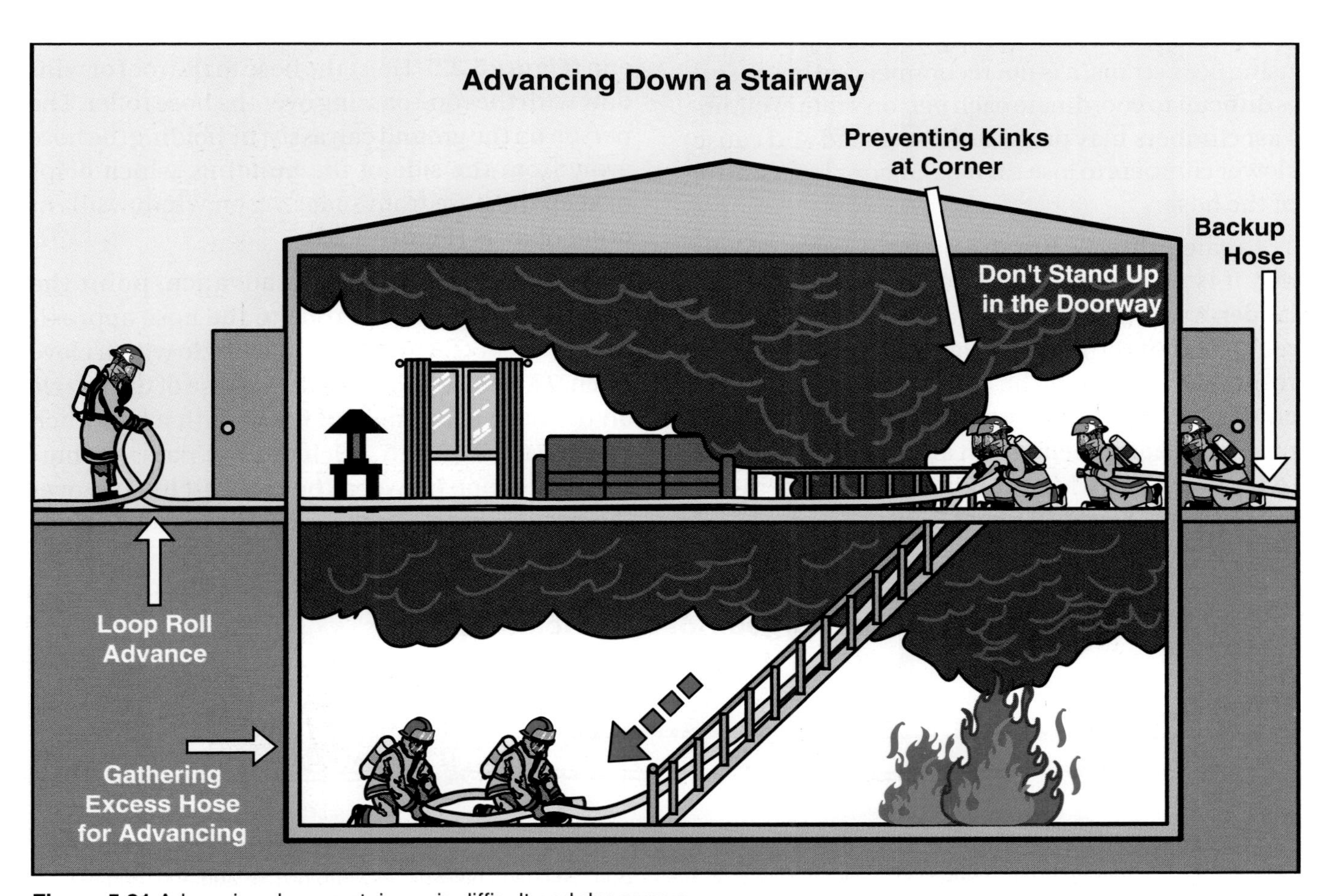

Figure 5.21 Advancing down a stairway is difficult and dangerous.

advancing it. As stated before, the person carrying the nozzle drapes the hose over one shoulder so that the nozzle hangs at the person's back. This position frees both hands for climbing the ladder. Position additional persons 10 feet (3 m) apart (approximately one ladder fly section) with each person carrying the hose using a strap over the shoulder so that the hands are free for climbing. When straps are not available, each person can loop the hose on the shoulder so that the hands are free. When the group wishes to advance the hose without moving the handlers up the ladder, each person takes a secure position and simply passes the hose along hand-over-hand.

If it becomes necessary to move charged hose up a ladder, more people are needed to handle the combined weight of the hose and water. Ladder loading also increases because of the weight of the charged hoseline. If the hose is over 2 inches (50 mm) in diameter, increase spacing between handlers to more than the standard 10 feet (3 m). All persons on the ladder must secure themselves and pass the hose upward hand-over-hand. While it is possible to carry a charged hose up a ladder using hose straps, it is not recommended because it is difficult to coordinate each person's rate of climb. Fast climbers may push the hose ahead and cause slower climbers to lose their balance or lose control of the hose.

To safely direct a fire stream from a ground ladder, it is necessary to secure the hoseline to the ladder. A safe aboveground fire-fighting operation requires that the hoseline, ladder, and emergency responder directing the stream be secured. The emergency responder, equipped in full personal protective equipment, must also take precautions to ensure safety while working on the ladder with the hoseline. See **Skill Sheet 5-30.**

Hoseline Hoists

Fire hose can also be advanced up the outside of a building by hoisting it with a rope. A charged line can be hoisted with difficulty, and the potential for damage to the nozzle and other equipment exists. Whenever possible, hoist an uncharged line because it is easier and far more maneuverable than a charged-line hoist advancement. Regardless of whether a charged or uncharged line is advanced, always protect the hoseline by using an edge protector to prevent unnecessary abrasion when the hose passes over a roof edge, parapet wall, or windowsill.

To hoist an uncharged hoseline, fold the first 6 feet (1.8 m) back over itself. If the nozzle is attached, fold the hose in a manner so that the nozzle shutoff handle (bale) is against the hose to prevent it from snagging on something as the hose is hoisted and allow the end to pass over the hose roller easily.

Tie the end of the rope, which has been lowered to the ground from above, with a clove hitch around the doubled hose near the nozzle. Then tie a half hitch halfway between the nozzle and the folded end **(Figure 5.22)**. Hoist the hose to the roof or window with the rope passing over the hose roller. The person on the ground can assist by holding the hose away from the side of the building, which helps to keep the hose from snagging on windowsills or other objects **(Figure 5.23)**.

For a charged-line hoist advance, point the nozzle upward. Tie the rope to the hose approximately 6 feet (1.8 m) below the nozzle with a clove hitch. Then tie a half hitch at the base of the nozzle on the hose coupling and finish with a half hitch over the tip **(Figure 5.24)**. If desired, pass the short section of rope between the two half hitches over the nozzle shutoff to secure it in the closed posi-

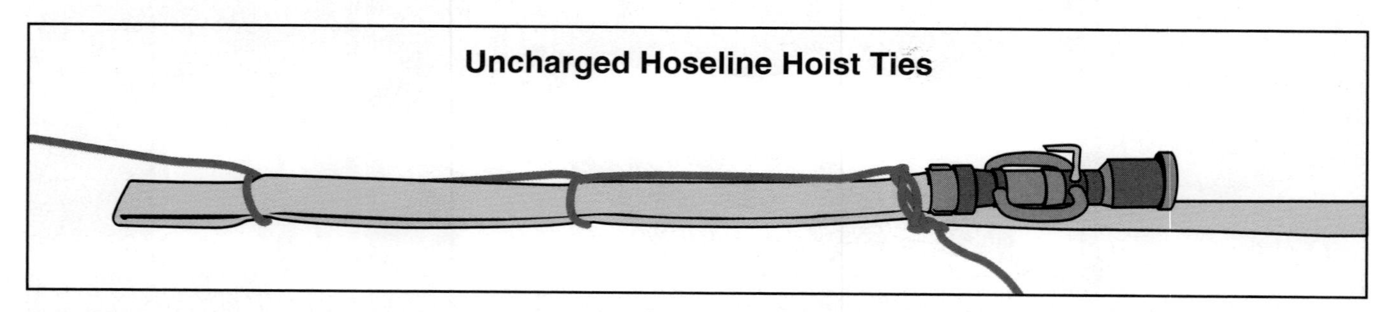

Figure 5.22 Uncharged hoseline hoist ties: Clove hitch near the nozzle, half hitch halfway between nozzle and end, and half hitch about 6 inches (152 mm) from the end.

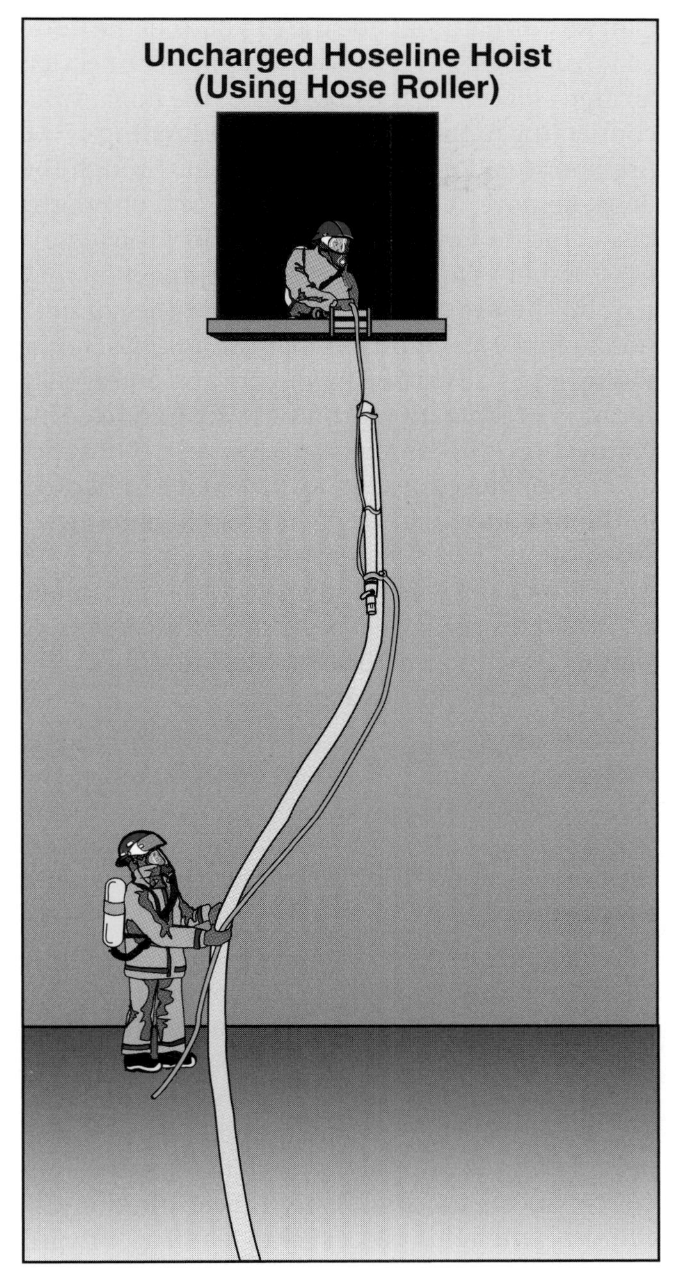

Figure 5.23 Uncharged hoseline hoist: Hoist hose and allow the rope to pass over the hose roller while the person on the ground assists by holding the hose away from the side of the building.

tion. Hoist the hose to the roof or window in the same manner as done with an uncharged hoseline with the rope passing over the hose roller. Two or more persons may be necessary to hoist the hose, depending upon the size and total length of hose.

When hose hangs on the outside of a building, support it along its length to relieve pressure on the couplings. These couplings must not only withstand the internal water pressure but also the weight of the hose as it hangs from the building. If the hose is not being advanced up a ladder, the use of hose straps or ropes to secure the hose to the building is recommended. Another method is to "thread" the hose in and out of windows every other floor. This method has become less attractive because of the architectural spacing of windows in modern buildings and use of nontempered glass windows that do not open. It is, however, an option on old structures.

Booster Lines

Booster hose is usually stored on reels and unlike other fire hose is at least partially charged with water when in storage on the reels. For this reason, booster hose can be fairly heavy when fully extended. A booster hose reel has an adjustable brake that prevents the reel from inadvertently turning when the pumping apparatus is in motion. It also prevents the reel from spinning after the hose is deployed, which would cause the remaining hose to uncoil on the reel. If the brake is applied too firmly against the reel, deploying the booster hose becomes difficult.

To advance booster hose directly from the reel, simply grasp the nozzle firmly in both hands and walk toward the destination. If moving a long distance (over 50 feet [15 m]), it may be necessary for other persons to assist by pulling the additional

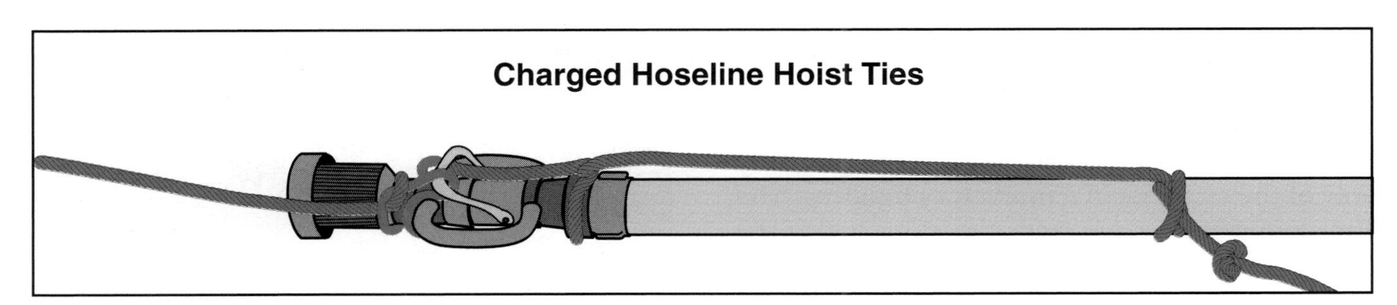

Figure 5.24 Charged hoseline hoist ties: Clove hitch 6 feet (1.8 m) below the nozzle, half hitch at the base of the nozzle, loop through the nozzle bale, and half hitch over the tip.

hose some distance behind the person with the nozzle. This distance varies with the diameter of booster hoseline, total distance deployed, and terrain encountered en route to the destination. It is also helpful to have a person located at the pumping apparatus to help pull the hose from the reel.

From Standpipes

Fighting fires in large buildings with a number of floors or large floor areas poses special problems for fire-suppression activities. In particular, the advancing of hoselines within these structures is a complicated operation that requires coordination among all operational elements. Conduct hoseline deployment either by advancing straight from the pumping apparatus (which is cumbersome with exceeding long hose deployments) or from standpipe connections located within the building. The most efficient method for hose deployment in a structure is from the standpipe connections that are located in stairwells or at other locations spaced throughout a large floor space. Regardless of the standpipe location, the fire-suppression crew must carry an adequate amount of hose from the pumping apparatus to the interior connection to complete the fire-suppression assignment.

Using a hose pack (also known as *high-rise pack* or *standpipe pack*) is the easiest way to make this connection. The manner in which the hose pack is arranged is found in the fire and emergency services organization's standard operating procedures, and the arrangement may be set up in the form of folds or bundles that are easily carried on the shoulder. Specially designed hose packs complete with nozzles, tools, and fittings may be prepared to facilitate these procedures. The command officer must be prepared to support the fire-suppression team with the resources required to safely deploy from the standpipe to the fire, and the driver/operator must provide the appropriate engine pump pressure to properly supply the nozzle. See **Skill Sheet 5-31** for details.

The fire-suppression crew along with other supporting elements advances to the fire floor by way of the stairway in a multistory building. The fire-suppression crew stops one floor below the fire floor or on the intermediate landing between the floors and connects to the standpipe at this location. This connection is usually located in the stairway directly outside the fire door to the floor. Old structures may have the standpipe connection located adjacent to the doorway to the floor. Avoid connecting to the standpipe on the landing of the fire floor to allow ease of movement through the door. Remove the hose outlet cap and check the discharge for foreign objects. Connect the female hose coupling to the outlet. This connection may require the use of a reducer to allow the connection from the standard 2½-inch (65 mm) discharge to the hose size carried by the fire and emergency services organization. Another way to make this connection is to connect a gated wye so that a second option hoseline can be deployed as a backup to the first hoseline. Often, a fire and emergency services organization may deploy a 2½-inch (65 mm) attack hoseline as a large hoseline to make an initial fire attack with the intent of extending smaller hoselines further into the building's interior once fire control is achieved.

Once the standpipe connection is complete, spread out or "flake" the hose up the stairs to the floor above the fire and then back down to the fire floor door. This procedure allows the fire-suppression team to easily move the hose onto the fire floor.

When deploying hose in a large area building, it is extremely important for the command officer to constantly monitor changing fire conditions. No safe area exists around the standpipe connection, so these connections are made in a variety of smoke and heat conditions. It is extremely important to provide ample support to the fire-suppression crews who are advancing from these interior discharges. Often it is necessary to leapfrog forward with hoselines stretched from the exterior to a standpipe connection and so forth until the fire is located and suppression can begin.

WARNINGS

- Do not advance into a structure fire without inspecting for fire progress.
- Do not overextend into a structure fire beyond the capabilities of the forces available.
- Pull ceilings and determine fire extension and development.
- Never work under a fire.

Roll-a-Loop

Usually no way is available to move the nozzle forward a short distance after a hoseline has been fully extended except to add a section of hose at the nozzle end. Adding a hose section takes considerable time because it requires shutting down the hoseline. A simple method, however, for advancing the nozzle a short distance is possible if the hose has been laid with curves in the line. A hoseline that is not absolutely straight provides slack that can move the nozzle forward. This movement can be easily done by simply straightening the hose progressively from the water source and moving toward the nozzle. As the slack hose accumulates, it tends to form an *S* shape. Lay one segment of the *S* over to form a large loop, stand the loop upright, and roll the slack hose toward the nozzle in a manner similar to rolling a hoop **(Figure 5.25)**. As the slack is removed, the hose lies in a straight line and the nozzle can be moved forward.

This procedure is limited to hoselines less than 4 inches (102 mm) in diameter. This size hose or larger sizes weigh enough that attempting to "roll" them could constitute a major physical risk to fire and emergency service responders performing the operation. Additionally, severe damage to the couplings or other objects can occur should the hose roll into them during the procedure.

Figure 5.25 Roll-a-loop advance: Lay one segment of the S over to form a large loop, and then roll it toward the nozzle to advance the slack hose.

Hoseline Operations

The methods of applying water through a nozzle attached to an attack hose vary with the size of hose used, the fire or emergency situation, and with the type of attack (see IFSTA's **Essentials of Fire Fighting**). An important point to remember with all methods, however, is that the nozzle must be controlled at all times. The nozzle operator keeps one hand on the shutoff valve whenever water is flowing through the nozzle **(Figure 5.26).** This control permits immediate shutdown of the nozzle if the hoseline becomes uncontrollable because of a pressure surge, loss of balance, or other problem.

Many methods of operating a hoseline are available. The method that is comfortable for a particular fire and emergency services responder is often contingent upon the responder's strength, general physical stature, and personal preference. The following sections describe the various methods of hoseline operations that are popular

Figure 5.26 Hoseline operation: Keep one hand on the shutoff valve whenever water is flowing through the nozzle.

and available for use: one-person methods, two-person methods, portable monitor connections, and hoseline control.

One-Person Methods

When one person is required to work a handline nozzle unassisted, some means must be provided for bracing the hoseline. Position the hose so that it extends straight back for at least 10 feet (3 m) behind the nozzle person. Stand facing the objective with the feet spread at least shoulder-width apart for good balance. Grasp the hose directly behind the nozzle with one hand and the shutoff with the opposite hand. The hose may be further anchored by placing a foot upon the hose.

Water flowing through any hoseline larger than 1 inch (25 mm) in diameter usually causes significant nozzle reaction. Use a hose strap to gain additional control of the hose. Attach the strap to the hose, and then place the strap over the shoulder so that the body absorbs the reaction force of the nozzle **(Figure 5.27)**.

If the hose needs to be moved during the fire attack or other emergency operation, close the nozzle and move the hose to the new location. Then straighten the hose behind the nozzle and operate it in the same manner as before.

Another means of controlling a hoseline when working alone is to reduce the water flow with the nozzle shutoff valve. Reducing the water flow also reduces the nozzle reaction, thus making the hose easier to handle. When using an adjustable flow nozzle, advise the driver/operator and select a lower flow setting to accomplish the same purpose.

When a large hoseline is positioned outside a structure for extended periods (for example, an exposure protection line), one person can safely and effectively operate the nozzle alone if the hose is properly arranged. Take approximately 25 feet (8 m) of the hose immediately behind the nozzle and form a loop. Pass the nozzle beneath the loop so that the loop rests on the end of the hose approximately 2 feet (0.6 m) behind the nozzle. This arrangement allows upward and downward movement of the stream and some movement side to side. Secure the loop by tying the hose at the crossover point with a hose strap. Kneel or sit on the hoseline at the crossover point and operate the nozzle **(Figure 5.28)**. One person can also control a hoseline for similar purposes if it is securely tied to a fixed object such as a telephone pole or parking meter.

Two-Person Methods

As stated earlier, the preferred practice when operating any hoseline during a fire attack or other emergency operation is to have at least two or more persons on the hoseline. This practice provides increased safety in case one person becomes disabled or an unforeseen change in the emergency conditions occurs. It also enables the persons to make use of the full fire flow potential of the nozzle.

Figure 5.27 Hoseline operation: Use a hose strap to counteract the backward force from the nozzle.

Figure 5.28 Operating large hoseline (one person): Kneel or sit on the hoseline at the crossover point and operate the nozzle.

The person at the nozzle holds it in much the same manner as when working alone: one hand on the hose directly behind the nozzle and the other hand on the nozzle shutoff valve. The nozzle person does not, however, brace the hose with a leg. The responsibility for anchoring the hose is given to the backup person so that the nozzle person has more freedom of movement than when working alone. The backup person remains on the same side of the hose and grasps it with both hands. This person cradles the hose against the inside of the closest leg and braces the hose against the front of the body and hip **(Figure 5.29).** The backup person thus takes most of the backward force of the nozzle, allowing the person on the nozzle the freedom of movement to easily direct the stream and adjust its pattern.

As in the one-person method, two people can better control the hose by using hose straps. Each person attaches hose straps and loops the straps in such a manner so that each person shares the backward force from the nozzle. This method is used primarily during exterior fire-suppression operations.

Figure 5.29 Operating hoseline (two persons): Both emergency responders stay on the same side of the hoseline; the backup person braces the hose against the front of the body and hip.

Another method may be used when a nozzle is equipped with handles. One person grips the shutoff with one hand and the handle with the other hand. The other person grasps the near handle with both hands **(Figure 5.30).** Using hose straps increases stability.

When advancing and operating a hoseline during interior fire-suppression operations, crew members should stay as low as possible to retain visibility and reduce their exposure to the heat that is present higher in a room. This procedure often means that crew members must crawl into the structure and operate the nozzle from prone positions. Nozzle crew members must practice these maneuvers to remain proficient.

WARNING

Stay as low as possible to reduce heat exposure when advancing hose into a structure.

The combination fire attack uses the steam-generating technique of ceiling-level attack combined with a direct attack on burning materials near floor level. The nozzle may be moved in a *T,*

Figure 5.30 Operating hoseline (two persons): Grasp the handles to provide stability.

Z, or *O* pattern, starting with a solid, straight, or penetrating fog stream directed into the heated gases at the ceiling level and then dropping down to attack the combustibles burning at the floor level **(Figure 5.31).** The *O* pattern of the combination attack is probably the most familiar application method. When performing the *O* pattern, direct the stream at the ceiling and rotate with the stream edge reaching the ceiling, wall, floor, and opposite wall.

Care must be exercised when using this method because much of the water is directed at smoke and does not extinguish the fire, which can cause additional water damage and disturbance of the thermal layers at the ceiling. Backup fire and emergency service responders assisting the responder at the nozzle should allow enough room to easily direct the nozzle while providing hoseline support for advancing or retreating the hoseline.

Portable Monitors

Portable monitors or deluge sets are capable of discharging large volumes of water when high-volume nozzles are used. A monitor can be an effective tool when combating an intense fire, but it can also pose a tremendous hazard to fire and emergency service responders and others if it is not deployed and secured in a proper manner. Depending upon the nozzle tip size or flow rate, the nozzle reaction can be significant. The device can be propelled in the opposite direction of flow causing great damage and possibly serious injury. A common method for controlling the monitor is to assign at least one person to stay at the device to control the flow and direction of the stream.

Occasionally, it becomes necessary to leave a monitor unattended, which could occur during operations where it is deployed to cool a heavily involved flammable liquids tank fire or other incident where extreme hazard precludes the presence of fire and emergency service responders. One of the best ways to secure an unattended portable monitor is to lace the hoselines supplying water to the monitor in such a manner so that they counteract the nozzle reaction forces.

CAUTION

Never move a monitor while it is charged and flowing water. The reaction force can damage equipment and severely injure fire and emergency service responders operating it.

Hoseline Control

Special situations often rise when a hoseline does the unexpected — it breaks and the loose end is flailing in the street. Tremendous quantities of water are being lost. The simplest control method is to

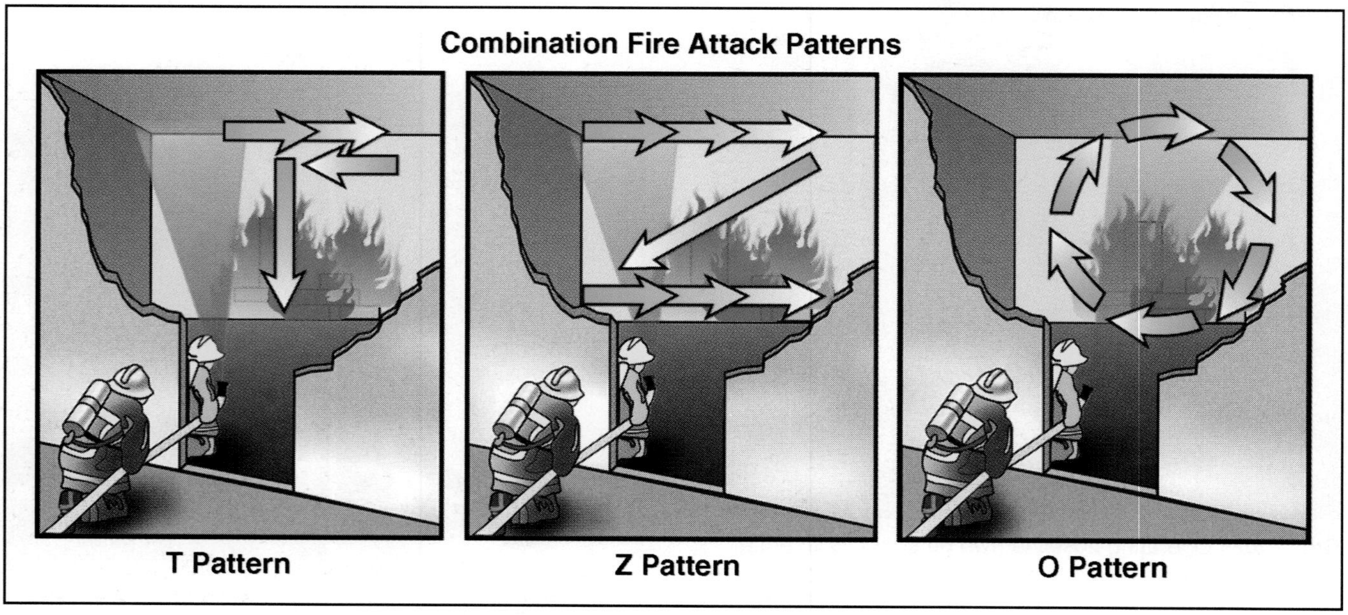

Combination Fire Attack Patterns

T Pattern

Z Pattern

O Pattern

Figure 5.31 Combination fire attack: Move the nozzle in *T*, *Z*, or *O* patterns.

close the discharge valve that controls water flow into the hose. Another method is to apply a hose clamp on the coupling closest to the burst section. If a clamp is not available, however, it is sometimes possible to shut down the line by kinking the hose. This procedure is outlined in **Skill Sheet 5-32.**

Should the nozzle "get away" from a fire-suppression team and become a "wild line," a potentially life-threatening situation exists. The charged hose and heavy nozzle can quickly do major damage and cause severe injuries to fire and emergency services personnel. When possible, turn off the water supply. If turning off the water cannot be accomplished quickly, retrieve the wild line by "crawling" up the line (advance cautiously) toward the nozzle. Most often the nozzle lashes back and forth on the ground; however, occasionally it may jump and come back. Watch the nozzle when retrieving it to avoid being struck. Once the line is controlled and the water turned off, inspect the nozzle for damage to the shutoff or tip. Inspect the coupling also because there may be a separation between the coupling and the hose.

Summary

Basic fire hose skills allow fire and emergency service responders to efficiently operate as members of fire-suppression teams. The simple tasks of rolling hose and attaching couplings, although outwardly simple in concept, must be practiced to achieve a high level of performance to ensure that these procedures can be accomplished without hesitation or fault when needed. Reading or viewing fire hose handling and nozzle evolutions does not complete the training requirements for fire and emergency service responders to become adept at these skills. Adequate hands-on training is essential to master these techniques. It is always easier to efficiently perform skills that are often employed because they become routine.

Because some of the hose-handling skills described in this chapter may be used only once or twice in a fire and emergency services responder's career, it is essential to supplement extensive training in these skill areas to be prepared to use them during fire-suppression or other emergency activities. Many times a certain fire-fighting technique is seldom used because of a lack of understanding its application, personal safety concerns, or the possibility of problems during the evolution. Thus fire and emergency service organizations become reluctant to employ the procedure and it becomes a "lost art" — one that could be of great benefit during certain limited circumstances. Avoid this attitude and constantly train in all fire-suppression or other emergency activities, even those that appear to be simple or are not often employed.

Connecting Fire Hose Foot-Tilt Method

Skill Sheet 5-1

Step 1: Stand facing the two couplings so that one foot is close to the male end.

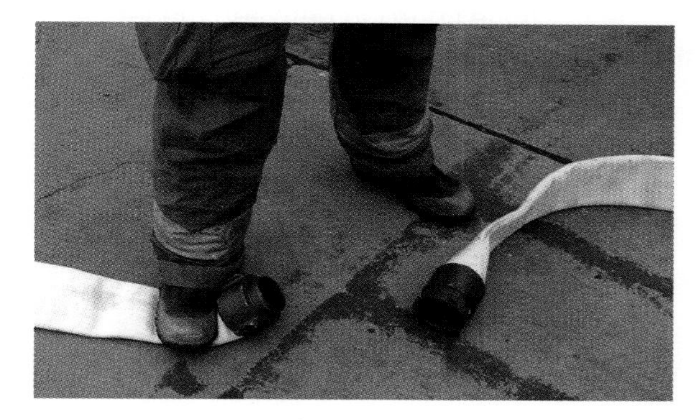

Step 2: Place a foot on the hose directly behind the male coupling.

Step 3: Apply foot pressure to tilt the male coupling upward.

NOTE: Position feet well apart for balance.

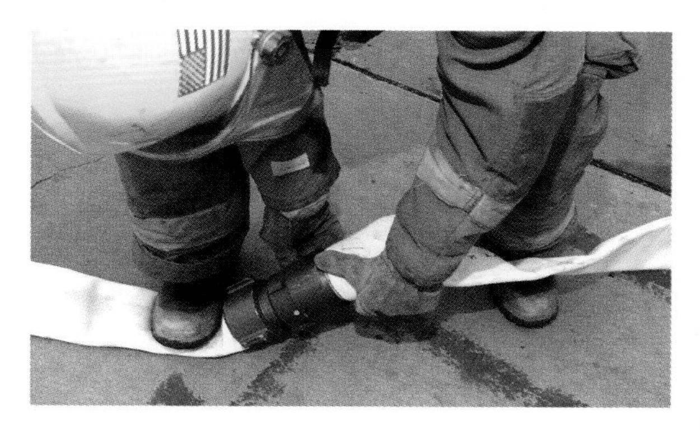

Step 4: Grasp the female end by placing one hand behind the coupling and the other hand on the coupling swivel.

Step 5: Bring the two couplings together, and turn the swivel clockwise with the thumb to make the connection.

5-2 Skill Sheet — Connecting Fire Hose
Between-the-Feet Method

Step 1: Cradle the male coupling between the feet so that the coupling is tilted up.

Step 2: Pick up the female coupling of the opposite hose.

Step 3: Align the female coupling with the upturned male coupling.

Step 4: Turn the swivel clockwise to make the connection.

Connecting Fire Hose Across-the-Leg Method

Skill Sheet 5-3

Step 1: Grasp the female coupling in one hand with the hose straight behind.

Step 2: Bend the knee slightly on the corresponding side.

Step 3: Lay the hose across the thigh (see photo).

Step 4: Pick up the male coupling in the opposite hand, and bring the two couplings together.

5-3 Skill Sheet

Step 5: Turn the swivel clockwise to make the connection.

Connecting Fire Hose Two Persons — Threaded Couplings

Skill Sheet 5-4

Step 1: ***Firefighters 1 and 2:*** Face each other while holding either the male or female connection with hoses extended behind each.

Step 2: ***Firefighter 1:*** Hold the male coupling with the threads outward, toward Firefighter 2.

Step 3: ***Firefighter 2:*** Align the couplings using the Higbee indicator and turn the swivel clockwise to make the connection.

5-5 Skill Sheet — Connecting Fire Hose: Two Persons — Sexless Couplings

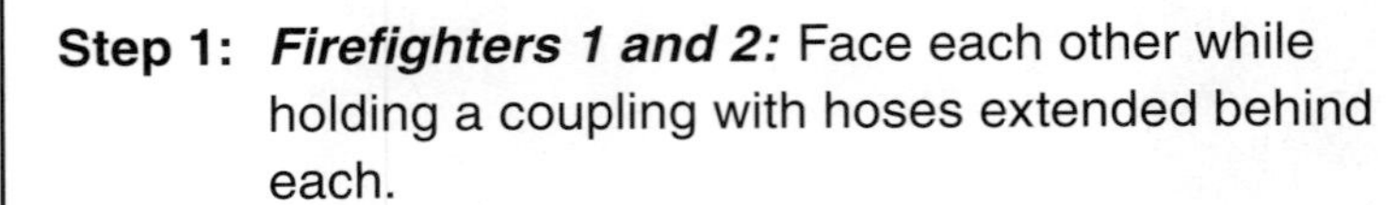

Step 1: ***Firefighters 1 and 2:*** Face each other while holding a coupling with hoses extended behind each.

Step 2: **Firefighter 1:** Hold one of the swivels so that it cannot turn (the sexless or Storz coupling has a swivel on each coupling).

NOTE: NFPA 1963 requires a locking mechanism to keep the couplings from disconnecting.

Step 3: ***Firefighter 2:*** Align the couplings and insert the lugs.

Step 4: ***Firefighter 2:*** Complete the connection by turning the swivel one-third turn clockwise to engage and lock the lugs securely.

Connecting Fire Hose Fixed Male Fitting

Skill Sheet 5-6

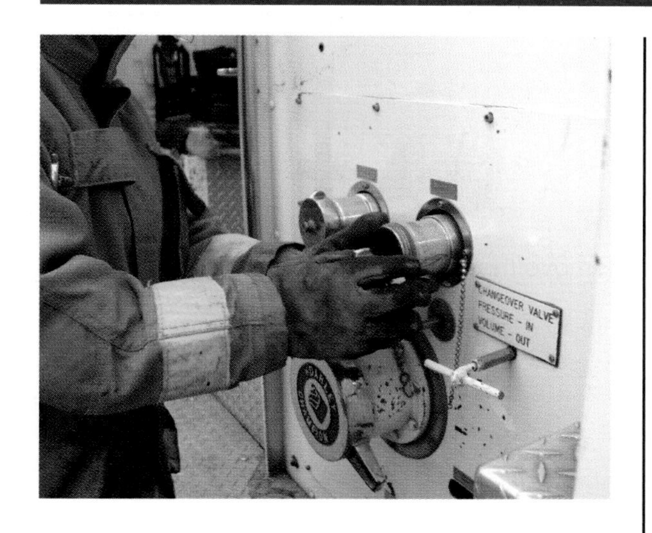

Step 1: Remove the cap on fixed fitting. Place the female coupling against the male threads of the discharge fitting.

Step 2: Align the coupling with the male fitting.

Step 3: Turn the swivel clockwise to tighten the coupling and complete the connection.

Connecting Fire Hose
Fixed Female Fitting

Step 1: Remove the male plug from fitting.

Step 2: Align the male coupling with the female fitting.

Step 3: Turn the swivel counterclockwise to tighten the coupling and complete the connection.

Attaching a Nozzle

Method 1 — Two Handed

Skill Sheet **5-8**

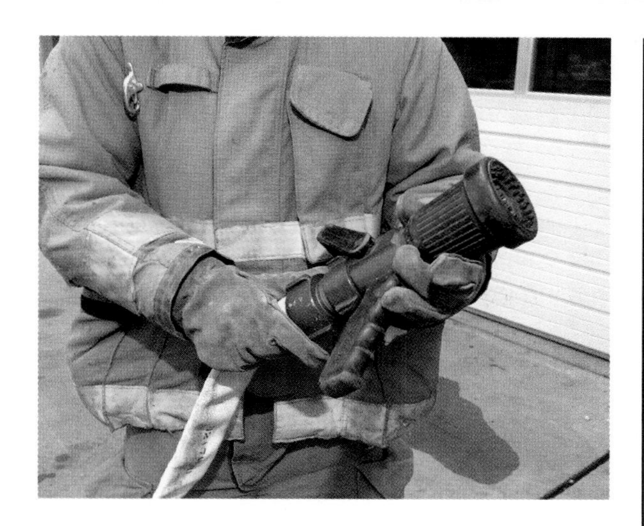

Step 1: Hold the nozzle firmly in one hand and the male coupling in the opposite hand.

Step 2: Align the nozzle threads against the coupling threads.

Step 3: Turn the nozzle clockwise to complete the connection.

5-9 Skill Sheet — Attaching a Nozzle Method 2 — With Foot

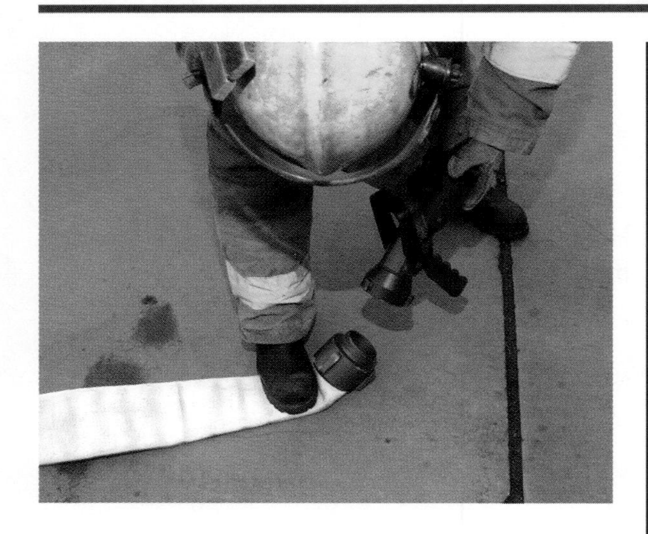

Step 1: Place one foot on the hose directly behind the male coupling and apply pressure to tilt the coupling upward.

Step 2: Hold the nozzle in both hands, and place it against the upturned male coupling.

Step 3: Turn the nozzle clockwise to make the connection.

Loosening a Tight Connection Knee-Press Method

Skill Sheet 5-10

Step 1: Grasp the coupled hose (which is lying on the ground) behind the female coupling, and push it up and forward to stand the male coupling below it on end.

NOTE: The principle of this one-person method is that compression of the hose gasket permits the swivel to turn more easily.

Step 2: Bend the hose over the top of the female coupling and as close to the coupling as possible.

5-10 Skill Sheet

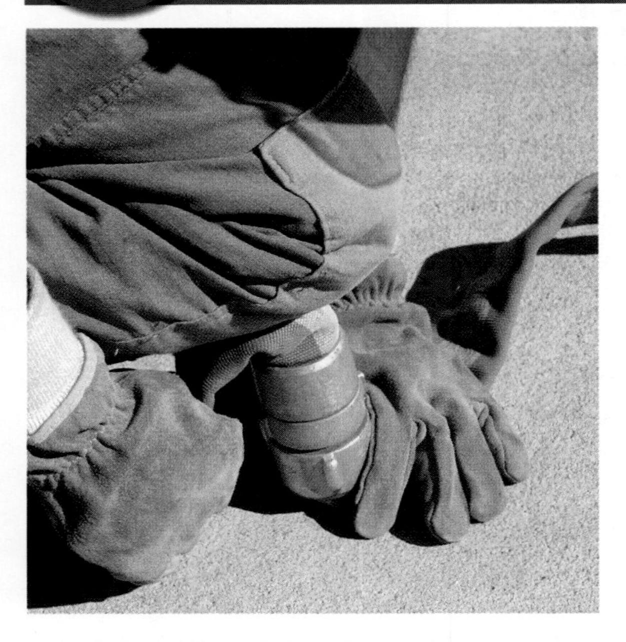

Step 3: Place one knee on the hose above the shank of the female coupling, keeping the upper leg in a vertical plane in alignment with the couplings.

Step 4: Lean forward and apply body weight to the connection.

Step 5: Quickly snap the swivel in a counterclockwise direction at the same time weight is applied to loosen the connection.

Loosening a Tight Connection Two Persons — Stiff-Arm Method

Skill Sheet 5-11

Step 1: ***Firefighters 1 and 2:*** Face each other with the hose coupling in between, and position feet about shoulder-width apart.

Step 2: ***Firefighters 1 and 2:*** Take a firm, two-handed grip on the coupling (Firefighter 1 holds the male shank and Firefighter 2 holds the female swivel).

Step 3: ***Firefighters 1 and 2:*** Interlock the fingers around the coupling to allow a tighter grip if desired.

Step 4: ***Firefighters 1 and 2:*** Hold the arms rigidly straight, and start with the body leaning slightly to the right. Apply counterclockwise force by leaning to the left while keeping the arms locked straight.

NOTE: This levering action produces a counterclock-wise force on the couplings. The weight of the body supplies most of the force for loosening the connection.

5-12 Skill Sheet Loosening a Tight Connection Two-Spanner-Wrench Method

Step 1: Place one spanner wrench across the lugs on the swivel of the female coupling.

Step 2: Place a second spanner wrench across the lugs on the shank of the male coupling.

Step 3: Hold the spanner wrench on the male coupling firmly in place.

Step 4: Rotate the spanner wrench on the female coupling counterclockwise to loosen the connection.

5-13 Skill Sheet Rolling Fire Hose

Straight Roll (Street Roll)

Step 1: Lay the hose straight and flat on a clean surface.

Step 2: Roll the male coupling over onto the hose to start the roll.

Step 3: Form a coil that is open enough at the center to allow the fingers to be inserted.

Step 4: Continue rolling the coupling over onto the hose, keeping the edges aligned, to form an even roll.

Step 5: Keep the hose edges aligned on the remaining hose to make a uniform roll as the roll increases in size.

Step 6: Lay the completed roll on the ground, and use a foot to tamp any protruding coils down into the roll.

5-14 Skill Sheet — Rolling Fire Hose

Donut Roll — Method 1

Step 1: Lay a section of hose flat and in a straight line.

Step 2: Pick up the male coupling, carry it to the opposite end, and place it next to the female coupling.

NOTE: This step forms a loop in the hose with two parallel segments that should lie flat, straight, and without twists.

Step 3: Walk to the looped end and face the couplings.

Step 4: Start the roll about 30 inches (762 mm) from the loop on the male coupling side by forming a bight (lift the hose enough to create a bend).

NOTE: The 30-inch (762 mm) measurement is for 2½-inch (65 mm) hose; for 1½-inch (38 mm) hose, measure in about 18 inches (457 mm) from the looped end.

Step 5: Roll the bight toward the male coupling to form a small coil, leaving enough space in the center to insert the fingers for carrying the finished roll.

5-14 Skill Sheet

Step 6: Continue rolling the hose toward the male coupling, keeping the edges aligned to form an even roll and pulling the female side back a short distance to relieve the tension if the hose behind the roll becomes tight.

Step 7: Draw the female coupling end around the male coupling to protect the coupling threads and complete the roll.

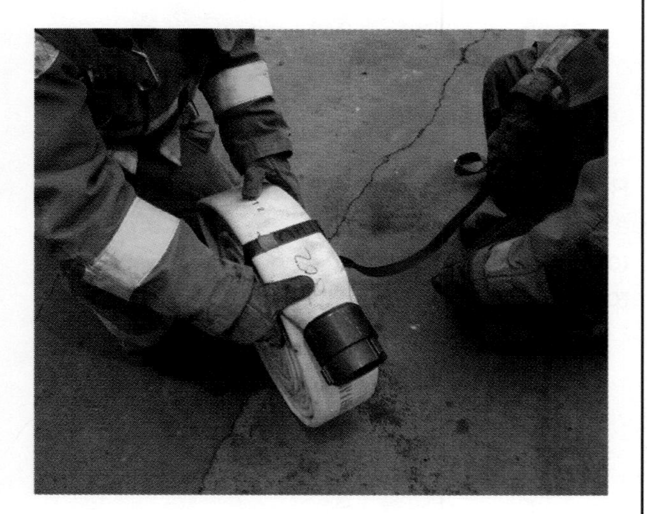

Step 8: Secure hose roll with a hose strap for storage in an apparatus.

Rolling Fire Hose

Skill Sheet 5-15

Donut Roll — Method 2

Step 1: Lay a section of hose flat and in a straight line.

Step 2: Lift the hose and form a bight at a point 5 or 6 feet (1.5 m or 1.8 m) from the hose midpoint.

5-15 Skill Sheet

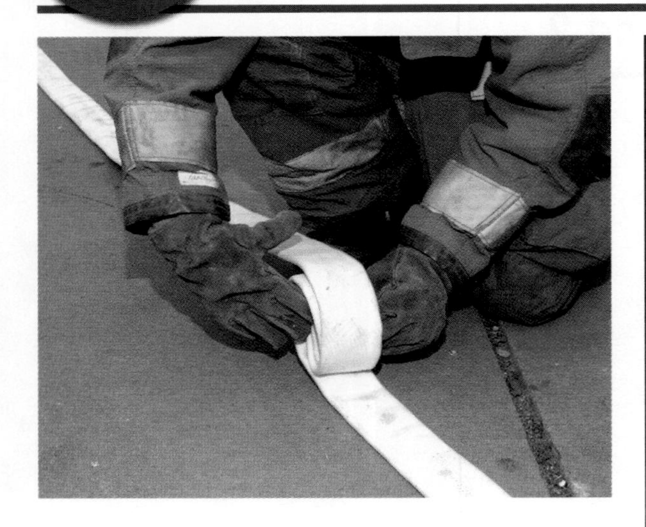

Step 3: Roll the bight toward the male coupling to form a small coil, leaving enough space in the center to insert a hand for carrying the finished roll.

NOTE: When the roll is complete, the male coupling will be inside the roll. The female coupling will be about 3 feet (1 m) ahead of the male coupling.

Step 4 Continue rolling the hose, building the roll upon the male half of the outstretched hose and aligning the edges to make a uniform roll.

NOTE: The female half of the hose is dragged forward as it is pulled into the roll.

Step 5: Draw the female coupling end around the male coupling end to protect the coupling treads and complete the roll.

Rolling Fire Hose

Donut Roll — Two Persons

Skill Sheet 5-16

NOTE: Either firefighter may perform the first three steps.

Step 1: Lay a section of hose flat in a straight line.

Step 2: Pick up the male coupling and carry it to the opposite end so that the hose is doubled back on itself.

Step 3: Place the male coupling on the hose 3 to 4 feet (0.9 m to 1.2 m) short of the female coupling (see photo).

Step 4: ***Firefighter 1:*** Straddle the hose at the looped end, and face the couplings.

Firefighter 2: Take a position 4 to 5 feet (1.2 to 1.5 m) in front of and facing Firefighter 1.

Step 5: ***Firefighter 1:*** Pick up the looped end and roll the loop over to form a small coil, leaving enough space in the center to insert a hand for carrying the finished roll (see photo).

Step 6: ***Firefighter 1:*** Roll the hose upon the doubled hose, keeping its edges aligned to make a uniform roll.

Step 7: ***Firefighter 2:*** Pull the slack hose back as it appears ahead of the roll; maintain alignment of the doubled-hose segments.

Step 8: ***Either Firefighter:*** Draw the female coupling end around the male coupling end to protect the coupling threads and complete the roll.

Twin Donut Roll

Step 1: Lay a section of hose flat, without twisting, to form two parallel lines from the loop end to the couplings (couplings should now be next to each other).

Step 2: Start the roll by folding the loop end over and upon the two hose lengths.

5-17 Skill Sheet

Step 3: Roll both lengths simultaneously toward the coupling ends to form a twin roll with a decreased diameter.

Step 4: Insert a strap through the center of the roll for carrying purposes.

Rolling Fire Hose

Self-Locking Twin Donut Roll

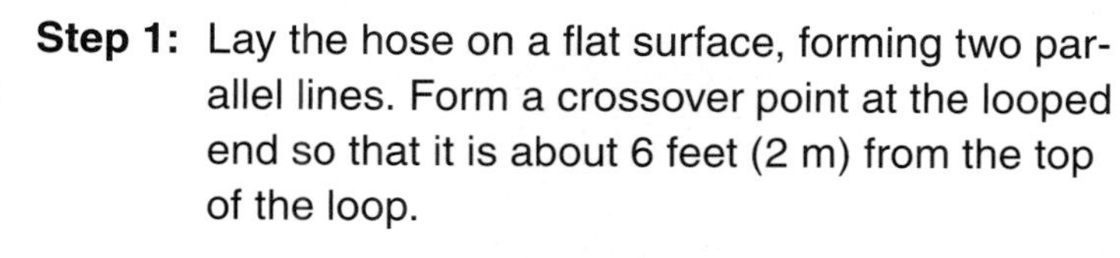

Step 1: Lay the hose on a flat surface, forming two parallel lines. Form a crossover point at the looped end so that it is about 6 feet (2 m) from the top of the loop.

NOTE: The size of the large loop formed by the crossover determines the length of the shoulder sling for carrying.

Step 2: Face the coupling ends, bring the top of the loop forward toward the couplings, and place it flat on top of the crossover to form a butterfly loop.

NOTE: The butterfly's "wings" form the shoulder loop for carrying.

Step 3: Hold the loop and crossover segments together, and roll the hose forward to form two tight coils on the parallel segments.

Step 4: Complete the twin roll so that the couplings lie across the top of each roll.

5-18 Skill Sheet

Step 5: Adjust the loops to form a shoulder sling by pulling one side of the loop through the roll so that it becomes larger than that on the opposite side.

Step 6: Slip the long loop through the short loop just behind the couplings and tighten snugly. Carry over shoulder.

All Skill Sheet 5-18 photographs are courtesy of Illinois Fire Service Institute.

Carrying Fire Hose

Skill Sheet

Shoulder-Loop Carry From the Ground

NOTE: This method requires one firefighter for each section of hose. The first 8 steps are performed by the first firefighter.

Step 1: Deploy the desired amount of hose in a straight line.

Step 2: Take a position at one end (either male or female coupling) of the hose, facing the other end.

Step 3: Pick up the end of the hose and place it over the shoulder so that the coupling hangs behind the body at waist height (see photo).

Step 4: Step forward to form a 3-foot (0.9 m) loop (bight) in the hose.

Step 5: Hold the hose on the shoulder with one hand, stoop, and pick up the looped section of hose at a point closest to the feet.

5-19 Skill Sheet

Step 6: Lift the loop and lay it on the shoulder without twisting or turning it over.

NOTE: Expect the shouldered loop to have a half-twist in it.

Step 7: Adjust the loop so that the bottom of the loop hangs at about mid-calf.

Step 8: Step forward approximately 15 feet (5 m) so that a second firefighter may load shoulder loops in the same manner.

NOTE: Repeat the loading process with additional fire-fighters until the entire length of hose is shoulder loaded and ready to carry.

Carrying Fire Hose

Accordion Shoulder Carry From the Ground

Skill Sheet **5-20**

Step 1: Deploy the desired amount of hose in a straight line.

Step 2: Take a position at one end (either male or female coupling) of the hose, and face the other end.

Step 3: Pick up the end of the hose and place it over the shoulder so that the coupling hangs behind the body at waist height (see photo).

Step 4: Hold the hose in front of the body, walk slowly forward, and form a loop that hangs at knee height.

5-20 Skill Sheet

Step 5: Continue walking, guiding the hose back over the same shoulder to form a loop that hangs at about knee height behind the body.

Step 6: Continue walking slowly down the hoseline, forming alternating loops in front of and behind the body until the entire hose section is picked up.

NOTE: Additional firefighters may shoulder accordion loads in the same manner for longer hoselines.

Carrying Fire Hose

Skill Sheet 5-21

Accordion Shoulder Carry From an Accordion-Loaded Hose Bed

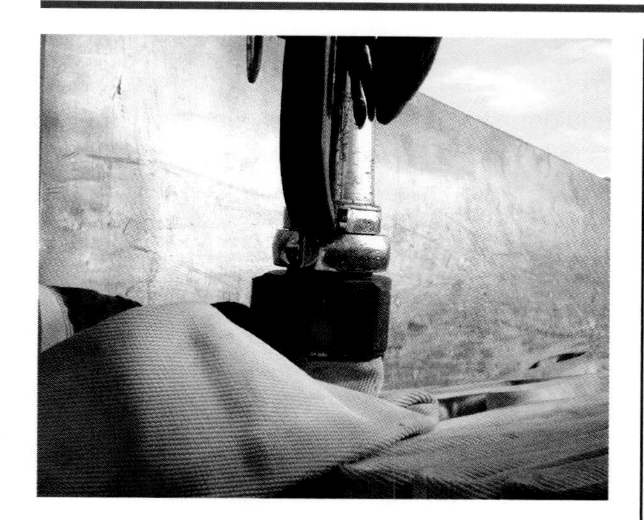

Step 1: ***Firefighter 1:*** Attach a nozzle to the end of the hose if desired.

Step 2: ***Firefighter 1:*** Face the hose bed, grasp the nozzle or coupling with one hand and the first section of hose (up to the next coupling) with the other hand.

Step 3: ***Firefighter 1:*** Pull the hose folds in one smooth motion about half way out of the hose bed, rotate the gathered folds 90 degrees so that the fold with the nozzle is on the bottom of the stack, and pivot away from the hose bed, placing the folds on one shoulder.

NOTE: Make sure that the hose is flat on the shoulder with the nozzle or coupling in front of the body.

5-21 Skill Sheet

Step 4: ***Firefighter 1:*** Hold the bundle tightly on the shoulder, step down from the pumping apparatus tailboard, and walk forward about 15 feet (3 m) to clear the folds from the apparatus.

Step 5: ***Firefighter 2:*** Step to the tailboard and face the load. In one smooth motion, grasp the running length of the hose (up to and including the next coupling), rotate the folds 90 degrees so that the running length is on the bottom, and pivot away from the hose bed, placing the folds on the shoulder.

Step 6: ***Firefighter 2:*** Hold the bundle tightly on the shoulder, and walk forward about 15 feet (3 m) so that another firefighter may shoulder an accordion load.

Step 7: ***Additional Firefighters:*** Load the hose in the same manner until the required amount of hose is removed from the bed.

Step 8: ***Last Firefighter:*** Disconnect the hose from the hose bed.

Carrying Fire Hose

Skill Sheet 5-22

Accordion Shoulder Carry From a Hose Load Other Than an Accordion Load

Step 1: ***Tailboard Firefighter:*** Attach the nozzle to the end of the hose.

NOTE: The Tailboard Firefighter remains at the rear of the pumping apparatus to help other firefighters load their shoulder bundles.

Step 2: ***Firefighter 1:*** Take a position at the pumping apparatus tailboard facing in the direction of travel.

Tailboard Firefighter: Place the initial fold of hose over Firefighter 1's shoulder so that the nozzle can be held at chest height.

Step 3: ***Firefighter 1:*** Hold each layer of hose to prevent it from slipping off the shoulder.

Tailboard Firefighter: Bring the hose from behind the back over Firefighter 1's shoulder so that the rear fold ends at the back of Firefighter 1's knee.

5-22 Skill Sheet

Step 4: ***Tailboard Firefighter:*** Make a front fold that ends at knee height, and bring the hose back over Firefighter 1's shoulder. Continue making knee-high folds until an appropriate amount of hose is shoulder loaded.

Step 5: ***Firefighter 1:*** Move forward approximately 15 feet (3 m) to provide room for Firefighter 2 to receive a shoulder load.

Firefighter 2: Take a position at the tailboard, facing the direction of travel.

Step 6: ***Tailboard Firefighter:*** Load hose onto Firefighter 2's shoulder in the same manner as before, making knee-high folds until about one section of hose is loaded.

Firefighter 2: Move forward approximately 15 feet (3 m) to provide room for Firefighter 3 to receive a shoulder load.

Step 7: ***Tailboard Firefighter:*** Repeat the loading process with each additional firefighter until the desired length of hose is loaded.

Step 8: ***Tailboard Firefighter:*** Uncouple the hose from the hose bed and hand the coupling to the last firefighter.

5-23 Skill Sheet — Carrying Fire Hose

Horseshoe Shoulder Carry From the Ground

Step 1: Lay a section of hose in a straight line. Pick up the male coupling, carry it to the opposite end, and place it approximately 12 inches (305 mm) to the side and even with the female coupling.

Step 2: Go to the opposite end and pick up the hose in the center of the loop. Carry the loop back between the couplings, and lay it on the ground even with the couplings.

Step 3: Go to the center of the folded hose, place one foot against the hose as a pivot, and swing the folded ends around and even with the couplings.

Step 4: Pick up the bundle at the center and place it on the shoulder.

CAUTION

During the lift, the hose may swing, increasing the possibility of injury to the back. To avoid back injury, lift with the knees and not with the back.

Step 5: Proceed to deployment location.

5-24 Skill Sheet — Carrying Fire Hose

Horseshoe Underarm Carry From the Ground

Step 1: Lay the section of hose in a straight line. Pick up the male coupling, carry it to the opposite end, and place it approximately 12 inches (305 mm) to the side and even with the female coupling.

Step 2: Go to the opposite end and pick up the hose in the center of the loop. Carry the loop back between the couplings, and lay it on the ground even with the couplings.

Step 3: Go to the center of the folded hose, and swing the folded ends around and even with the couplings.

Step 4: Go back to the center of the folded hose, again place one foot against the hose as a pivot, and swing the looped end around to the couplings so that the couplings are in the center of the bundle.

Step 5: Kneel beside the completed bundle, compress the folds together, and stand the bundle on edge. Pick up the bundle and cradle it under one arm.

CAUTION

To avoid back injury, always remember to lift with the legs and not the back.

5-25 Skill Sheet — Dragging Fire Hose

Single Section — Method 1

Step 1: Lay a section of hose straight. Stand to the side of one end of the hose, pick up the coupling with the hand nearest the hose, and then walk to the midpoint of the hose section.

Step 2: Pick up the hose at the midpoint with the other hand, and loop it over the shoulder nearest the hose.

Step 3: Step to the other side of the hose, permitting the hose to cross the chest, and advance to the next coupling.

Step 4: Pick up the second coupling with the free hand. Walk forward to drag the hose.

5-26 Skill Sheet Dragging Fire Hose

Single Section — Method 2

Step 1: Lay a section of hose straight. Pick up a coupling and place it over the shoulder so that the hose drapes across the chest.

Step 2: Walk to the other coupling, pick it up, and place it over the opposite shoulder so that the hose crosses the chest in the opposite direction.

Step 3: Turn, step inside the loop that has been formed, and walk back to the end of the loop.

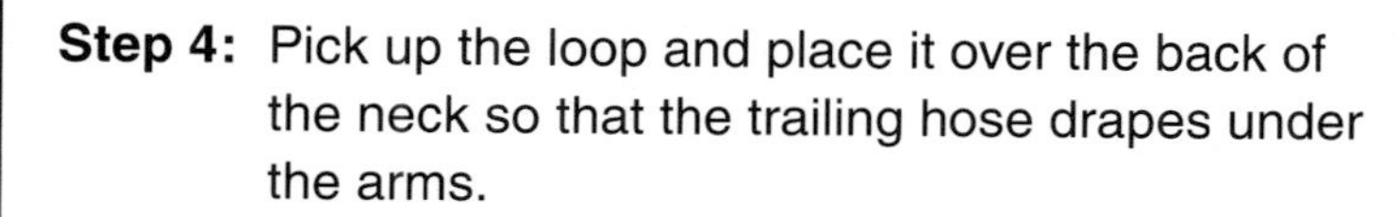

Step 4: Pick up the loop and place it over the back of the neck so that the trailing hose drapes under the arms.

Step 5: Walk forward dragging the hose to the destination.

5-27 Skill Sheet Dragging Fire Hose Multiple Sections — Method 1

Step 1: ***Firefighter 1:*** Take a position at the pumping apparatus tailboard, facing toward the direction of travel (away from the apparatus).

Tailboard Firefighter: Attach the nozzle if one is required, and hand it to Firefighter 1.

NOTE: For this procedure, the Tailboard Firefighter stays at the pumping apparatus tailboard to help the firefighters load the hose to be dragged.

Step 2: ***Firefighter 1:*** Lay the hose over the shoulder and hold the nozzle in front of the body. Walk forward slowly (see photo).

Tailboard Firefighter: Pull off a section of hose and then call for Firefighter 1 to stop and Firefighter 2 to step in on the same side of the hose as the previous person. Place the hose over Firefighter 2's shoulder with the coupling held in front about chest high.

Step 3: ***Tailboard Firefighter:*** Give a verbal signal for the two firefighters to walk forward and drag the hoseline.

Step 4: ***Tailboard Firefighter:*** Shoulder load firefighters and pull more hose in the same manner, stopping the advancing group when the coupling of the last hose section appears.

Step 5: ***Tailboard Firefighter:*** Uncouple the last hose section from the hose bed, and either connect the hose to a discharge valve or carry it to the destination, serving as the last carrier in the group.

Dragging Fire Hose Multiple Sections — Method 2

Skill Sheet 5-28

Step 1: ***Firefighter 1:*** Attach the nozzle to the end of the hose if a nozzle is required.

Step 2: ***Firefighter 1:*** Pull enough hose from the hose bed to allow placing the nozzle on the ground approximately 25 feet (8 m) behind the tailboard and on the side toward where the hose will be carried.

Step 3: ***Firefighter 1:*** Pull more hose from the bed and make a fold that starts at the tailboard and ends even with the nozzle. Lay this first fold beside the nozzle fold on the side away from where the hose will be carried.

Step 4: ***Firefighter 1:*** Continue to pull hose from the hose bed, making more folds in the same manner, and laying each progressively away from the nozzle in an accordion fashion.

5-28 Skill Sheet

Step 5: ***Firefighter 1:*** Uncouple the hose from the hose bed when the appropriate amount of hose has been pulled and either connect it to the pumping apparatus or (if the hose is needed at another location) lay the last coupling alongside the last fold.

Step 6: ***Firefighter 1:*** Stand at the nozzle, facing away from the folds. Pick up the nozzle, lay the hose over the shoulder, and hold the nozzle securely to the chest with one hand.

Step 7: ***Firefighter 1:*** Pick up a fold with the free hand and walk forward approximately 25 feet (8 m).

Step 8: ***Firefighter 2:*** Move in and stand at the end, facing away from the next two folds. Pick up a fold in each hand and walk forward approximately 25 feet (8 m) as Firefighter 1 moves forward the same distance.

Step 9: ***Additional Firefighters (as needed):*** Repeat this process until the hose has been dragged to its desired location.

5-29 Skill Sheet — Advancing Hoselines

Uncharged Hoseline

Step 1: Pick up the nozzle end of the hoseline and drape it over one shoulder and across the front of the body. The nozzle is kept to the front of the body.

Step 2: Reach back with the other hand, grasp a fold in the hoseline, and pull it forward under that arm while continuing to hold the hose and nozzle.

Step 3: Advance the hoseline, maintaining control of the nozzle and ensuring that the hoseline is not tangled or presenting a tripping hazard.

Advancing Hoselines

Securing Hoseline to a Ladder

Skill Sheet 5-30

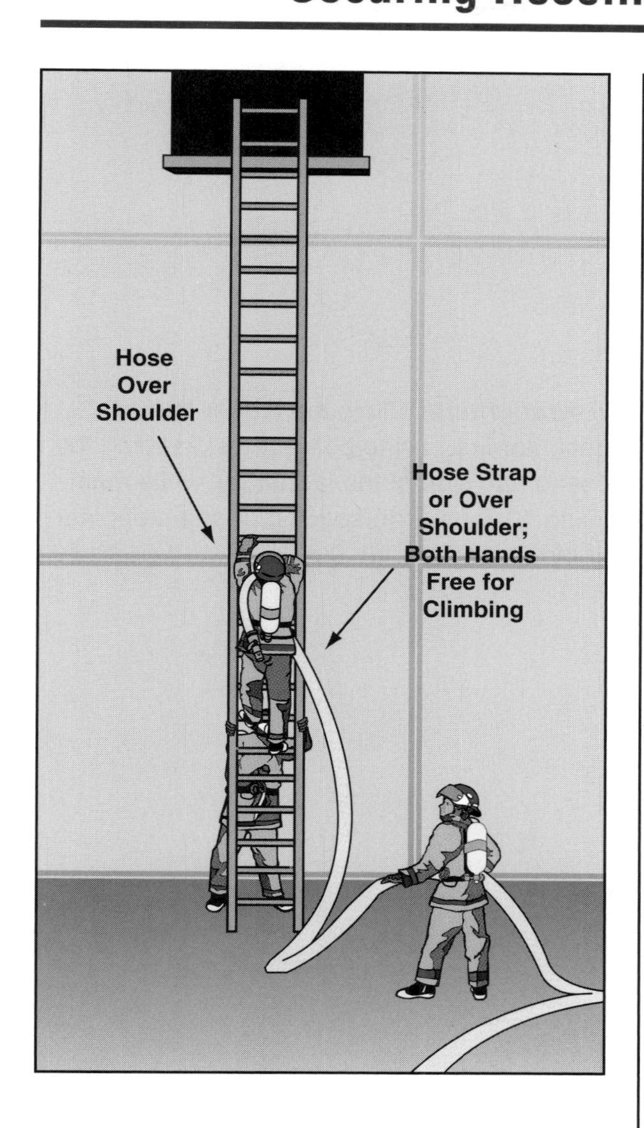

Step 1: ***All Firefighters:*** Take positions about 10 feet (3 m) apart on the same side of the hose, facing the nozzle.

NOTE: This procedure requires a team of four or five firefighters: One to secure and stabilize the ladder, one to remain on the ground and advance hose, and two or three to climb.

Step 2: ***Firefighter 1:*** Place the hoseline on either the right or left shoulder with the nozzle lying on the back.

Step 3: ***All Firefighters:*** Advance, allowing the hose to drape and hang to knee level in front. Leave about 10 feet (3 m) of hoseline between each firefighter.

5-30 Skill Sheet

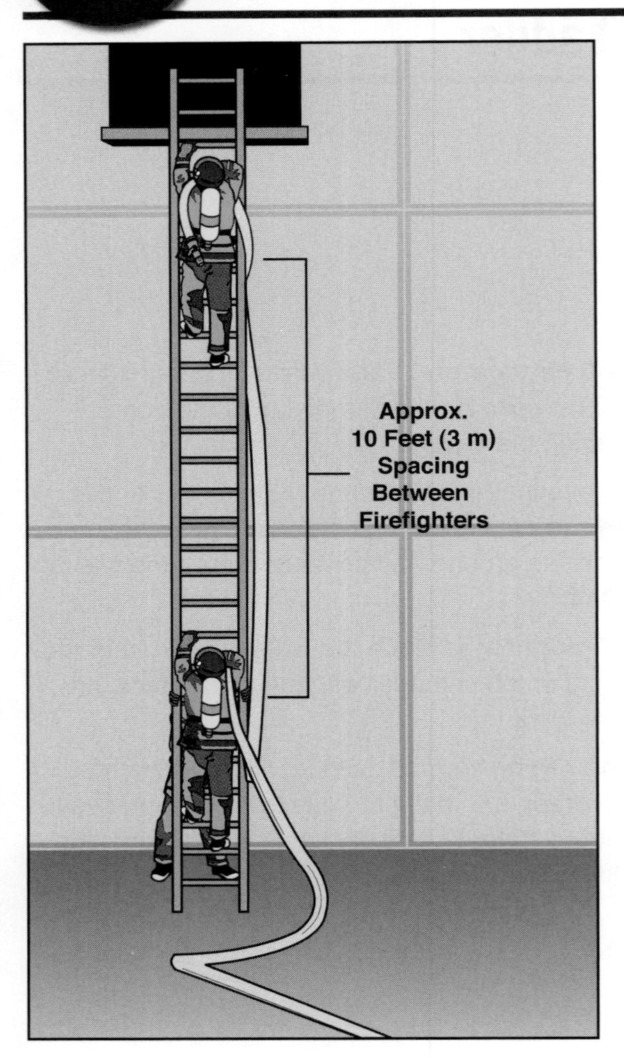

Step 4: ***All Firefighters:*** Climb the ladder (Nozzle Firefighter first) using both hands, keeping the hose to one side of the ladder rail while maintaining 10-foot (3 m) spacing or one firefighter per ladder section.

Step 5: ***All Firefighters:*** Once in position, leg lock into the ladder.

5-30 Skill Sheet

Step 6: ***Nozzle Firefighter:*** Pass the nozzle through the rungs at chest height, allowing it to hang down about 2 feet (0.6 m) on the inside of the ladder.

Step 7: ***Nozzle Firefighter:*** Tie the hose to the ladder with a clove hitch, rope hose tool (as shown being looped on the left and hooked on the right), or utility strap at a point two or three rungs below where the nozzle is hanging.

Step 8: ***Nozzle Firefighter:*** Step down two rungs and lock into the ladder once the hose and nozzle is securely attached to the ladder.

Step 9: ***Nozzle Firefighter:*** Adjust the hose strap as necessary when the hoseline is charged.

Step 10: ***Nozzle Firefighter:*** Move back to nozzle, lock in, and direct nozzle stream.

Advancing Hoselines From a Standpipe

Skill Sheet 5-31

Step 1: ***All Firefighters:*** Gather equipment and assemble as an entry team upon arrival at the scene.

Step 2: ***All Firefighters:*** Gain entry to the smoke tower stairs and advance to the standpipe connection, one floor below the incident floor.

5-31 Skill Sheet

Step 3: ***Firefighter 1:*** Remove the cap on the standpipe valve.

Step 4: ***Firefighter 1:*** Connect a gated wye to the standpipe valve, and connect the standpipe pack attack line to one side of the wye.

Step 5: ***Other Firefighters:*** Advance up to the incident floor while deploying attack hoseline up the stairs to the floor above the incident and then back down to the incident floor. Keep the deployed line to the outside of the stairs against the wall.

Step 6: ***Company Officer:*** Notify Firefighter 1 located at the standpipe connection to charge the hoseline when entry team is ready.

All Skill Sheet 5-31 photographs are courtesy of Rick Montemorra, Mesa (AZ) Fire Department.

5-32 Skill Sheet — Controlling Hoselines

Kinking the Hose to Remove a Burst Section

Step 1: Start at a point well ahead of the burst section, and straighten the hose progressively, moving from the water source toward the nozzle.

NOTE: As the slack hose accumulates, it will form an S shape.

Step 2: Lay one segment of the *S* back over itself to form a small loop. Stand the loop on edge.

Step 3: Press the loop downward to form a fold. Kneel on the bends that are formed in the hose. Allow the pressure of the water trapped in the loop to slowly release and collapse the hose.

NOTE: The hose is considered shut down at this point.

Step 4: Replace the burst hose section with another section of hose while the flow is stopped.

Chapter 6
Supply Hose Loads and Deployment Procedures

Chapter 6
Supply Hose Loads and Deployment Procedures

Fire hose is used in several general ways: (1) to draft water from a static water source to a fire pump, (2) to transport water or extinguishing agent from a pressurized source to a fire pump or series of fire pumps (in a supply capacity), and (3) to transport water or extinguishing agent from a fire pump to the nozzle(s) (in an attack capacity). This chapter describes the use of fire hose to supply water or extinguishing agent from a source to the location of an emergency incident where it is delivered to attack hoselines. A number of methods for loading supply fire hose on the pumping apparatus and then deploying it are available. Hose loads are designed so that fire hose can be deployed in an expedient manner during fire-suppression or other emergency operations. Various methods of selecting and making the appropriate fire hose connections to a fire hydrant are also presented. The hose loads and deployment methods described in the following sections are found in use throughout North America and can be adapted to the many types of pumping apparatus and hose bed configurations found in the fire and emergency services today.

NOTE: For reference purposes, the end of the hose bed closest to the pumping apparatus tailboard is the *rear;* the end closest to the apparatus cab is the *front.* A layer of hose in the pumping apparatus hose bed is a *tier* **(Figure 6.1).**

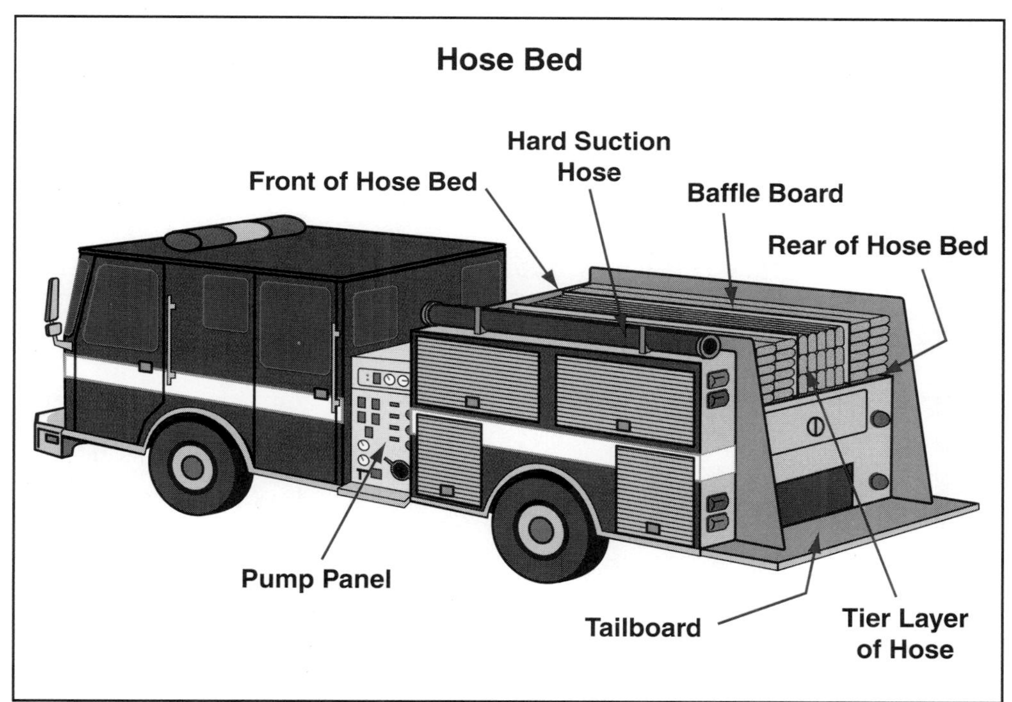

Figure 6.1 A hose bed showing the locations and terminology used during hose loading/deployment procedures.

Deploying Supply Fire Hose

Supply fire hose deployment and layout procedures often vary regionally from organization to organization. However, the basic methods of deploying supply fire hose outlined in this section remain fundamental regardless of location, size, or type of hose. The deployment of supply fire hose can go from a water source to the fire/emergency scene (forward lay), from the fire/emergency scene to a water source (reverse lay), or from a midpoint in the deployment to the water source with one pumping apparatus and to the fire/emergency scene with another pumping apparatus with the hose connected at the meeting point (split lay). These basic methods (forward, reverse, and split) are presented to provide the foundation for developing efficient fire hose deployment options that specifically meet an individual fire and emergency services organization's requirements. Combination and dual lays can be developed from these three fundamental configurations.

Some mention must be given to the direction in which supply hose (particularly supply hose with threaded couplings) is deployed before discussing the methods of loading hose. Threaded coupling hose must be arranged in the hose bed so that when hose is deployed, the end with the female coupling is toward the water source and the end with the male coupling is toward the fire/emergency scene. When hose is deployed in this manner, several options are available such as the following:

- At the water source, hose can be connected to the male threads of a pumping apparatus discharge valve or to the male threads of a hydrant.
- At the emergency incident, hose can be connected to the auxiliary intake valve of the pumping apparatus or it can be connected directly to nozzles and appliances (all of which have female threads).

A double female or double male coupling adapter can be used when it becomes necessary to connect two male couplings or two female couplings together. The use of sexless couplings simplifies the procedures and options available to fire and emergency service organizations when selecting the type of deployment that best serves their needs.

Forward Lay

With the *forward lay,* hose is laid from the water source to the fire/emergency scene **(Figure 6.2).** Hose beds configured for forward lays are arranged

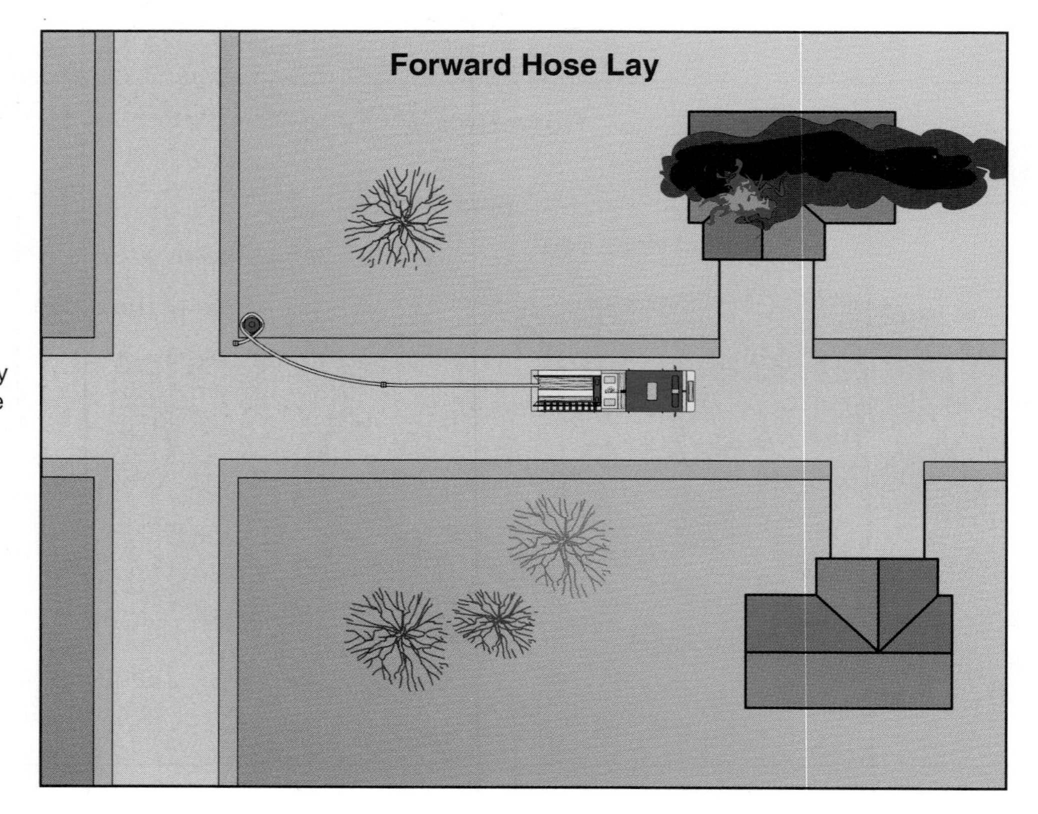

Figure 6.2 The forward hose lay proceeds from the water source to the emergency scene.

so that the female coupling is deployed first. This hose lay is often used when the water source is a hydrant or other pressurized source but the pumping apparatus must stay at the fire/emergency scene.

The primary advantage with this hose lay is that a pumping apparatus remains at the incident scene to provide additional equipment and tools for a fire-suppression or other emergency effort if needed. The pumping apparatus driver/operator also retains visual contact with the fire and emergency services crew and, therefore, can better react to changes in the fire/emergency operation than if the pumping apparatus is at the hydrant.

A disadvantage with the forward hose lay, however, is that if a long length of hose is deployed, it may be necessary for a second pumping apparatus to boost the pressure in the hoseline at the hydrant. Although it is not a bad idea to have a second pumping apparatus positioned at the water source, it complicates the operation and requires additional time to implement. The use of a hydrant valve (sometimes referred to as a *four-way valve*) allows the hydrant to be opened, charging the supply line to the fire/emergency scene, while allowing a second pumping apparatus to deploy a second supply line without shutting down the hydrant. Another disadvantage is that one member of the crew is temporarily unavailable for assignment at the actual incident scene because that person must stay at the hydrant long enough to make the connection and open the hydrant.

When deploying hose forward from the water source to the incident scene, the pumping apparatus may be used in several ways. The option chosen often depends upon the diameter of the hose being deployed. Supply hose delivers a volume of water directly proportional to its size. Therefore, when large diameter supply hose is being deployed, locate the pumping apparatus at the most tactically advantageous position at the incident scene (following the fire and emergency service organization's procedures and emergency response guidelines), connect the hose to an intake valve, and pump the water to the hoselines.

When a supply hose is deployed and the crew is involved in a quick-attack operation using preconnected attack hoselines, the driver/operator may wish to first pump water from the booster tank and then switch to the supply hose when it becomes charged with water. The critical point in this situation is that fire and emergency service crews could be placed in a dangerous situation if the pumping apparatus tank water becomes depleted before the supply line is charged. The option of starting a quick attack using water from the booster tank should be chosen only if the crew is reasonably certain it can achieve fire knockdown or control the emergency using only the water in the tank.

Another option when deploying supply hose forward is for the pumping apparatus to drop tools and hose at the incident scene and return to the hydrant to increase the water pressure by pumping the supply water to the attack hoselines **(Figure 6.3, p. 248).** When this option is used, the forward-deployed hose becomes an attack hose and must meet the specifications of attack hoselines. See Fire Hose Classifications section in Chapter 1, Fire Hose and Couplings. The procedures for an attack-line forward lay are described in Chapter 7, Attack Hose Loads, Finishes, Hose Packs, and Deployment Procedures. Although this alternative is available for implementation during fire-suppression/emergency scene activities, its use is usually limited to emergency operations that are over 500 feet (152 meters) from the hydrant where supply lines smaller than 3 inches (77 mm) are used.

The procedures for completing a forward supply hose lay are described in **Skill Sheet 6-1.** These procedures can be modified to accommodate the type of pumping apparatus, hose diameter, and kind of equipment used by the local fire and emergency services organization.

CAUTION

Fire hose classified as supply hose is designed with a maximum operating pressure of 185 psi (1 276 kPa) [12.75 bar]. If supply hose is used to directly supply attack hoselines, a pressure-relief device must be used to limit the pressure in that hoseline to 185 psi (1 276 kPa) [12.75 bar].

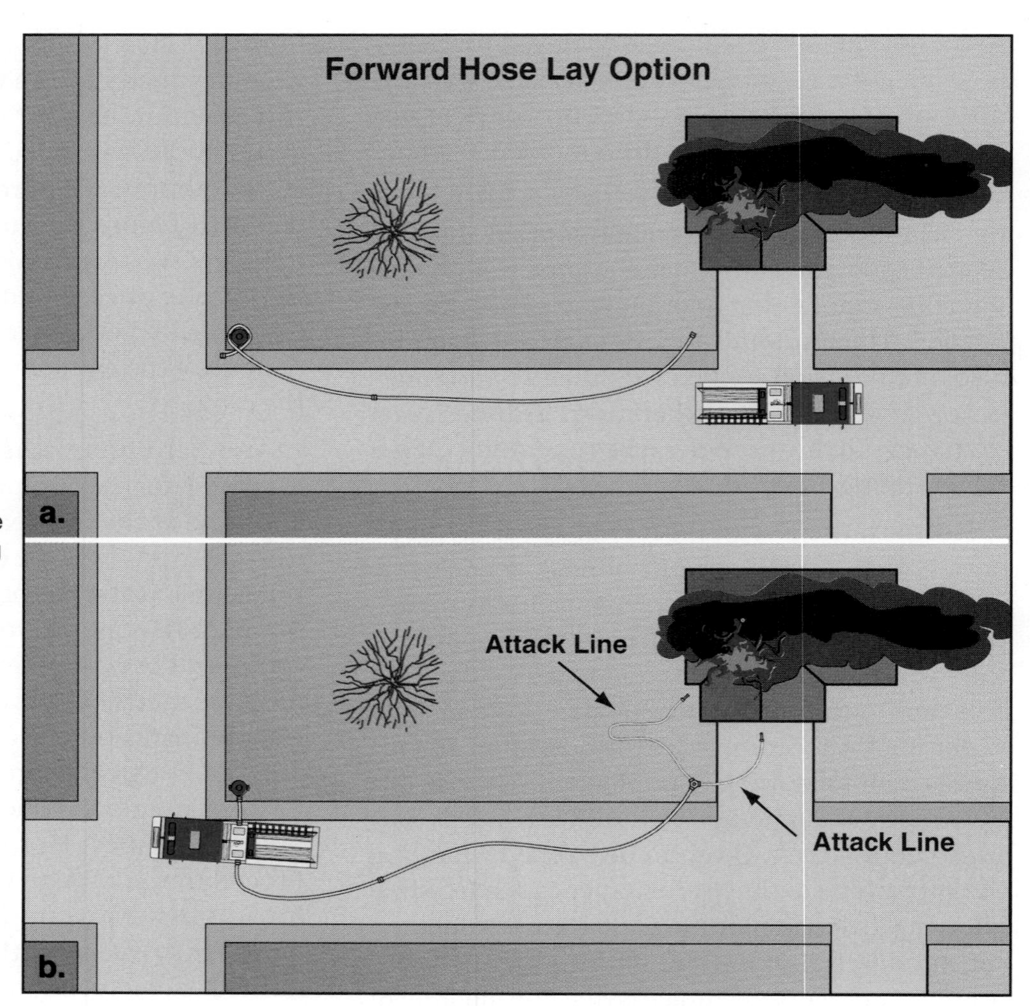

Figure 6.3 Forward hose lay option: (1) The pumping apparatus drops tools and hose at the incident scene and (2) returns to the hydrant to increase the water pressure by pumping the supply water to the attack hoselines.

Reverse Lay

The *reverse lay* deploys fire hose from the fire/emergency scene to the water source **(Figure 6.4).** Hose beds configured for reverse lays are loaded so that the male coupling is deployed first. The reverse method is most often used when hydrant spacing is very close (under 200 feet [61 m]), the driver/operator can see the incident scene, the water pressure is low (under 40 psi [276 kPa {2.76 bar}]), and the supply line is over 500 feet (152 m).

This method may also be used when a pumping apparatus is available to support the operation but is somewhat distant from the incident scene, requiring an extended response time to get to the scene. An advantage of the reverse lay is that personnel on the first pumping apparatus on the scene can size up the incident before committing to the deployment of a supply line. This situation is especially true when operating in rural or developing suburban areas where water sources are limited. A reverse hose lay is also the most expedient way to deploy hose if the pumping apparatus that deploys the hose must stay at the water source such as when drafting or increasing hydrant pressure to the supply line.

A disadvantage with the reverse lay, however, is that essential equipment, including attack hose, tools, and spare self-contained breathing apparatus cylinders, must be removed and placed at the fire/emergency location before the pumping apparatus can proceed to the water source. This situation causes some delay in the initial attack. The reverse lay also obligates one person, the driver/operator, to stay with the pumping apparatus at the water source, thus preventing that person from performing other essential fire/emergency scene activities (including being counted as a member of a backup crew to comply with Occupational Safety and Health Administration [OSHA] regulations [Title 29 *CFR* 1910.134]).

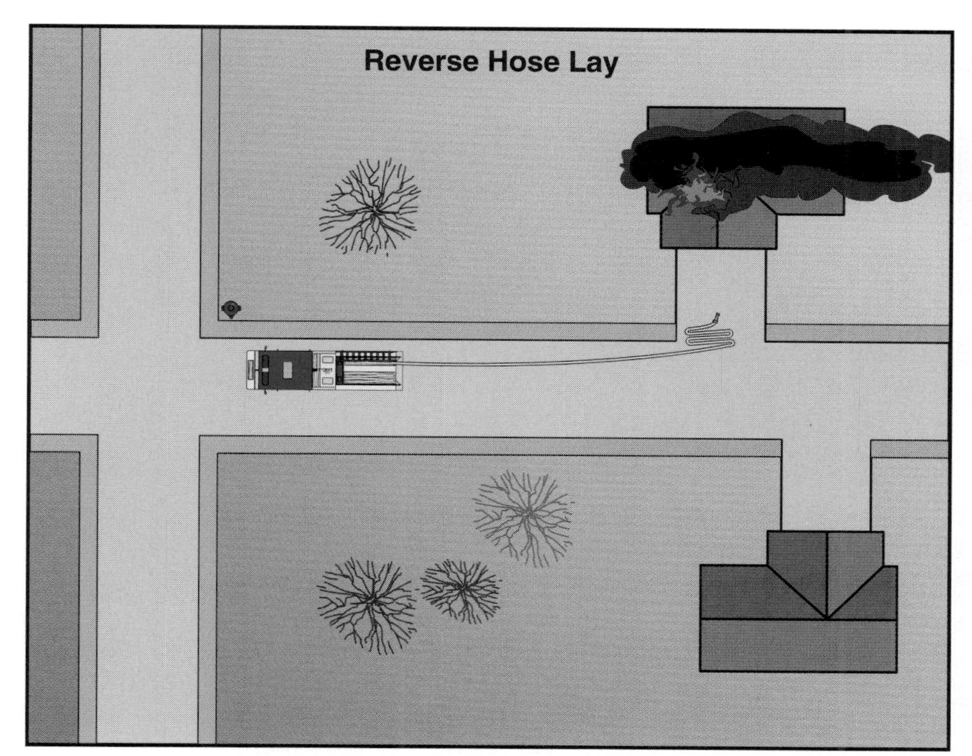

Figure 6.4 The reverse hose lay proceeds from the emergency scene to the water source.

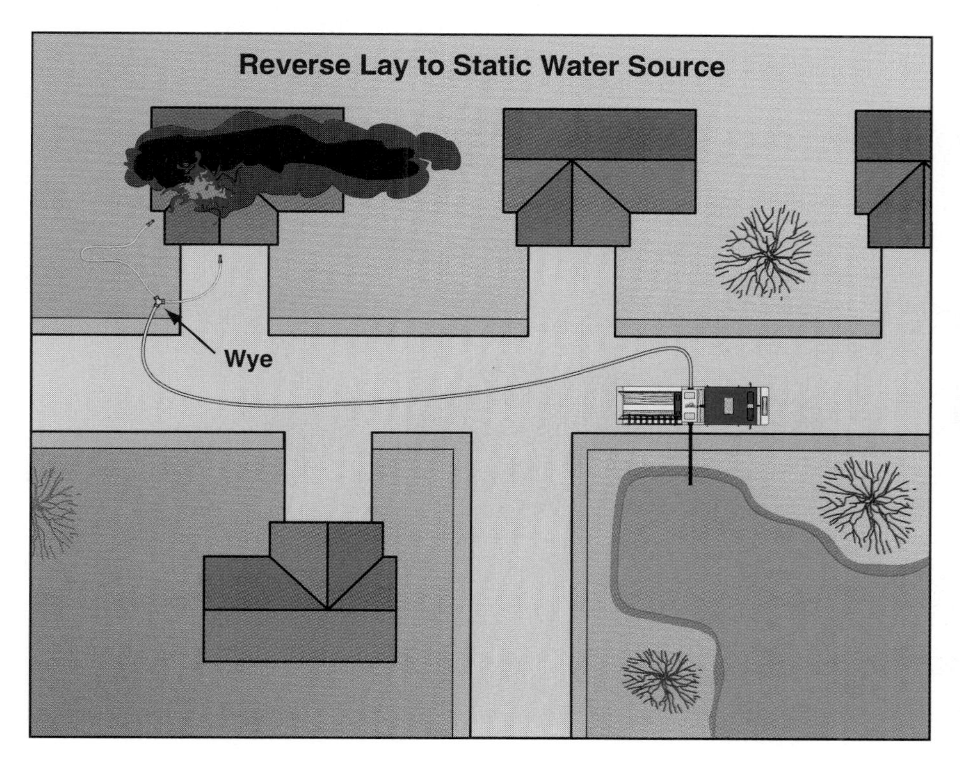

Figure 6.5 A reverse hose lay may be used when only one pumping apparatus is available and a static water supply is distant from the fire scene.

Deploying hose from the incident scene back to the water source is a standard method to begin developing a relay operation when using 2½-, 3-, or 3½-inch (65 mm, 77 mm, or 90 mm) supply hose. In most cases with this size hose, it is necessary to place a pumping apparatus at the hydrant to supplement hydrant pressure to the supply hose. It is, of course, always necessary to place a pumping apparatus at a static water source implementing a drafting operation **(Figure 6.5).** The reverse lay is the most direct way to accomplish this operation.

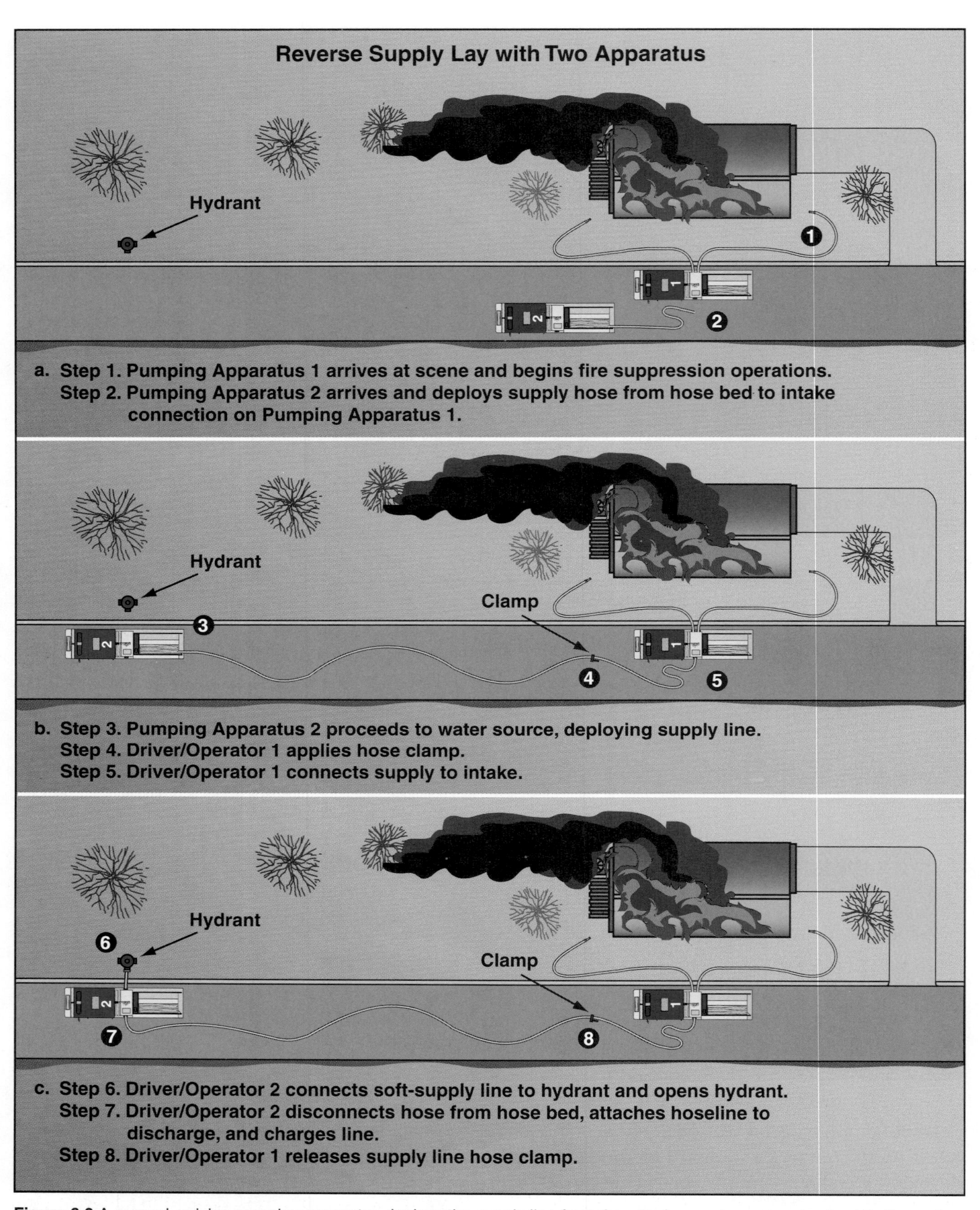

Figure 6.6 A second-arriving pumping apparatus deploys the supply line from the attack pumping apparatus back to the water source.

A classic supply-line evolution involving two pumping apparatus — an attack pumping apparatus and a water supply pumping apparatus — calls for the first-arriving pumping apparatus to go directly to the scene to start an initial attack on the fire/emergency, while the second-arriving pumping apparatus deploys the supply line from the attack pumping apparatus back to the water source **(Figure 6.6).** This operation is relatively simple because the second pumping apparatus needs only to connect its just-deployed hose to a discharge, connect a suction hose, and begin pumping.

When it is anticipated that additional water may be required, a variation of this operation is to have the first pumping apparatus complete a forward lay and begin fire-suppression/emergency activities, while the second pumping apparatus deploys a second hoseline using a reverse lay from the fire to the hydrant. Both hoselines are then pumped back to the fire by the second pumping apparatus. This variation is especially useful when hose diameters are less than 3½ inches (90 mm), restricting the quantity and distance water can be pumped.

When deploying a supply hose reverse, it is not necessary to use a four-way hydrant valve on the hydrant as long as the pumping apparatus is connected to the hydrant and the supply hose. A four-way hydrant valve can be used, however, if it is expected that the pumping apparatus will later disconnect from the supply hose and leave the hose connected to the hydrant. This situation may be desirable when the demand for water diminishes to the point that the second pumping apparatus can be made available for response to other emergency incidents. As is the case with a forward lay, using the four-way valve in a reverse lay provides the means to switch from pump pressure to hydrant pressure without interrupting the water flow.

The reverse lay is also used when the first-in pumping apparatus arrives at a fire/emergency scene and must work alone for an extended period. In this case, the reverse-deployed hose becomes an attack hoseline. It is often connected to a reducing wye so that two smaller hoses can be used to make a two-directional attack on the fire/emergency incident (the attack-line reverse lay is described in greater detail in Chapter 7, Attack Hose Loads, Finishes, Hose Packs, and Deployment Procedures).

Skill Sheet 6-2 describes the procedures for deploying fire hose from a second pumping apparatus using a reverse lay from a pumping apparatus positioned at the scene to a hydrant. This operation can be modified to accommodate most types of pumping apparatus, hose, and equipment. It can also be easily incorporated with other hose lays to allow a fire and emergency services organization a flexible response to a variety of water-flow needs as demanded by the incident situation.

Split Lay

The *split lay* is a hoseline deployed in part as a forward lay and in part as a reverse lay by two pumping apparatus from a point between the incident and the water source **(Figure 6.7, p. 252).** This hose lay usually starts at a road/street corner, driveway, or highway intersection. A second pumping apparatus can then make a reverse hose lay to the water supply source from the point where the initial line was laid by the first pumping apparatus. Care must be taken to avoid making the hose lay too long for the pump, hose size, and required gallons per minute (liters per minute) delivery. See **Skill Sheet 6-3** for the procedures for deploying supply hoselines with a split hose lay.

Basic Hose Lay Variations

Dual and combination lays, as well as the configuration of the hose bed, can incorporate any of the three basic hose loads. These variations can be deployed in a forward lay, reverse lay, or split lay.

Dual Lay

The term *dual lay* can refer to any one of a number of ways to deploy multiple supply hoselines: two or more same-sized hoselines deployed as supply hoselines by one pumping apparatus. Dividing a hose bed into two or more separate sections or compartments allows the driver/operator a number of options for laying multiple supply hoselines of the same diameter. Depending upon how the hose beds are loaded (for forward or reverse hose lays), hoselines can be deployed in the following ways:

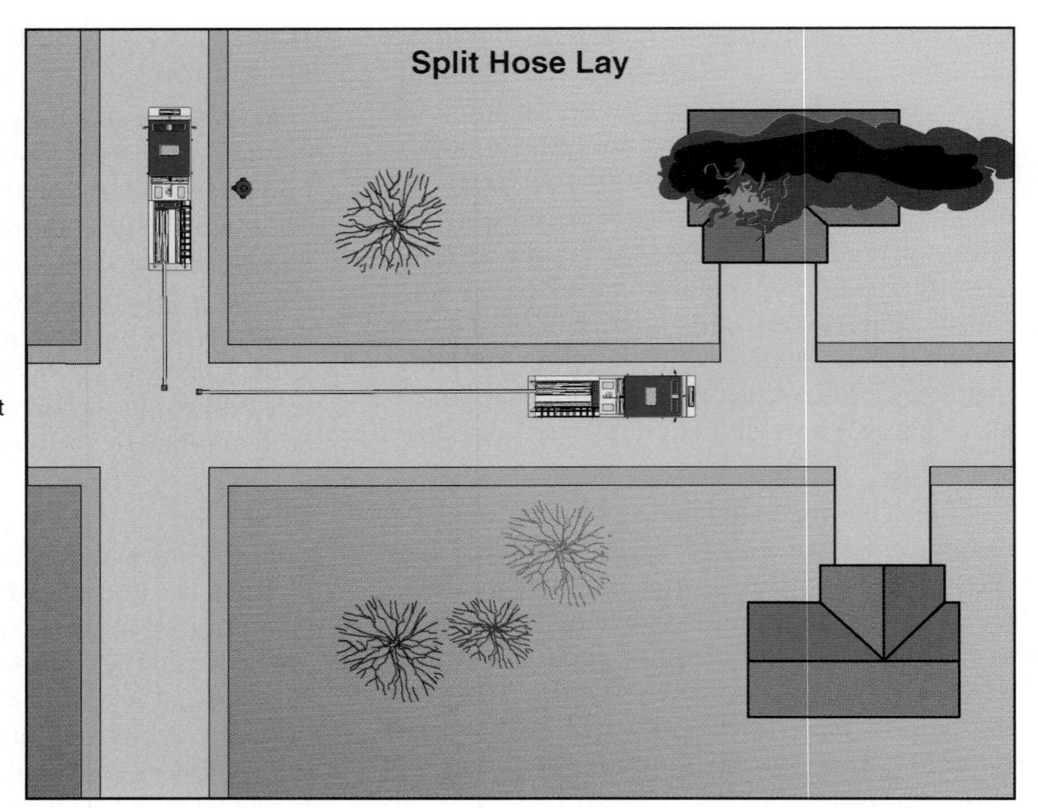

Figure 6.7 A split hose lay is performed by two pumping apparatus, beginning at a point between the incident and the water source.

- Two lines deployed forward (from water source to emergency scene) **(Figure 6.8a)**
- Two lines deployed reverse (from emergency scene to water source) **(Figure 6.8b)**
- Forward lay followed by reverse lay **(Figure 6.8c)**
- Reverse lay followed by forward lay **(Figure 6.8d)**

Combination Lay

A *combination lay* is when two different-sized hoselines are deployed from one pumping apparatus and laid parallel to each other. Clearly, there are many options when the hose bed is divided. One of the most versatile arrangements is one in which one section of the hose bed contains large diameter hose (over 3½ inches [90 mm]) while the other portion of the hose bed contains 2½- or 3-inch (65 mm or 77 mm) diameter hose that can be used for either supply or attack modes. These arrangements are often described as *combination loads* (see Selecting/Making Hose Loads section).

A pumping apparatus configured in this manner can deploy large diameter supply hose when the fire/emergency situation requires that the pumping apparatus deploy its own supply line and work alone (deploying it forward so that the pumping apparatus stays at the incident scene). It can use its small diameter hose as a supply line at fires/emergencies with less demanding water-flow requirements while at the same time having large diameter hose ready to be charged for large fire-suppression/emergency needs. A divided hose bed with two hose sizes, therefore, offers the command officer a number of choices when determining the best way to commit limited resources **(Figure 6.9, p. 254)**.

Selecting/Making Hose Loads

Three basic hose-load choices are available when a standard hose bed is used: accordion, horseshoe, and flat. Each of these hose loads may be employed in the hose bed of a single vehicle to accommodate special situations anticipated within a jurisdiction. Fire hose can also be carried in a reel load that does not require a hose bed. The following sections discuss hose-load selection, the various load types, and hose-loading guidelines .

The basic three loads can be used in a standard hose bed or in a bed that has been partitioned with dividers or baffle boards. Partitioning the hose bed

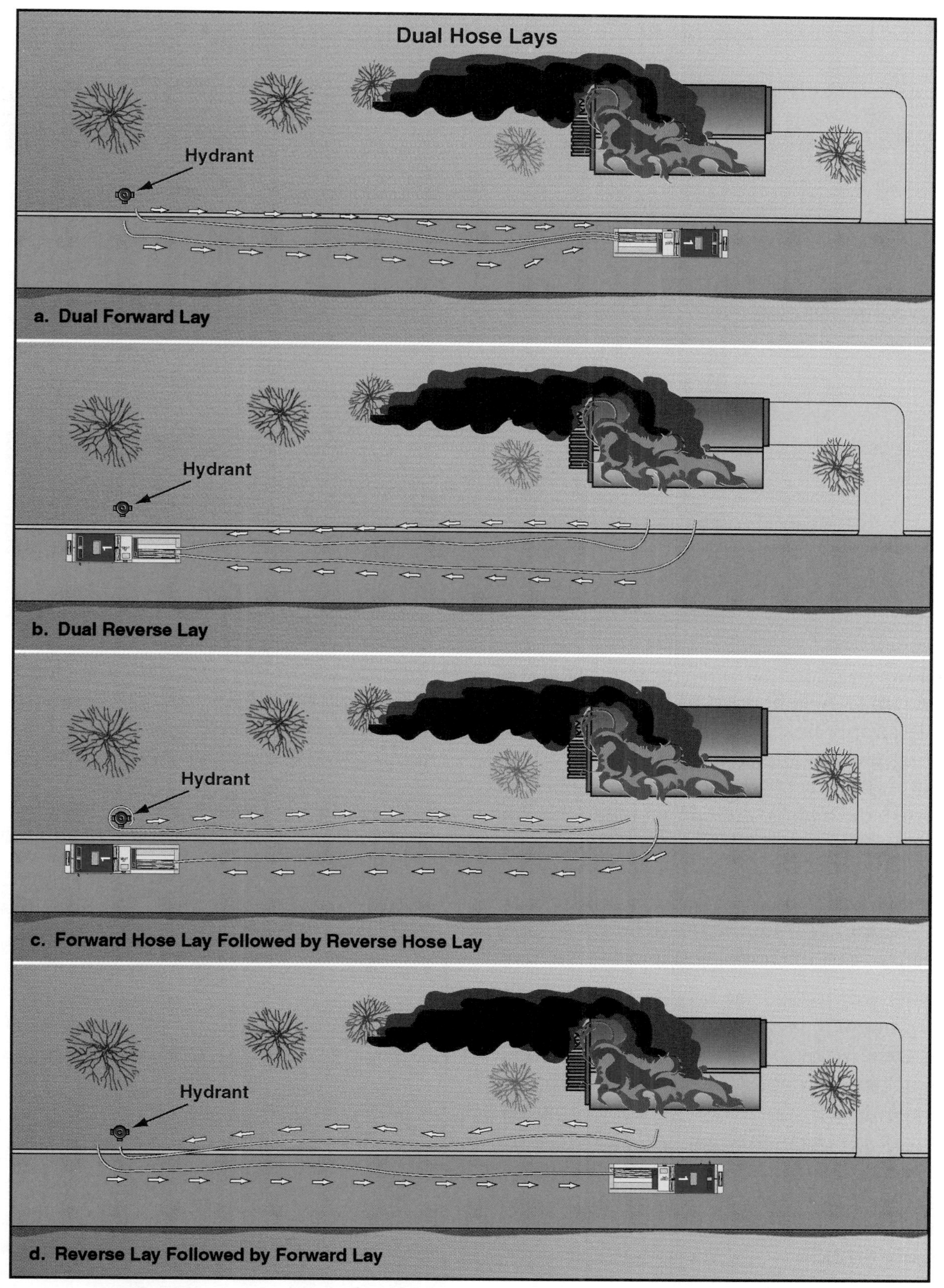

Figure 6.8 (a) Dual forward hose lay (two lines of the same size deployed forward from water source to incident). (b) Dual reverse hose lay (two lines of the same size deployed reverse from incident to water source). (c) Forward hose lay followed by reverse hose lay (two lines of same size deployed forward from hydrant; then pumping apparatus turned around and deployed one line reverse to the hydrant). (d) Reverse hose lay followed by forward hose lay (one line deployed reverse from incident to water supply; then a forward lay is deployed from the water source to incident).

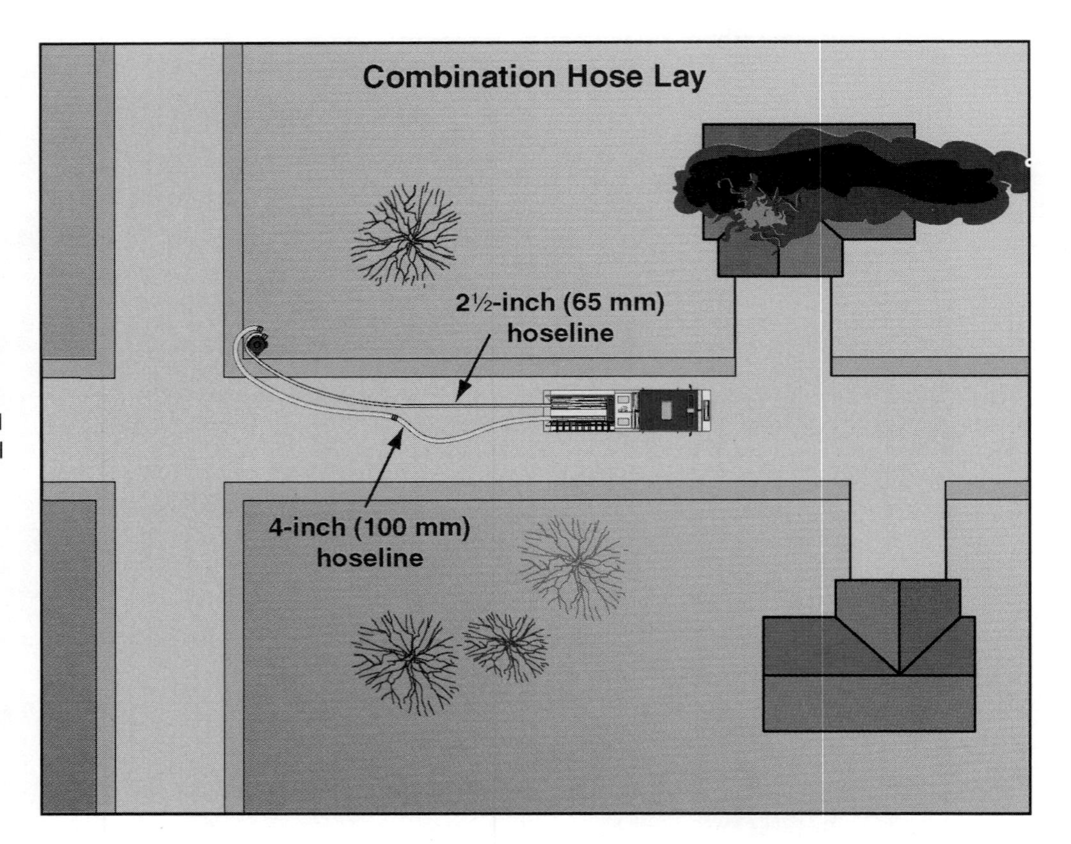

Figure 6.9 A combination supply hose deployment consists of two different sized hoselines deployed in parallel lines.

into separate sections permits a number of options for deploying single hoselines such as deploying multiple supply hoselines in each direction or deploying parallel hoselines concurrently. **Figure 6.10** shows a typical divided hose bed. Note that hose in the beds is interconnected so that the entire complement is deployed in a single-length hose lay. If dual hoselines are deployed, the couplings between the beds are first disconnected so that the couplings from the top of both beds can be pulled to start the hose lay **(Figure 6.11).**

Another way to load the hose in a divided bed is to load one side for a forward lay and the other side for a reverse lay (combination load). This type of loading permits deploying hose in either direction as required by the fire/emergency situation. When loading hose in this manner, the beds must be interconnected with a double male or double female adapter if it becomes necessary to deploy the entire hose complement in a single-line hose lay **(Figure 6.12).**

Regardless of the type of hose load selected, start by placing the male coupling first in the hose bed if it is being configured for deploying hose in a forward lay. In this way, the female end of the hoseline will leave the hose bed first when the hose is deployed. For a reverse lay, start with the female coupling so that the male end of the hose leaves

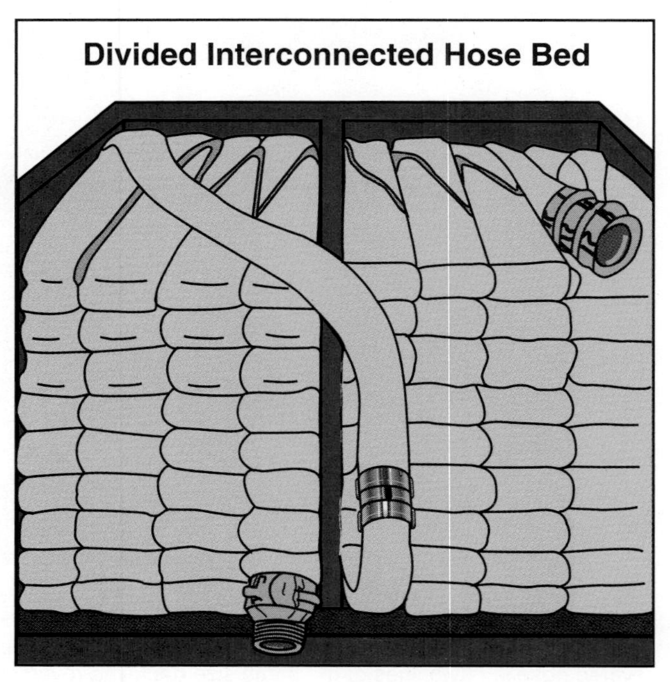

Figure 6.10 A divided interconnected hose bed allows the deployment of either a single supply hoseline or dual supply hoselines.

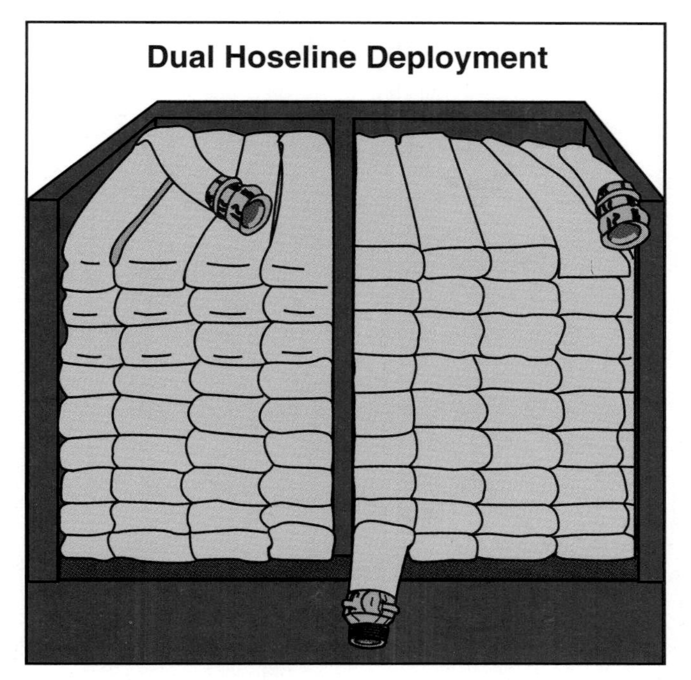

Figure 6.11 If dual hoselines are deployed, the couplings between the beds are first disconnected so that the couplings from the top of both beds can be pulled to start the hose lay.

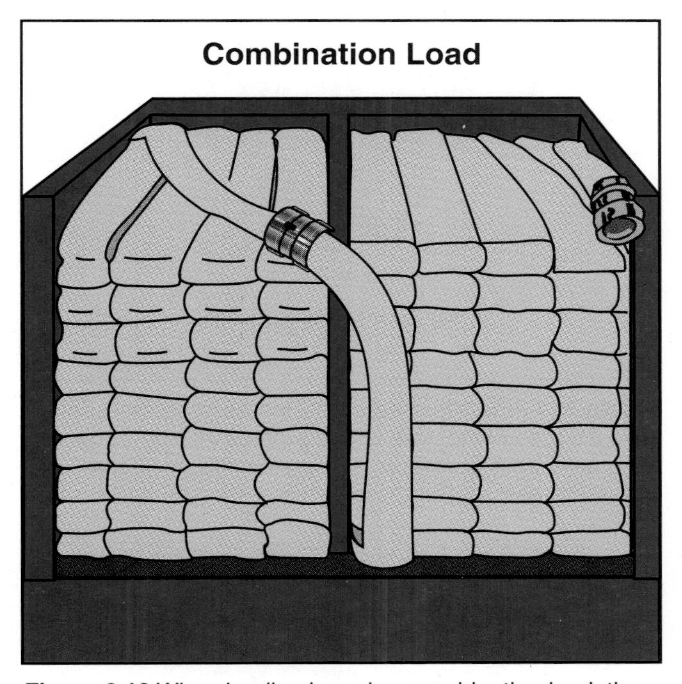

Figure 6.12 When loading hose in a combination load, the hose beds must be interconnected with a double male or double female adapter if it becomes necessary to deploy the entire hose complement in a single-hoseline lay.

the bed first. Place the first coupling to the rear of the bed. The coupling can be placed on either side if the bed is not divided.

Hose-Load Selection

Choosing hose loads for a pumping apparatus depends on the particular needs of the fire and emergency services organization. A hose load and deployment procedure that works for one organization may not work as well for another because conditions vary from community to community. Hose loads must be designed to accommodate special needs and conditions present within each jurisdiction.

One of the first steps in deciding which hose load to use is to define the mission of the pumping apparatus and its crew. This step is a primary factor in selecting not only the hose load but also in determining the deployment methods that best suit the most demanding situations the unit will encounter. Loads and deployment procedures are designed so that hoselines can be placed into service in the shortest time possible with the fewest number of personnel.

Loads and deployment procedures are also designed for a maximum number of deployment options to meet the water-supply needs in any kind of situation. For example, in a small two-station fire department with one pumping apparatus at each station, the mission of the first-arriving unit is most often to conduct initial fire-suppression and mitigation activities. In this case, the hose might be loaded so that the pumping apparatus can deploy a supply line from the hydrant to a fire as it approaches the emergency scene. The use of a four-way hydrant valve allows the water supply from the hydrant to be established before the arrival of a second pumping apparatus. The second-arriving unit, which often arrives minutes later, may be designated as the *water-supply unit*. This pumping apparatus not only pumps water to the first-arriving pumping apparatus through the first-deployed line but may also deploy additional lines if the situation demands more water. When pumping apparatus expect to operate in either an attack mode or a water-supply mode, they are loaded with similar hose-bed configurations.

Another factor to consider when designing hose loads is the size of the pumping apparatus fire pump. Pump size has a direct bearing on the manner in which hose is loaded and deployed and on the size of hose used. A pumping apparatus with a large-capacity fire pump carries a hose load of sufficient diameter and length to deliver a volume of water equal to the maximum output of the pump. Three basic ways to maximize water deliveries are as follows:

- Carry large diameter fire hose in at least one section of the hose bed.
- Carry large amounts of small diameter fire hose (under 3½ inches [90 mm]) in a divided hose bed so that multiple hoselines can be deployed.
- Carry fire hose on two units where one pumping apparatus is equipped with attack hose and the second pumping apparatus is equipped as a supply hose tender carrying sufficient quantities of supply hose to meet local conditions.

The use of large diameter hose makes sense for a number of reasons. Large diameter hose takes up less space in the hose bed than an equivalent amount of small diameter hose. Equivalency in this case means knowing the amount of small diameter hose needed to deliver the same amount of water the same distance and at the same pressure as a single length of large diameter hose. Small diameter hose must be deployed in multiple lengths to accomplish this equivalency. For example, it takes at least four 2½-inch (65 mm) hoselines to deliver 1,000 gpm (3 785 L/min) of water at the same pump pressure as a single 5-inch (125 mm) hoseline of the same length **(Figure 6.13).** Conversely, the pump pressure required to move 1,000 gpm (3 785 L/min) through two 100-foot (31 m) 2½-inch (65 mm) hoselines moves the same amount of water 750 feet (229 m) through a single 5-inch (125 mm) hoseline.

One of the most significant advantages of large diameter hose is that it can not only deliver large amounts of water great distances with a relatively low loss of pressure but it can also deliver small amounts of water with virtually no pressure loss. For this reason, the incident commander should not restrict the use of large diameter hose to only large, high-volume fires or other emergencies. **Appendix D,** IFSTA Friction Loss Calculations (English System), and **Appendix E,** IFSTA Friction Loss Calculations (International System), describe the relationships between fire hose diameter, fire hose length, fire flow, and subsequent expected friction losses.

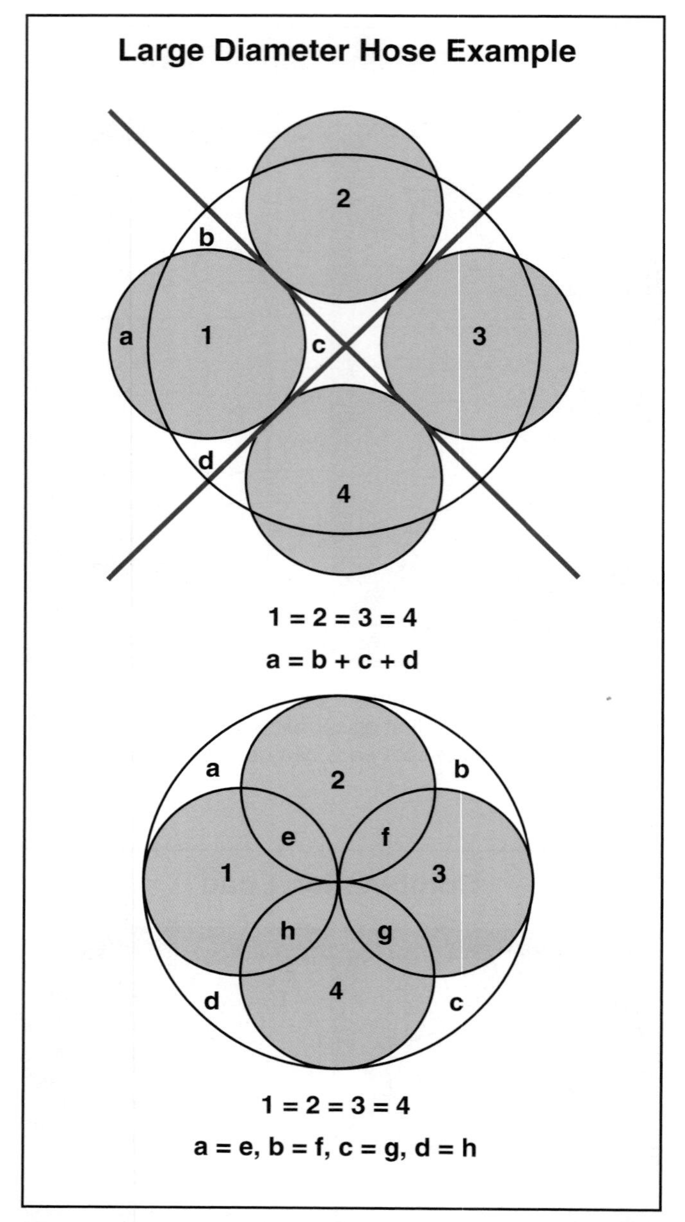

Figure 6.13 When the diameter of the supply hose is doubled, the area of the hose opening increases approximately four times, while the pump pressure remains the same in both instances.

Carrying supply hose of 2½ or 3 inches (65 mm or 77 mm) has some advantages however. A pumping apparatus loaded in this manner can use the entire complement of hose for either attack or supply hoselines if the hose meets the NFPA attack fire hose standards. Proponents of small diameter hose argue that having the flexibility to use this hose in either capacity outweighs the disadvantage of having to deploy multiple hoselines to supply maximum water demands. NFPA 1901, *Standard for Automotive Fire Apparatus* (2003), recommends that a minimum of 800 feet (244 m) of 2½-inch (65 mm) or larger hose be carried. To effectively use the larger pumping apparatus fire pumps that are available today (1,000 gpm [3 785 L/min] and larger) and to provide the maximum fire flows they can produce, use large diameter supply hose.

Carrying both supply and attack hose on two separate pumping apparatus that respond simultaneously is a practice that works well when sufficient apparatus and personnel are available. The greatest disadvantage of such a system is its high cost, both initially to implement the system and then on a continuing basis. The high initial cost for a pumping apparatus makes this practice almost impossible for a fire and emergency services organization with limited funds. With modern pumping apparatus designed to respond to a wide variety of emergencies, it is most common to find both attack and supply hoselines on the same pumping apparatus. The maximum use of the pumping apparatus is realized with this hose configuration, allowing supply-line deployment and attack-line evolutions to function independently without an additional vehicle.

The nature of the water source is an important factor to consider when determining hose loads. If a pumping apparatus works in a municipality that has a well-designed water system with hydrants spaced at regular intervals along each main, the need for carrying large amounts of hose is less than that of a rural community where long hose lays must be made from a distant water source. However, it is important to remember that in a large fire/emergency situation, more than one hydrant may be needed to supply the volume of required fire flow. Therefore, it is important to carry enough supply hose on the pumping apparatus for these situations.

Another important factor in determining the size of hose and type of load for a pumping apparatus is whether that unit may be required to work independently for an extended time. In jurisdictions where a second-arriving pumping apparatus is more than 10 minutes away, pumping apparatus must be configured to be self-sufficient to deploy a high-volume supply hoseline with little need for another pumping apparatus to increase the pressure from the hydrant.

Accordion Load

The *accordion load* derives its name from the manner in which the hose appears after loading: Hose is laid progressively on edge in folds that lie adjacent to each other (accordion shape) **(Figure 6.14).** An advantage of this load is its ease of loading: Its simple design requires only two or three persons (although four persons are best) to load it in a matter of minutes. Another advantage is that hose for shoulder carries can be easily taken from the load by simply picking up a number of folds and placing them on the shoulder (see Chapter 5, Basic Methods of Handling Fire Hose, Accordion-Shoulder Method section). See **Skill Sheet 6-4** for the procedures for loading the forward accordion load and **Skill Sheet 6-5** for loading the reverse accordion load procedures.

A disadvantage of this load is that the hose folds contain sharp bends, which require that the hose be reloaded periodically if it is not used on a regular basis. Reloading relocates the bends within each length of hose in order to prevent damage to the lining and the outer cover. Another disadvantage is that the hose tends to wear along its edges. This wear is due to the combined effect of the vibration caused by the motion of the pumping apparatus and the weight of the layered hose on itself. This load is not recommended for large diameter supply

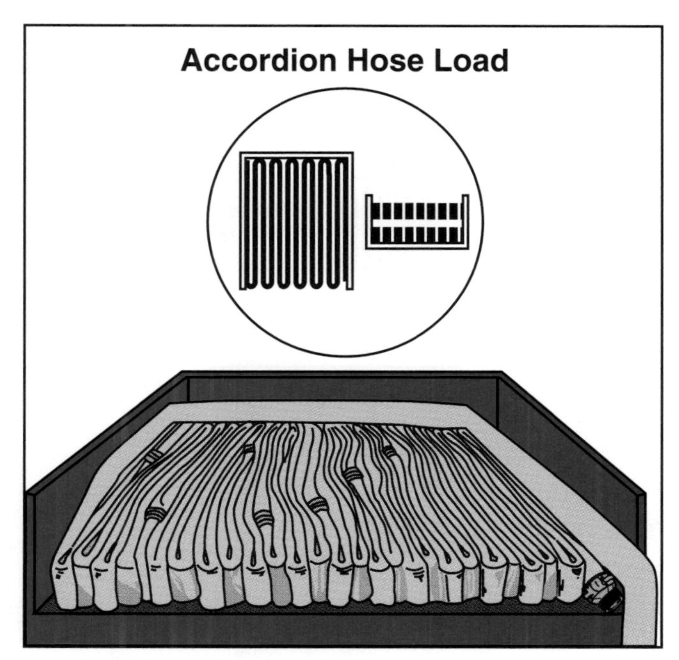

Figure 6.14 Accordion hose load: Hose loaded progressively on edge in folds that lie next to each other.

hose because the remaining folds tend to collapse into the flat position as the hose deploys, which could cause them to tangle.

Horseshoe Load

The *horseshoe load* is also named for the way it appears after loading — it resembles a horseshoe **(Figure 6.15).** Like the accordion load, it is loaded on edge, but in this case the hose is laid around the perimeter of the hose bed in a *U*-shape or horseshoe configuration. Each length is progressively laid from the outside of the bed toward the inside so that the last length is at the center of the horseshoe. The primary advantage of the horseshoe load is that it has fewer sharp bends in the hose than the accordion or flat loads; the bends occur only at the rear of the hose bed.

A disadvantage of the horseshoe load occurs most often in wide hose beds. The hose sometimes deploys in a wavy or snakelike (serpentine pattern) lay in the street as the hose is pulled alternately from one side of a bed and then the other. Although this serpentine pattern occurs to some extent with all hose loads, it is most pronounced with the horseshoe load. It is essential for the command officer and the driver/operator to realize that between 10 and 15 percent of the total length of the hose is "lost" because of the wavy pattern during deployment of this hose load. Another disadvantage with the horseshoe load is that folds for a shoulder carry cannot be obtained as easily as they can with an accordion load. In this case, two persons are required to make the shoulder folds for the carry (see Chapter 5, Basic Methods of Handling Fire Hose, Accordion-Shoulder Method section).

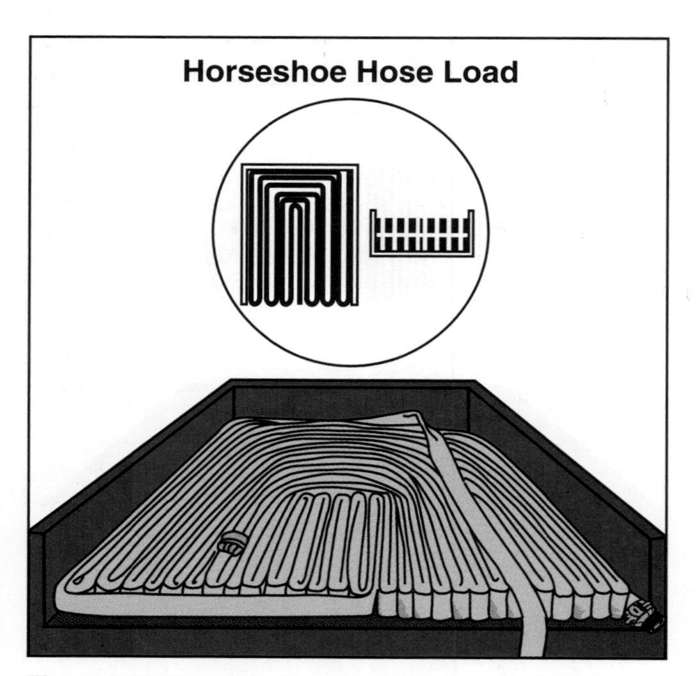

Figure 6.15 Horseshoe hose load: Hose loaded on edge and laid around the perimeter of the hose bed.

Like the accordion load, the hose is loaded on edge, which can promote wear on the hose edges. The horseshoe load is not recommended for large diameter supply hose because as the hose deploys, the hose remaining in the bed tends to collapse, causing the hose to twist and then tangle as it deploys.

In a single-hose bed, the horseshoe load may start on either side. In a divided hose bed, lay the first length against the partition with the coupling hanging an appropriate distance below the hose bed. Gauge this distance on the anticipated height of the adjacent hose bed so that the coupling can be connected to the last coupling of the load on the opposite side (crossover) and laid on top of the load. This procedure allows for easy disconnection of the couplings when the load must be split to deploy dual lines. With a combination load (one side loaded for a reverse lay and the other side loaded for a forward lay), use an adapter to connect identical couplings. See **Skill Sheet 6-6** for the procedures for loading the forward horseshoe load and **Skill Sheet 6-7** for loading the reverse horseshoe load procedures.

Flat Load

As the name implies, a *flat load* is where the hose is laid so that its folds lie flat rather than on edges **(Figure 6.16).** Of the three supply hose load types, the flat load is the easiest to load. It is suited for any size of supply hoseline, including large diameter supply hose where it is the loading method of choice. Hose loaded in this manner is less subject to wear from pumping apparatus vibration during travel than other loads.

A disadvantage of this load is that the hose folds contain sharp bends (like the accordion load), which require that the hose be reloaded periodically to relocate bends within each length to prevent damage to the hose lining. Another disadvantage of the load is that it requires two persons to make folds for shoulder carries. Loading supply hose in a forward flat load is described in **Skill Sheet 6-8.**

In a single-hose bed, the flat load may be started on either side, ensuring that the coupling is brought to the front of the compartment. Large diameter supply lines are usually loaded in sections that are 100 feet (31 m) in length. Additionally, the use of sexless couplings for large diameter supply hoselines continues to grow in popularity. Because of the size of the hose and couplings, it is important to place the couplings in the hose bed in a carefully designed plan that first allows the hose to deploy easily without snagging and second allows all of the hose needed for the hose bed to be loaded into the bed **(Figure 6.17)**. It is necessary to load the hose in a manner that places the couplings in the front of the bed so that the coupling will not form a lump in the bed, which disrupts the easy deployment of the hose.

When the hose bed is divided to provide the option for a dual lay, place the first length against the partition with the coupling hanging an appropriate distance below the hose bed. Again, gauge this distance on the anticipated height of the adjacent hose bed so that the coupling can be connected to the last coupling of the load on the opposite side (crossover) and laid on top of the load. This procedure allows for easy disconnection of the couplings when the load is deployed as a dual lay.

With a divided load where the option of either a forward or reverse lay is desired, use an adapter to connect identical threaded couplings. This procedure is not necessary when the large diameter supply hose is provided with sexless couplings, which eliminates the need to use adapters to complete the connections.

The flat load can be adapted for reloading large diameter supply hose directly from the street/road after an emergency incident by allowing the pumping apparatus to straddle the hose while it drives slowly forward as the hose is progressively loaded into the bed. Use a hose wringer or roller to expel the air and water from the hose as it is placed in the hose bed. NFPA 1500, *Standard on Fire Department Occupational Safety and Health Program* (2002), permits reloading large diameter hose on moving apparatus once the fire and emergency services organization has written and implemented standard operating procedures that ensure the safety of those performing this task.

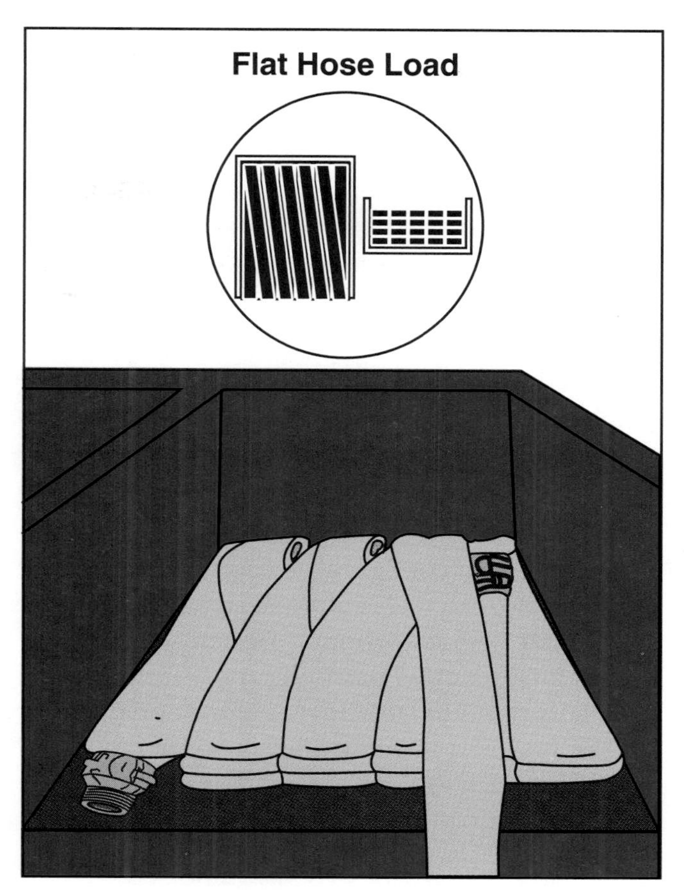

Figure 6.16 Flat hose load: Hose loaded so its folds lie flat rather than on edge.

Figure 6.17 Large diameter hose and couplings require that the hose load be designed to permit the maximum amount of hose to be loaded while ensuring that it is deployed in a safe and effective manner. *Courtesy of Sam Goldwater.*

Reel Load

Another way to load hose, especially large diameter supply hose, is to put it on a large pumping-apparatus-mounted reel **(Figure 6.18)**. An advantage to the reel load is that there are no folds in the hose. For this reason, the hose does not need to be reloaded periodically to relieve stress on bends.

A disadvantage is that only one hoseline can be deployed at a time unless multiple reels are mounted on the apparatus body. Another disadvantage is that travel vibration causes the hose to sometimes work loose and hang loosely under the reel, which can cause the reel to jam or the hose to tangle as it deploys.

An apparatus configured to carry large diameter supply hose on a reel is often designated as a *hose wagon* because its primary mission is to deploy supply hose for other pumping apparatus. The reel is usually mounted on the rear of the apparatus or on the rear of a towed trailer in some industrial uses. The reel is either hydraulically or electrically powered to aid in reloading **(Figure 6.19)**.

When reloading large diameter supply hose onto a reel at the emergency scene, the simplest method is the same as that used for reloading the flat load:

Figure 6.18 Large diameter hose reel mounted in a pumping apparatus compartment. *Courtesy of Sam Goldwater.*

Figure 6.19 Large diameter hose reel mounted on a towed hose reel trailer. *Courtesy of Sam Goldwater.*

Straddle the hose with the pumping apparatus and progressively load it as the vehicle moves forward slowly. Another method used in the station for loading large diameter hose from a storage roll involves the use of a hose-loading table with a rotating top. The roll is simply placed on the table and unrolled as it is loaded onto the reel.

Hose-Loading Guidelines

Hose loading on pumping apparatus is not an emergency operation, although if it is not well planned and conducted properly, the entire fire-suppression/emergency operation can be negatively affected. When fire hose is needed at an emergency, the proper hose load permits efficient and effective emergency operations. Regardless of the hose load selected, observe the following general hose-loading guidelines:

- Check fire hose coupling gaskets and the swivel action before making a connection.
- Keep the flat sides of the hose on the same plane when the couplings are attached. It is not important to align the lugs of the hose couplings.
- Tighten couplings hand-tight while connecting hose. Do not use wrenches or undue force.
- Remove wrinkles from fire hose when it must be bent to form a loop in the hose bed by pressing with the fingers so that the inside of the bend is folded smoothly.
- Make a short fold or reverse bend (called a *Dutchman*) in the hose during the loading process with the coupling facing the rear of the hose bed so that the coupling does not have to be turned around to deploy from the bed. This procedure prevents the coupling from catching on the walls of the hose bed compartment or cover **(Figure 6.20)**.
- Load large diameter supply hose (3½-inch [90 mm] or larger) so that all couplings are located at the front of the hose bed. This procedure saves space and allows the hose to lie flat. Lay couplings in a manner that does not require them to turn over when the hose pays out of the bed.
- Pack hose in a manner that allows it to deploy freely from the hose bed. Packing hose too tightly causes the couplings to snag as the hose pays out of the bed. The hose needs to be loose enough to easily insert a hand between the folds.

Figure 6.20 A Dutchman fold/bend is used in a hose load to allow the coupling to be pulled straight out of the hose bed without having to pivot and turn it.

WARNING

All fire and emergency service personnel must be off a pumping apparatus and in a safe position after loading hose before the driver/operator moves the pumping apparatus.

Selecting/Connecting a Hydrant

The success or failure of a fire or other emergency operation requiring the use of water or water-based extinguishing agents often depends upon the proper selection and connection of the fire hydrant. To perform this operation successfully, the incident commander must recognize and address the following limitations:

- Long distance from the emergency scene requires a long supply line deployment
- Too close to the incident places personnel and equipment in a dangerous position and subjects them to harm or damage
- Hydrant does not possess adequate residual pressure

The hydrant connection itself must be kept as simple as possible. A hydrant that is overgrown with brush, obviously damaged, or equipped with nonconforming threads and couplings will adversely effect the completion of the connection. The company officer or driver/operator must evaluate these factors to ensure that the process of selecting and connecting the hydrant can be accomplished with ease and efficiency.

Hydrant Selection

In jurisdictions where the primary water source is a water system with hydrants, it is usually best to use the hydrant closest to the fire/emergency scene. This hydrant choice minimizes the length of hose deployed and thus minimizes the friction loss in the hoseline. Because some communities may not have a modern water utility system or the system is undersupplied due to population growth, alternative "stronger" (high-pressure) hydrants or water sources may be required to sustain or provide higher water flow rates during emergencies.

Some fire and emergency service organizations prefer to go directly to the incident scene to conduct an operational size-up before committing the pumping apparatus to deploying a supply hose. In an organization where this practice is a standard procedure, pumping apparatus hose loads are usually designed to deploy reverse hose lays. In this case, it is not unusual for a pumping apparatus to deploy a hoseline to the closest hydrant beyond the fire/emergency scene.

Regardless of the method adopted by a fire and emergency services organization, all deployment procedures are designed to be flexible enough to modify should the unexpected occur. Hydrants fail for many reasons, the most common ones being a lack of maintenance or mechanical damage from an outside source. Water mains are sometimes shut down for expansion of the water system, repair, or routine maintenance. In each of these instances, the fire and emergency services organization must be prepared to identify, locate, and operate alternative hydrants or water sources.

Situations exist when the closest hydrant may not be the best choice. In some cases, it may be necessary to select a more distant hydrant. For example, if the closest hydrant has a significantly smaller flow than a more distant hydrant (perhaps because the closest hydrant is on a smaller water main than the distant hydrant), it may be better to deploy a hoseline to the hydrant with the greater flow. It may also be better to choose a distant hydrant on the same side of the street/road as the fire/emergency scene rather than a closer hydrant on the opposite side of the street/road or when the scope of the incident may expand and involve the hydrant. This position is especially important when large diameter supply hose is used because deploying it across a street/road impedes the passage of incoming pumping apparatus and other emergency vehicles operating at the emergency scene.

CAUTION

A more distant hydrant may be a better choice if the closest hydrant is located very close to an intensely burning structure. Use of the closest hydrant in this case would unnecessarily expose the pumping apparatus and fire and emergency services personnel to extreme heat from the fire.

Regardless of the potential traffic problem posed by large diameter supply hoselines, their use makes hydrant selection much easier. A pumping apparatus loaded with large diameter supply hose can provide an adequate water supply and initiate fire-suppression activities or other emergency operations as a single unit. When sexless couplings are used, the hose load can be deployed as either a forward or reverse lay without considering the need for thread adapters. A hydrant or other water source that has reduced or minimized water pressure and water flow can still be used because of the low friction loss inherent with large diameter supply hose.

Hydrant Connection

When a pumping apparatus deploys hose forward from a hydrant, the hose is anchored to the hydrant so that as the apparatus proceeds to the incident scene, the hose is pulled from the bed and laid in the street. Anchoring the hose in this manner is safer than having a person stand in the street holding the end of the hose. If a coupling snags in the hose bed or if the hose tangles, the hydrant will be able to better withstand the sudden jerk of the hose than a person in a freestanding position. Anchoring the hose to the hydrant or other solid object also allows the pumping apparatus to deploy hose without needing to leave someone at the hydrant. In this case, the hose is deployed and left uncharged until a second pumping apparatus and crew arrive to make the hydrant connection and charge the supply line.

Numerous terms for making a hydrant connection are used throughout the fire and emergency services (see sidebar). This manual will use the term *make a hydrant.*

Fire Hydrant Connection Terms

Many local terms for making the fire hydrant connection exist. Some of the more commonly used phrases are as follows:

- Make a hydrant
- Catch a hydrant
- Dress a hydrant
- Snub a hydrant
- Address a hydrant
- Wrap a hydrant
- Hook a hydrant
- Twist a hydrant

Many other similar phrases, too numerous to be listed here, are used throughout the fire and emergency services. These phrases all describe the same procedure.

Communication is essential to the success of any supply-hose-deployment operation. The command officer must continually monitor and access the situation and clearly convey decisions to fire and emergency services personnel. As a pumping apparatus approaches an incident, the command officer and driver/operator must work in concert to decide the most advantageous method of ensuring a water supply. Using response maps, computer-aided dispatch information, as well as personal knowledge of the location, a hydrant can be selected. Once the hydrant has been chosen, this information must be given to other members of the responding fire-suppression/emergency team. This communication can be accomplished face to face, over headphones, by hand signals (for example, two upstretched arms), or other methods outlined in a fire and emergency service organization's standard operating procedures.

The emergency responder assigned to pulling hose and anchoring it to the hydrant waits for a direct order from the command officer before proceeding. An emergency responder does not act automatically whenever the pumping apparatus stops at a hydrant because the officer in charge may have the pumping apparatus pause at a hydrant to make an assessment of the fire/emergency situation from that vantage point. If the command officer chooses to use another hydrant, premature action by an emergency responder could have a detrimental effect on the fire-fighting/emergency operation.

The command officer also communicates whether the supply hose is to be charged as soon as an emergency responder makes the connection. An option here is for the emergency responder's action to be predetermined by policy. Some fire and emergency service organizations specify that whenever a forward lay is made, the hydrant is charged as soon as possible. This action, of course, requires that someone in the crew take immediate action when the pumping apparatus stops at the incident scene. That person is responsible for either applying a hose clamp to the hose near the pumping apparatus or disconnecting the hose from the hose bed and immediately connecting it to an intake.

The procedure for anchoring and connecting hose to a hydrant starts when the driver/operator stops the pumping apparatus with the tailboard approximately 10 feet (3 m) past the hydrant. The command officer then gives the appropriate command for an emergency responder to begin the hose-deployment operation. The procedures for making the connection to a hydrant using a four-way valve are described in **Skill Sheet 6-9.**

Summary

Fire-suppression/emergency operations depend upon the rapid and efficient deployment of supply fire hose. Although pumping apparatus carry some water on board, it is a limited quantity that cannot be counted upon to sustain fire-suppression/emergency activities for more than 1 or 2 minutes. This situation is true for both rural and urban fire-fighting/emergency situations. Water must be supplied from a source, whether from an improved pressurized system in a city or a rural water shuttle system originating at a pond or stream. In either case, supply fire hose provides the conduit for the water to travel from a source and arrive at the fire pump. The hose size, type, and length are determined locally according to circumstance and need. Each fire and emergency services organization must evaluate the fire-suppression/emergency challenges that it faces and select the supply hose and loading methodology that best meets these needs.

This chapter outlined the various types of fundamental hose lays (forward, reverse, and split) that can be selected by a fire and emergency services organization for use on its pumping apparatus. Depending upon unique local conditions and needs, one of these types can be implemented to best meet the fire-suppression/emergency requirements of the community. Additionally, a variety of supply hose loads (including the accordion, horseshoe, flat, and reel) were described. These loads are illustrated in the skill-sheet section to ensure that consistent and safe procedures are employed when preparing the hose bed. The final part of the chapter was devoted to procedures associated with deploying a supply hoseline from a hydrant.

Regardless of the type of hose load or method of deployment selected, each fire and emergency services organization must continue to evaluate the changing needs of the community they serve with a constant eye to modifying and improving the water supply system that is employed. Changing local conditions, as well as new hose technology, necessitate that each fire and emergency services organization study the available hose deployment methods and loads to continually improve the service that it delivers.

Deploying Supply Hoselines Forward Hose Lay

Skill Sheet 6-1

Step 1: ***Driver/Operator:*** Stop the pumping apparatus at the selected hydrant.

Hydrant Firefighter: Take tools, devices, and other equipment necessary to complete the connection. Pull enough supply hose to reach and wrap around the hydrant. Place a loop of hose around the hydrant or other effective anchor to secure it (see photo).

NOTE: Wrap the hose around the hydrant in a manner that restrains the hose when the pumping apparatus moves away from the hydrant.

Step 2: ***Hydrant Firefighter:*** Signal the driver/operator to proceed and deploy the hose to the incident.

Step 3: ***Hydrant Firefighter:*** Flush the hydrant.

6-1 Skill Sheet

Step 4: ***Hydrant Firefighter:*** Connect supply hose to hydrant.

Step 5: ***Driver/Operator:*** Proceed to the fire and position the pumping apparatus as directed. Apply a hose clamp and signal the hydrant firefighter to charge the line.

Step 6: ***Driver/Operator:*** Disconnect the supply hoseline from the hose bed.

Step 7: ***Driver/Operator:*** Connect the hose to the fire pump intake valve, release the hose clamp, and open the pump intake valve.

6-2 Skill Sheet — Deploying Supply Hoselines Reverse Hose Lay

Step 1: ***Supply Driver/Operator:*** Stop the supply pumping apparatus so that its hose bed is slightly past the intake valve of the attack pumping apparatus.

Supply Firefighter: Pull sufficient hose to reach the intake valve, and anchor the hose (always remaining in view of the driver's side mirror).

Step 2: ***Supply Driver/Operator:*** Deploy the hose to the water source when signaled that the hose is secured and anchored.

Step 3: ***Attack Driver/Operator:*** Apply a hose clamp to the hose near the intake of the attack pumping apparatus.

Step 4: ***Supply Driver/Operator:*** Position supply pumping apparatus at the hydrant, and connect the hose to the hydrant and the intake side of the pump.

Step 5: ***Supply Driver/Operator:*** Pull the remaining length of the last section of hose coming from the hose bed, disconnect the couplings, and return the male coupling to the hose bed, while protecting its threads.

Step 6: ***Supply Driver/Operator:*** Connect hydrant supply line to the pump intake and open hydrant valve.

6-2 Skill Sheet

Step 7: ***Supply Driver/Operator:*** Connect the deployed supply hose to a discharge valve, and charge the hose back to the attack pumping apparatus.

Deploying Supply Hosellines Split Hose Lay

Skill Sheet 6-3

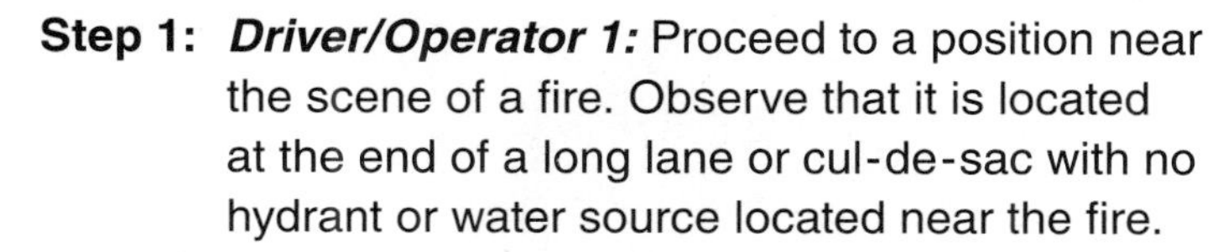

Step 1: ***Driver/Operator 1:*** Proceed to a position near the scene of a fire. Observe that it is located at the end of a long lane or cul-de-sac with no hydrant or water source located near the fire.

Step 2: ***Firefighter 1:*** Deploy the supply hose from the hose bed, and anchor it near the street intersection.

Step 3: ***Driver/Operator 1:*** Proceed to the fire, and deploy hose in a forward lay once it is secure.

Step 4: ***Driver/Operator 2:*** Stop at the intersection where the first hoseline was deployed.

Firefighter 2: Deploy the supply hoseline near the first hoseline, ensuring that the hose does not snag in hose bed.

Step 5: ***Driver/Operator 2:*** Proceed to a water source, deploying supply hose in a reverse lay, and set up a hydrant or drafting operation.

6-3 Skill Sheet

Step 6: ***Firefighter 2:*** Connect the hoselines and signal Driver/Operator 1 that the connection is complete.

Step 7: ***Driver/Operator 1:*** Signal Driver/Operator 2 to charge the line.

Loading Supply Hose
Forward Accordion Load

Skill Sheet 6-4

Step 1: Place the male coupling in the front of the hose bed in either the right or left corner.

Step 2: Lay the first length of hose in the bed on its edge against the outer wall.

Step 3: Bend the hose back on itself at the rear of and even with the rear of the bed. Then extend the hose to the front of the bed.

Step 4: Fold the hose back on itself and bring it to the rear of the bed. Stagger the folds at the edge of the bed so that every other bend is approximately 2 inches (50 mm) shorter than the edge of the bed. (This stagger may also be done at the front of the bed if desired.) Repeat the process until the first tier of hose is complete.

6-4 Skill Sheet

Step 5: Angle the hose up to begin the next tier upon reaching the opposite side of the hose bed.

Step 6: Make the first fold of the second tier directly over the last fold of the first tier. Begin this transition at the front of the bed with the first bend at the rear of the bed.

Step 7: Continue the second tier in the same manner as the first, laying hose in a progressive manner across the hose bed until the tier is complete.

Loading Supply Hose
Reverse Accordion Load

Skill Sheet 6-5

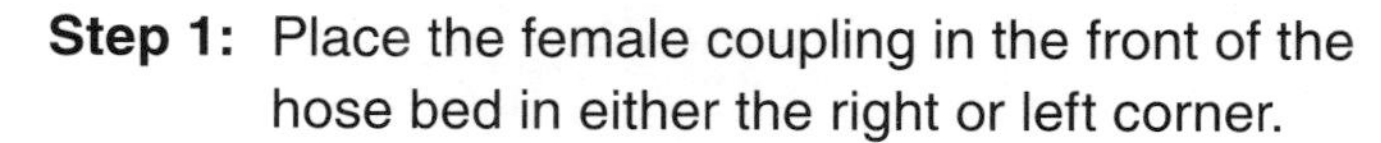

Step 1: Place the female coupling in the front of the hose bed in either the right or left corner.

Step 2: Lay the first length of hose in the bed on its edge against the outer wall.

Step 3: Bend the hose back on itself at the rear of and even with the rear of the bed. Then extend the hose to the front of the bed.

Step 4: Fold the hose back on itself and bring it to the rear of the bed. Stagger the folds at the edge of the bed so that every other bend is approximately 2 inches (50 mm) shorter than the edge of the bed. (This stagger may also be done at the front of the bed if desired.) Repeat the process until the first tier of hose is complete.

Step 5: Angle the hose up to begin the next tier upon reaching the opposite side of the hose bed.

Step 6: Make the first fold of the second tier directly over the last fold of the first tier. Begin this transition at the front of the bed with the first bend at the rear of the bed.

Step 7: Continue the second tier in the same manner as the first, laying hose in a progressive manner across the hose bed until the tier is complete.

6-6 Skill Sheet — Loading Supply Hose
Forward Horseshoe Load

Step 1: Place the male coupling in a front corner of the hose bed, and lay the first length of hose on its edge against the wall.

Step 2: Make the first fold at the rear even with the edge of the hose bed.

Step 3: Load the hose to the front, and then place it around the perimeter of the bed so that it comes back to the rear of the bed and then along the opposite side.

Step 4: Bring the hose to the front, and then place it around the perimeter of the bed so that it comes back to the rear of the bed along the opposite side. Continue in this manner until the first tier is filled to the center.

Step 5: Bring hose up from the center of the first tier and then take it to the rear of the bed to begin the second tier. Complete the second tier in the same manner as the first tier was done.

6-7 Skill Sheet — Loading Supply Hose

Reverse Hoseshoe Load

Step 1: Place the male coupling in a front corner of the hose bed, and lay the first length of hose on its edge against the wall. Make the first fold at the rear even with the edge of the hose bed.

Step 2: Lay the hose to the front, and then place it around the perimeter of the bed so that it comes back to the rear of the bed along the opposite side.

Step 3: Make a fold at the rear in the same manner as before, and then lay the hose back around the perimeter of the hose bed inside the first length of hose.

Step 4: Lay succeeding lengths of hose progressively inward toward the center of the tier until the entire space is filled. If desired, stagger the folds so that every other bend is approximately 2 inches (51 mm) inside adjacent bends.

Step 5: Start the second tier by extending the hose from the last fold directly over to a front corner of the bed, laying it flat on the hose of the first tier.

Step 6: Make the second and succeeding tiers in the same manner as the first. Lay the crossover length flat on the second tier, but lay it to the opposite corner from that in the first tier. Make crossovers in succeeding tiers to alternate corners.

6-8 Skill Sheet — Loading Supply Hose Forward Flat Load

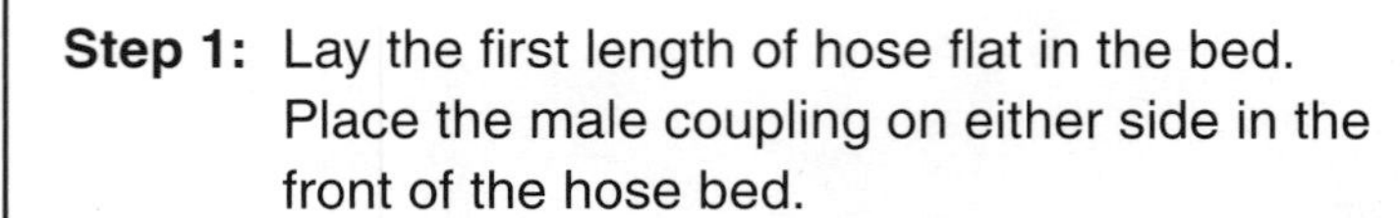

Step 1: Lay the first length of hose flat in the bed. Place the male coupling on either side in the front of the hose bed.

Step 2: Fold the hose back on itself at the front of the hose bed.

Step 3: Lay the hose back to the rear on top of the first length.

Step 4: Fold the hose so that the bend is even with the rear edge of the bed.

Step 5: Lay the hose back to the front of the bed angling it to make the front fold adjacent to the previous fold. Provide a space across the front of the hose bed, depending upon the type and size of the couplings.

Step 6: Continue to lay the hose in folds across the bed to the opposite side to complete the first tier.

Step 7: Build the second and succeeding tiers in the same manner as the first. Place the folds progressively across the hose bed. Set the folds at the back of the bed back 2 inches (51 mm) every other tier to allow room for the folds.

6-9 Skill Sheet — Connecting Supply Hose to Hydrant Four-Way Valve

NOTE: Pumping Apparatus 1 arrives at the hydrant.

Step 1: ***Hydrant Firefighter 1:*** Take the hydrant tools and pull enough hose to wrap around the hydrant.

Step 2: ***Hydrant Firefighter 1:*** Wrap the hydrant with the hose, and signal Driver/Operator 1 to deploy the hose.

Step 3: ***Hydrant Firefighter 1:*** Flush the hydrant before attaching the four-way hydrant valve.

Step 4: ***Hydrant Firefighter 1:*** Unwrap the hose from the hydrant, and attach the four-way valve to the appropriate hydrant outlet after Pumping Apparatus 1 has deployed several sections of hose.

Step 5: ***Hydrant Firefighter 1:*** Connect supply hose to a four-way valve if it is not already attached.

NOTE: Pumping Apparatus 1 continues to the scene deploying supply hose.

Step 6: ***Driver/Operator 1:*** Apply a hose clamp on the supply line 20 feet (6 m) behind Pumping Apparatus 1.

Step 7: ***Driver/Operator 1:*** Give the signal to charge the line.

Step 8: ***Hydrant Firefighter 1:*** Open the hydrant fully when signaled to do so. Return to Pumping Apparatus 1 and tighten the couplings; push hose to the curb.

6-9 Skill Sheet

Step 9: ***Driver/Operator 1:*** Uncouple the supply hose from the bed.

Step 10: ***Driver/Operator 1:*** Connect the hose to the pump, and release the hose clamp.

NOTE: Pumping Apparatus 2 arrives at the hydrant.

Step 11: ***Driver/Operator 2:*** Stop Pumping Apparatus 2 at the hydrant.

Step 12: ***Hydrant Firefighter 2:*** Connect the intake sleeve to the large connection of the four-way hydrant valve and large pump intake.

Step 13: ***Hydrant Firefighter 2:*** Open the valve to permit water flow into the pump.

Step 14: ***Hydrant Firefighter 2:*** Connect a discharge line to the four-way valve inlet.

Step 15: ***Driver/Operator 2:*** Apply proper pressure to the discharge line to support Pumping Apparatus 1 through the original supply line.

Step 16: ***Driver/Operator 2:*** Switch the four-way valve from hydrant to pumping apparatus supply where necessary. Charge other supply lines as needed.

Chapter 7

Attack Hose Loads, Finishes, Hose Packs, and Deployment Procedures

Chapter 7
Attack Hose Loads, Finishes, Hose Packs, and Deployment Procedures

When attack fire hose is used to move water from a fire pump to nozzles for application on a fire or other emergency situation, it is most often deployed manually. As with supply hose, the way attack hose is loaded depends upon a number of factors, including the number of fire and emergency service responders available, type of pumping apparatus, and diameter of hose. No matter which arrangement is used, however, attack fire hose must be loaded so that it can be pulled from the pumping apparatus and advanced in the most expedient manner. This chapter explores the methods of loading and deploying fire attack hose for use during fire-suppression or other emergency operations along with information on preconnected attack hose loads, attack hose load finishes, and hose packs.

The four following basic attack load methods are used to arrange attack fire hose for deployment from the pumping apparatus:

- ***Accordion, horseshoe, or flat load*** — Allows the use of the load as either a supply or attack hoseline. The use depends upon the conditions that are found upon the arrival of the pumping apparatus at the scene.
- ***Attack hose load finish*** — Attaches to the end of a hose bed load, allowing the pumping apparatus to proceed to a water source after deploying attack fire hose at the scene of the incident and then using a reverse lay to the water source.
- ***Preconnected quick-attack hoseline load*** — Connects hoseline to a pump discharge outlet at the time the hose is loaded.
- ***Portable hose pack*** — Preassembles hose for high-rise, low-rise, forestry service, warehouse, parking lot, hose station, and shipboard applications.

Attack Hose Lays

When a pumping apparatus crew must work alone for an extended period of time, which can occur when other pumping apparatus come from distant locations, a water supply hose must be deployed. This situation often requires the first-arriving pumping apparatus to position at the water source to draft or boost hydrant pressure. The hose deployed between the emergency scene and the water source, although supplying water to the emergency scene, is considered attack hose and must meet the stringent attack hose requirements defined in NFPA 1961, *Standard on Fire Hose* (2002). The hose can be laid forward or reverse, but in either case, enough must be manually deployed and positioned at the emergency scene to reach the most remote portions of a fire-involved or exposed structure.

A disadvantage when placing the first-arriving pumping apparatus at the water source is that fire-fighting tools and equipment must be removed from the apparatus at the emergency scene before it proceeds to the water source. Using this attack-hose evolution causes some delay in the initial-attack operation because the supply hose connections can only be accomplished after the equipment is removed from the pumping apparatus, the supply hose deployment is complete, and all hose connections are made at the water source. A way to expedite this task, however, is to clearly outline this process in a standard operating procedure (SOP) and assign specific pieces of equipment to each fire and emergency services responder. This procedure ensures that all tools necessary for fire-suppression or other emergency tasks are removed from the pumping apparatus in the shortest possible time at the scene.

Equipment that is essential to the fire-suppression or other emergency operation must include attack and supply fire hose, self-contained breathing apparatus (SCBA) and spare cylinders, nozzles, forcible-entry tools, wyes or siameses, axes, ladders, pike poles, and other items as designated in the fire and emergency services organization's SOPs. A forward large diameter supply line deployment in concert with preconnected attack hoselines provides for adequate water supplies while allowing all equipment and resources to be at the incident scene.

Forward Lay: Single Attack Hose

When removing attack hose from the pumping apparatus, pull it in the same way as is done during a forward supply hose deployment, with one additional step: Pull enough additional attack hose to reach the rear of the structure or emergency scene once the pumping apparatus deploying the hoseline arrives at the scene. If the hose comes entirely from the supply hose bed, pull it out of the bed in a manner that prevents it from tangling when it is advanced toward the emergency scene. Additionally, ensure that the hose is attack-hose certified and capable of sustaining the pressures expected of this hose type.

The procedure for forward-laying an attack hose from a flat-loaded hose bed is similar. After sufficient hose and equipment is removed from the pumping apparatus, it returns to the hydrant or other water source to charge and pump the hoseline from that location. This procedure is normally used when a second pumping apparatus that can provide additional support at the emergency scene is expected, but is delayed in arriving. Fire-suppression or other emergency operations can begin with the first pumping apparatus providing pressurized water from a remote source until the second pumping apparatus arrives. **Skill Sheet 7-1** describes the forward lay, single attack hose operation.

Reverse Lay: Single Attack Hose

When deploying an attack hose in a reverse lay, pull enough hose to reach all the way around the emergency incident. As with a forward attack lay, pull the hose from the apparatus bed in a manner that prevents it from tangling when it is advanced toward the emergency scene. **Skill Sheet 7-2** describes the procedure for deploying a reverse lay, single attack hose from the pumping apparatus.

Hose Load Finishes

Hose load finishes are arrangements of hose that are sometimes placed on top of a hose load and connected to the end of the supply hose load. It is a way to *finish* a hose load with attack fire hose that can be quickly pulled at the beginning of a forward or reverse lay supply line. The attack load that is most often used is a preconnected attack hoseline.

Finishes fall into two categories: those for forward lays and those for reverse lays. A finish for a reverse load allows the equipment to be removed from the pumping apparatus as a package and placed close to the emergency operation. Finishes for forward lays are usually designed to speed the pulling of hose for supply lines when making simple hydrant connections.

Forward Lay Finishes

A forward lay finish provides additional hose to reach from the place where the pumping apparatus starts the forward lay to the water source. This situation might be necessary when the pumping apparatus cannot easily get to a hydrant such as when the hydrant is set back on a narrow drive within an apartment complex or is some distance from an intersection. When a finish is used in a forward lay, the first step for laying the hose is to pull the finish from the hose bed. From this point on, the lay progresses as usual with a fire and emergency services responder anchoring the hose by holding it near the point where the finish is connected to the hose load.

One of the simplest finishes is a donut roll that is attached to the end of the supply load. Additional donut rolls can be added to the finish depending upon local SOPs **(Figure 7.1)**. Roll the donut so that the male coupling is on the outside of the roll. Then place the donut roll on top of the hose load, and connect the male coupling to the female coupling at the free end of the load. To pull the donut roll from the hose bed, simply grasp the female coupling and give a sharp tug to drop the roll from the hose bed to the ground. In many cases, this action is sufficient to secure the hose and allow the hose to deploy from the bed. After the pumping apparatus has deployed the hose down the street/road, the emergency responder anchoring/holding the hose grasps the female coupling and takes hydrant tools and hose to the hydrant to make the connection.

Although this finish is usually used as a supplement to a supply hoseline, it is included here because it is formed on top of the supply hose load. Used in a supply capacity, it is deployed first from the top of the hose bed (like the reverse lay finish). The forward lay finishes assist with the establishment of supply hoselines. Beginning where the pumping apparatus stops in the street/road to initially deploy a supply hoseline, a leading section of supply hose is prepared in a donut, two-donut, or rip-rap fashion, connected to the last section of loaded supply line, and placed on top of the supply hose load. This procedure allows the forward lay finish to be easily accessed at the rear of the hose bed so that an emergency responder can deploy it to the street/road.

Reverse Lay Finishes

The primary advantage of a reverse load hose finish is that the entire length of hose needed to reach the emergency scene from the street/road can be pulled in one motion. Most of the finishes described use an inline gate or gated wye that provides a better means of controlling incoming water than a hose clamp. One of the disadvantages of this particular load finish is that it does not lend itself well to deployment as a supply line. When used in this manner, the hose load finish must be disconnected and removed from the load to permit the deployment of the supply hoseline should the attack hoseline not be needed.

Three finishes are described in the sections that follow: Cisco, reverse horseshoe, and skid load. Each attack finish is designed for quick and easy deployment by one fire and emergency services responder. When pulled from the bed, the

Figure 7.1 Forward lay hose finishes: donut, two-donut, and rip-rap.

entire load is brought to the ground, and it does not interfere with the rest of the supply hose stored underneath.

Local emergency conditions and the capacity of the hose bed can assist in determining the desired length of each attack hoseline. When using small diameter attack hose, its length affects its overall performance. Friction loss is increased when small diameter hose is employed. Maneuverability is impeded due to weight and size when large attack hose is loaded. Each of the hose finishes described can be deployed from the hose bed by one emergency responder.

Cisco

The Cisco finish is made of two 100- to 150-foot (30 m to 45 m) lengths of hose, each connected to one side of a gated wye **(Figure 7.2).** Any diameter of attack hose can be used: 1½-, 1¾-, 2-, or 2½-inch (38 mm, 45 mm, 50 mm, or 65 mm). The small sizes require a 2½- by 1½-inch (65 mm by 38 mm) gated reducing wye; the 2½-inch (65 mm) hose requires a 2½- by 2½-inch (65 mm by 65 mm) gated wye. Two nozzles matching the diameter of the attack hose are also needed. The finished bundle folded in accordion fashion is stored on top of the hose load and tied together with straps or ropes. The procedure for loading the Cisco attack load finish for a reverse lay is described in **Skill Sheet 7-3.**

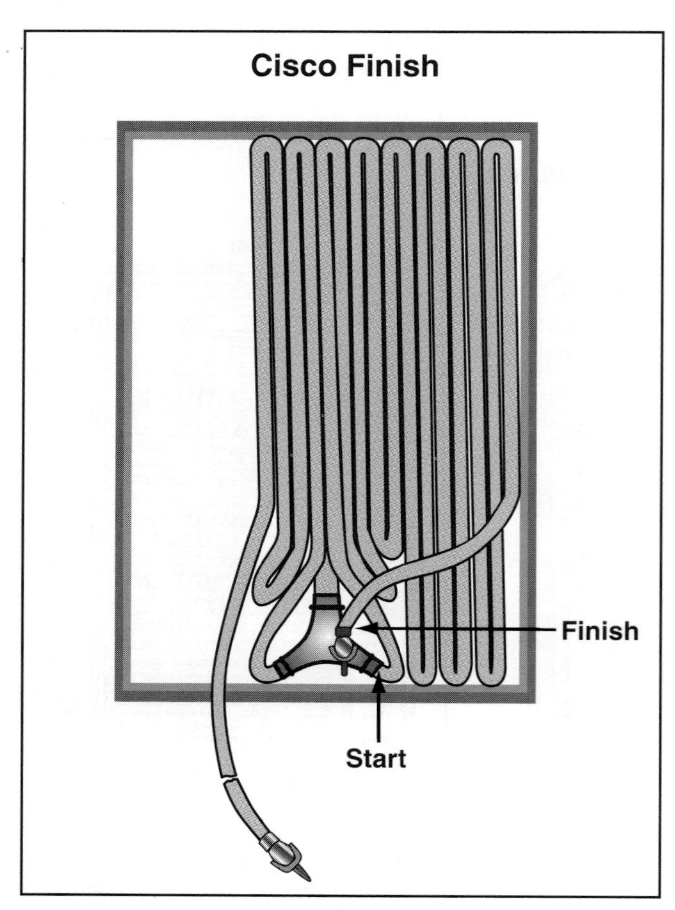

Figure 7.2 Reverse lay finish: Cisco.

To deploy the Cisco finish upon arrival at the scene of an emergency incident, the pumping apparatus must stop just beyond the involved or exposed structure. Following an evaluation and direction by the command officer, a fire and emergency services responder deploys the attack hoseline. This finish is deployed from the pumping apparatus and advanced by one or two emergency responders. Offensive interior fire-suppression operations must be conducted by teams of two emergency responders with a two-responder backup team available for immediate deployment for rescue. **Skill Sheet 7-4** outlines the procedures for deploying the Cisco load from the hose bed for a reverse lay.

Reverse Horseshoe

This finish is similar to the horseshoe load described in Chapter 6, Supply Hose Loads and Deployment Procedures, except that the *U* portion of the horseshoe is at the rear of the hose bed **(Figure 7.3).** The finish is made of two or three 50-foot (15 m) lengths of hose, each connected to one side of a wye. As with the Cisco finish, any size of attack hose can be used: 1½, 1¾, or 2½ inches (38 mm, 45 mm, or 65 mm). The small sizes require a 2½- by 1½-inch (65 mm by 38 mm) gated reducing wye; the 2½-inch (65 mm) hose requires a 2½- by 2½-inch (65 mm by 65 mm) gated wye. Two nozzles matching the diameter of the hose being used are also needed. The procedure for making a reverse horseshoe finish with 1½-inch (38 mm) hose for a reverse lay is described in **Skill Sheet 7-5.**

The reverse horseshoe finish can be pulled from the bed by one fire and emergency services responder. Once a second emergency responder arrives, this load can be advanced easily. The procedure for deploying and advancing this load and finish for a reverse lay is described in **Skill Sheet 7-6.**

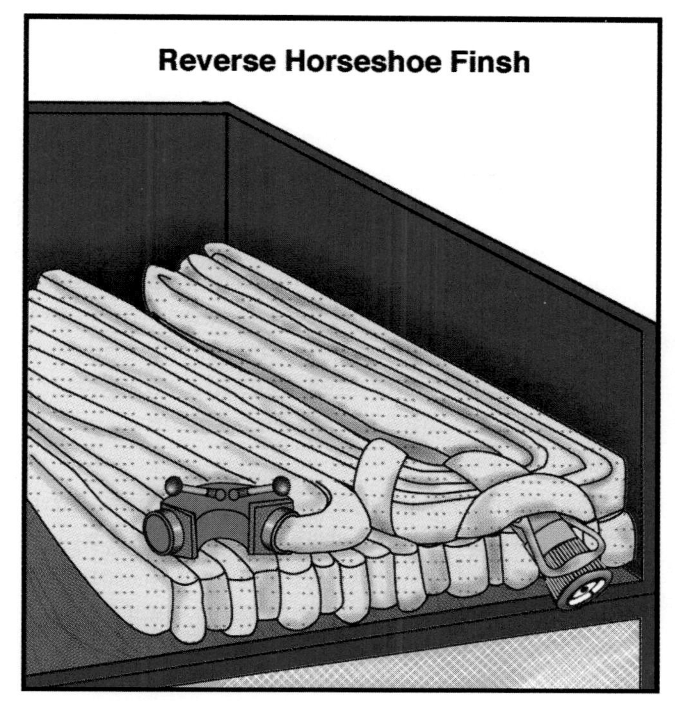

Figure 7.3 Reverse lay finish: Reverse horseshoe.

Skid-Load

Hose in this finish is loaded in accordion folds across the hose bed **(Figure 7.4).** The folds rest on two lengths of hose laid flat to act as skids (support) when the finish is pulled from the bed. Three sections of 2½-inch (65 mm) hose and a nozzle are used to make this load finish. Refer to **Skill Sheet 7-7** for a detailed procedure for loading this type of hose finish for a reverse lay.

The skid-load finish can be deployed by one fire and emergency services responder and advanced when a second responder arrives on the scene. Although one emergency responder can easily move the hose, teams of two or more responders must carry out fire-suppression or other emergency activities. The procedure for deploying a reverse lay load with a skid-load finish is described in **Skill Sheet 7-8.**

Preconnected Attack Hose Loads

Another way to carry attack hose is to preconnect it to a discharge valve and store it on the pumping apparatus in a manner that allows it to be easily deployed and advanced. The storage areas for these preconnected hoselines are places other than the main hose bed. Connecting the attack hose directly to a fire pump discharge valve and storing it in a dedicated compartment saves valuable time during emergency attack deployment. Another advantage of carrying attack hose outside the main hose bed is that there is no need to move it to the side when deploying supply hose, which was often the case before the advent of preconnected hosebed designs.

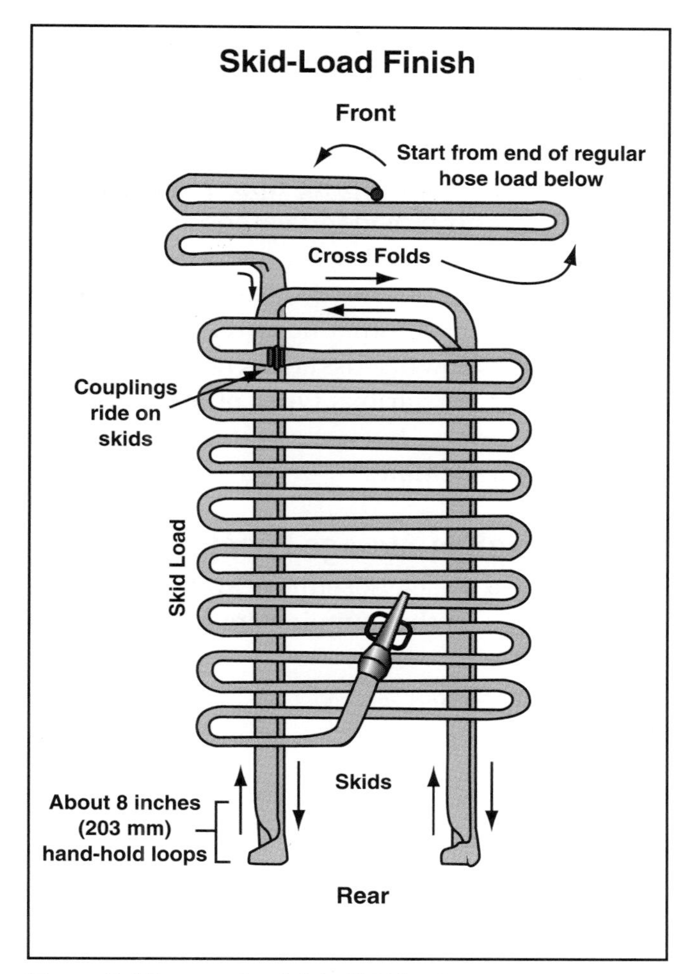

Figure 7.4 Reverse lay finish: Skid-load.

Any size of small diameter hose can be preconnected to the fire pump discharge valves. A typical quick-attack operation involves the pumping apparatus stopping near the emergency scene, fire and emergency responders donning self-contained breathing apparatus (SCBA) and pulling the preconnected hoses, and the driver/operator charging the hoses with water from the tank as soon as the hoselines are extended. This hose evolution does not restrict the deployment of a forward supply hose lay from the supply hose bed. The use of sexless couplings connecting large diameter supply hose with preconnected attack hoselines gives a flexible and efficient emergency operation plan for a fire and emergency services organization. A pumping apparatus can be assigned to easily provide attack hoselines, supply hoseline, or both during an operation without having to "dump" unwanted hose in the street/road while performing another hose-related task. The sections that follow discuss attack hoseline storage locations and the various types of preconnected hose loads (reverse horseshoe attack, flat, triple-layer, and minuteman).

Attack Hoseline Storage Locations

Preconnected attack hoselines can be carried in several places on the pumping apparatus such as the following:

- Longitudinal bed
- Transverse bed
- Tailboard compartment or hose tray
- Side compartment or bin
- Front bumper well
- Reel

Longitudinal Bed

Longitudinal hose beds for attack fire hose are generally located along the far left and right sides of the main hose bed **(Figure 7.5).** When space is limited, raised trays mounted over the main hose bed can be used to separate the attack hose from the supply hose underneath. A longitudinal hose bed is the traditional location to carry preconnected fire hose.

Years ago when firefighters were required to ride on the pumping apparatus tailboard when responding to fires or other emergencies, it was common to place hose in beds or raised trays that unloaded to the rear of the apparatus. This practice permitted the firefighters riding on the tailboard to immediately pull the preconnected hoselines as soon as the pumping apparatus arrived at an emergency scene. Although all fire and emergency service responders now ride in jump seats and crew cabs, there is still some advantage in loading attack hose so that it comes off the rear of the pumping apparatus.

Hose that deploys from the rear of the apparatus is in an ideal position when the pumping apparatus stops past the emergency scene (a standard practice with some fire and emergency service organizations). Loads that deploy from the rear are also located conveniently close for pulling and disconnecting the hose for a reverse lay deployment. The most notable disadvantage of this storage location is that the driver/operator cannot easily determine from the cab of the pumping apparatus which hoseline or hoselines have been deployed or if the hose has been completely pulled from the bed.

Transverse Bed

Transverse hose beds (also known as *mattydales* or *cross lays*) are most often found between the hose bed and the cab of the pumping apparatus, and they deploy to either side, depending upon need **(Figure 7.6).** Multiple transverse beds can be placed side by side or stacked. The primary advantage with loading preconnected hose into transverse hose beds is that it places the hose close

Figure 7.5 Longitudinal hose bed.

Figure 7.6 Transverse hose beds are usually located between the hose bed and the cab of the pumping apparatus and loaded so that they can be deployed to either side of the vehicle.

to seated fire and emergency service responders. This placement speeds a quick-attack operation because emergency responders do not need to go to the rear of the pumping apparatus to deploy an attack hose. The hose is also less prone to becoming entangled on the pumping apparatus, which occasionally happens when hose is pulled from rear beds and to the side of the apparatus. This placement also allows a separation between fire and emergency services personnel deploying supply lines from the rear of the main hose bed and those involved with advancing attack hose for fire suppression or other emergency activities. Another advantage with a transverse hose bed is that the driver/operator can more easily see the deployed hose from the pump control panel location.

Tailboard Compartment or Tray

Attack hose is sometimes carried at the rear of the pumping apparatus in a tailboard compartment **(Figure 7.7).** This location makes the hose

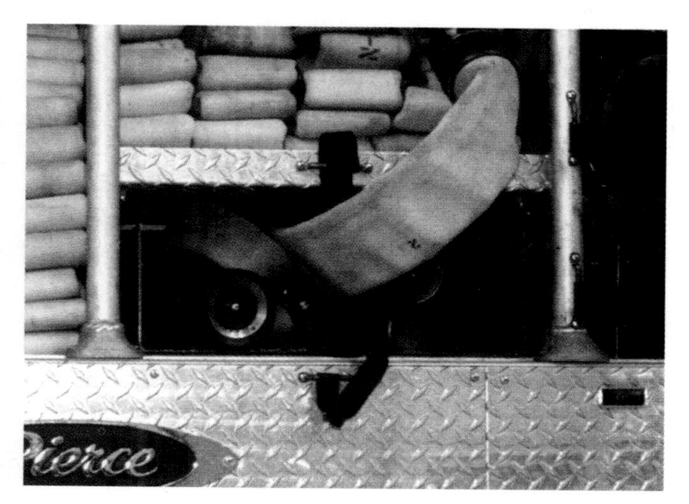

Figure 7.7 Tailboard compartment or hose tray. *Courtesy of Judy Halmich, Washington (MO) Fire Department.*

accessible for deployment while protecting it from exposure damage. Carrying hose in this way has some of the same advantages as carrying it in longitudinal beds. The attack hose deploys from the rear of the pumping apparatus, placing it in an ideal position when the apparatus stops and takes a position past the emergency scene. The load is conveniently close for deploying and disconnecting for a reverse lay but does not take space in the main supply hose bed.

The primary disadvantage is that the amount of hose that can be loaded is also limited because space is usually limited in the tailboard compartment. The configuration of this compartment is also limited as is the availability of equipment storage space on the rear step of the pumping apparatus. This situation also decreases the number of possible hose load options, and loading is sometimes a difficult task, allowing room for only one fire and emergency services responder to perform this task.

Side Compartment

Attack hose can be carried in side compartments and racks or bins mounted above the pumping apparatus side compartments. These locations give fire and emergency service responders easy access to the attack hoseline by placing it in a location next to the crew compartment. Upon arrival, the crew deploys the attack hose from the side compartment nearest the emergency scene. When pulling hose to the same side, the hose is less likely to snag on the pumping apparatus, and fire and emergency service personnel can work in an area separate from those who are pulling rear supply hoselines.

Like hose in tailboard compartments, however, space is limited, so the total length of hose that can be carried is also reduced to the capacity of the space available. Should this be the preferred method of attack hoseline storage, consider this factor during pumping apparatus design when preparing purchasing specifications.

Front Bumper Well

Attack fire hose carried in front bumper wells is especially well suited for an attack on small fires such as grass fires, trash or Dumpster® fires, and vehicle fires **(Figure 7.8).** The hose is easy to reach and easily reloaded. Space in the well is usually limited to 100 feet (30 m) of attack hoseline, which limits this type of load to no more than two sections of hose. This hoseline is sometimes referred to as a *trash line* and intended to be rapidly deployed, used, and then returned to storage.

Reel

Reels are most often used for booster hose but can be used for any size of attack hose **(Figure 7.9).** Some advantages with hose reels are that there are no bends in the hose, and it can be easily deployed and reloaded. A disadvantage with woven-jacket hose loaded on a reel, however, is that the entire length of hose must be deployed, disconnected from the remaining hose on the reel, and then attached to a fire pump discharge outlet before it can be charged with water. This evolution delays the fire-suppression or other emergency operation, especially if the pumping apparatus is located close to the emergency scene.

Reverse Horseshoe Attack Line Load

Two methods of loading the preconnected reverse horseshoe load are available **(Figure 7.10).** The first load is for large attack hose of 2½- or 3-inch (65 mm or 77 mm) diameter size, and it is best loaded into a wide hose bed. This load is usually deployed from the bed by a two-person team. The load is designed so that several persons can quickly deploy a large amount of attack hose.

The second reverse horseshoe load is for preconnected small attack hose (1½ or 1¾ inches [38 mm or 45 mm]), and it can be loaded into a narrow longitudinal bed or bin. Unlike the first load, one person can pull and extend this hose,

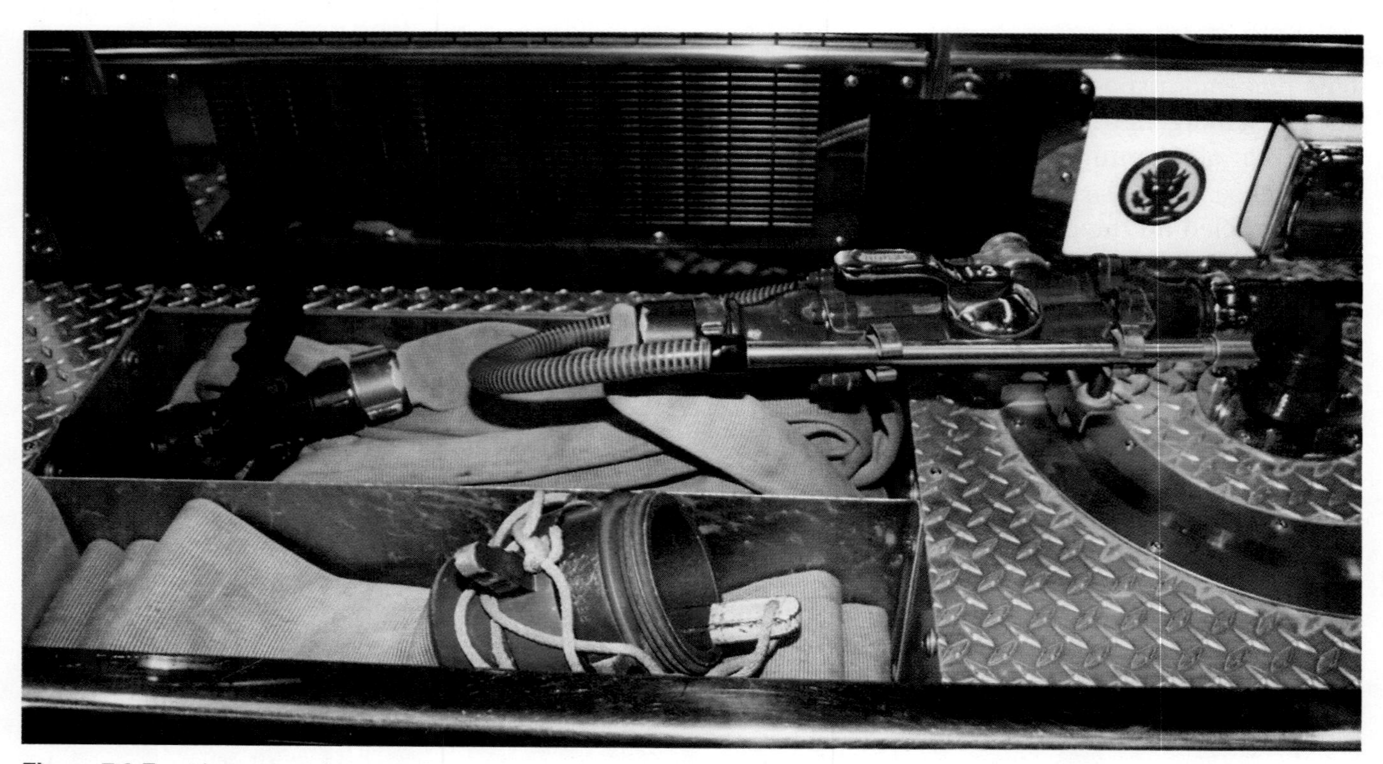

Figure 7.8 Front bumper well storage hose compartment. *Courtesy of Judy Halmich, Washington (MO) Fire Department.*

Figure 7.9 Hose reel with booster line. *Courtesy of Sam Goldwater.*

but a two-person team (with a backup team available) must conduct all interior structural fire-suppression activities.

WARNING

While one emergency responder can pull and extend a hoseline, all hoseline fire-suppression or other emergency activities must be conducted with a minimum of two emergency responders. For interior structural fire-suppression activities, an emergency backup team of two emergency responders must be available for immediate service also.

Large Attack Hose Method

Use 2½- or 3-inch (65 mm or 77 mm) hose with this preconnected reverse horseshoe load method, and put the load into a wide hose bed. This method allows one person to pull each tier and carry it. Therefore, load no more hose in a tier than a single fire and emergency services responder can carry (perhaps two sections of hose). If the load is connected to a discharge at the front of the bed, place an extra length of hose to the side of the first tier to provide enough slack to pull the tier clear of the bed. The procedure is described in **Skill Sheet 7-9**.

Carry the hose on the arm (as shown in **Skill Sheet 7-10**) so that it pays off as it is carried forward (in the same manner as the accordion shoulder carry described in Chapter 5, Basic Methods of Handling Fire Hose). The folds on either side of each tier provide a comfortable separation between emergency responders as they advance the hose.

Figure 7.10 Reverse horseshoe attack load in a removable tray. *Courtesy of Judy Halmich, Washington (MO) Fire Department.*

Small Attack Hose Method

The small attack hose load method is similar to the large attack hose method, except that it is for 1½- or 1¾-inch (38 mm or 45 mm) attack hose. The primary advantage with this load is that one emergency responder can pull the hose from the bed and deploy it around obstacles and up stairways without the hose snagging. It is usually loaded into a narrow longitudinal bed or bin so that each tier contains approximately one section of hose. Wider beds may accommodate as much as two sections of hose in each tier. Avoid packing the hose bed too tightly.

If the load is connected to a discharge at the front of the bed, place an extra length of hose to the side of the first tier to provide enough slack to pull the tier clear of the bed. **Skill Sheet 7-11** describes the procedure for loading preconnected small attack hose in a reverse horseshoe load. A single emergency responder can pull this load and deploy the hoseline when necessary. **Skill Sheet 7-12** describes the procedure for deploying and advancing this preconnected attack hoseline.

Flat Load

The flat load is adaptable for varying widths of hose and hose beds, and it is often used in transverse hose beds **(Figure 7.11).** This load is similar to the flat load for large supply hose (see Chapter 6, Supply Hose Loads and Deployment Procedures) with the exceptions that it is preconnected and hose loops are provided to assist in deployment. The hose loops are placed at regular intervals within the load so that equal portions of the load can be pulled at one time from the bed. The number of loops and the intervals at which they are placed depend upon the diameter and total length of the hose. The procedure for loading a preconnected attack hose flat load is described in **Skill Sheet 7-13.** The usual length of this load is 150 to 200 feet (45 m to 60 m) of 1½-, 1¾-, or 2-inch (38 mm, 45 mm, or 50 mm) hose loaded flat into a transverse hose bed.

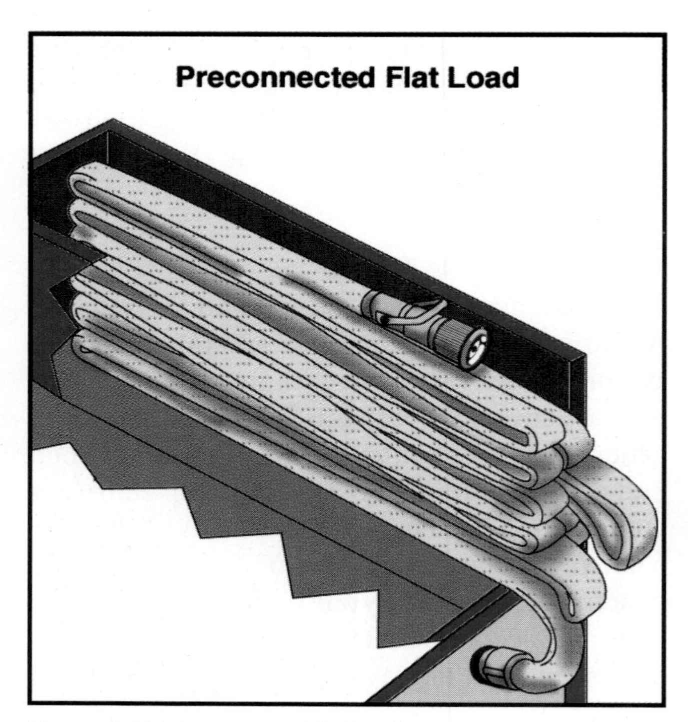

Figure 7.11 Preconnected flat load.

One fire and emergency services responder can pull and advance the flat load from a transverse hose bed. In beds loaded with an exceptionally large load of hose, two emergency responders may be required to fully extend the load. Although several different hose lengths and diameters can be successfully used with this hose load, it is best suited for use with 1½- to 2-inch (38 mm to 50 mm) diameter hose and usually limited to 150 feet (45 m). The procedure described in **Skill Sheet 7-14** is for the deployment of the preconnected attack hose flat load from a transverse hose bed by one fire and emergency services responder.

Triple-Layer Load

The triple-layer load gets its name because the load begins with hose folded in three layers **(Figure 7.12).** The three folds are then laid into the bed in an *S*-shaped fashion, which allows one fire and emergency services responder to deploy the load. The load is relatively cumbersome to place in the hose bed. Four to five emergency responders are usually required to keep the folds aligned as they are laid into the hose bed. To load preconnected attack hose for the triple-layer load, follow the instructions described in **Skill Sheet 7-15.**

A disadvantage with the triple-layer load is that the three layers, each of which may be 50 feet (15 m) long, must be completely removed from the bed

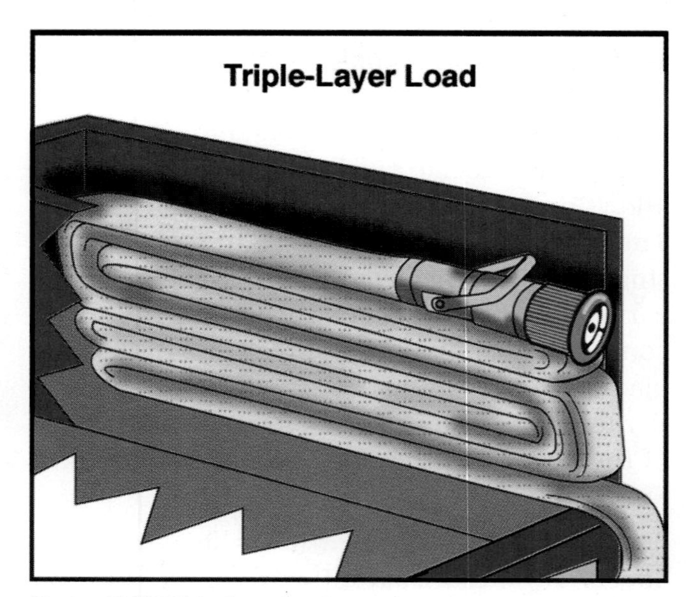

Figure 7.12 Triple-layer load.

before it is charged with water and the nozzle end of the hose can be advanced. If the pumping apparatus is positioned in a manner that places other equipment directly behind it, hose deployment is obstructed. Because this load is cumbersome when it is brought out of the bed, the most efficient method of deployment is with two emergency responders performing the operation. **Skill Sheet 7-16** describes the procedure for deploying preconnected attack hose from the triple-layer load.

Minuteman Load

The minuteman load is pulled directly from the bed to a shoulder of a single fire and emergency services responder **(Figure 7.13).** The hose pays off the shoulder as the hoseline is advanced. The primary advantage with this load is that it is carried on the shoulder completely clear of the ground, so hose will not snag on obstacles. This load is also particularly well suited for a narrow bed. A disadvantage encountered with this hose load is that it can be very awkward to carry when wearing an SCBA. This situation is especially true if the load is in a single stack, allowing hose to tip over and collapse on the shoulder if it is not held tightly in place (which could cause the hose to become entangled). **Skill Sheet 7-17** describes the minuteman loading procedure for preconnected attack hose in a double stack. The procedure for advancing the preconnected attack hose in a minuteman load is described in **Skill Sheet 7-18.**

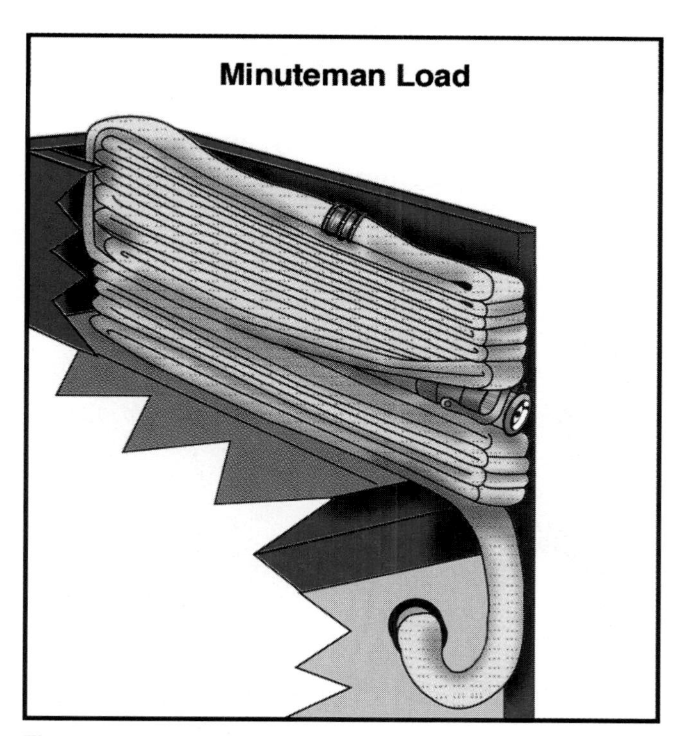

Figure 7.13 Minuteman load.

Hose Packs

A hose pack is an easily stored, self-contained fire-suppression assembly. It is easily stored in a side compartment on a pumping apparatus and provides fire and emergency service responders with a compact, preassembled hose system that can be transported into difficult locations with the least amount of effort.

The use of hose packs for specialized fire-suppression circumstances provides a coordinated and simplified method of bringing a large quantity of hose to a remote location. A hose pack eliminates the cumbersome activity of a direct hose advance from a hose bed. Hose packs are most beneficial when designed for use in conjunction with standpipes or hose stations and for deployment in unsupported nonstandpipe locations. Preconnected attack hoselines deployed from the pumping apparatus can be quickly extended by connecting a hose pack to the nozzle end of the hoseline and continuing the advancement. Wildland/interface packs allow remote pumping apparatus to support fire-suppression efforts by extending attack lines into areas that are only capable of being reached by foot (depending on the size, length, and type of hose as well as on the way the pack is designed).

Standpipe

One of the greatest problems in fighting fires in large buildings, particularly in multistoried buildings above the reach of elevating devices, is transporting water to the seat of the fire. Years ago, firefighters carried supply and attack hose into the buildings and up interior stairwells or exterior fire escapes to reach the fire floor. Eventually, however, new buildings were equipped with standpipe systems that allowed firefighters to only carry attack fire hose from the street to the floor where it was needed. A standpipe system may be vertical (for example, in high-rise buildings, in parking garages, and on ship decks) or horizontal (in shopping malls, factories, and warehouses).

Regardless of the type of application, the standpipe pack allows a fire-suppression crew to reduce the amount of supply and attack hose that must be carried and then deployed before fire-suppression activities can begin.

Although transporting hose to a standpipe connection is a much simpler evolution than deploying hose the entire distance from the pumping apparatus to the fire, it can still be a physically exhausting job. Fire and emergency responders must take not only hose and nozzles to the fire location but also SCBA, spare air cylinders, and other vital equipment for the operation. Lightweight hose is often packed into portable packs to facilitate their transport and deployment.

Many types of packs are available. Some are carried on the emergency responder's back, some are hand carried, and others are secured to two-wheeled handcarts. A typical hand-carried hose pack contains enough hose and equipment to allow the fire-suppression team to make all necessary connections for a self-sustained operation while deploying and advancing the hoseline from the standpipe to the location of the fire **(Figure 7.14).** The equipment contained within the pack must include all of the tools and fittings that permit the fire-suppression team to accomplish this evolution **(Figure 7.15).** A list that includes most of this equipment is as follows:

Figure 7.14 Standpipe hose pack.

- Hose pack carrying frame or harness that allows all equipment to be attached
- 100-foot (30 m) length of 1½-inch, 1¾-, or 2-inch (38 mm, 45 mm, or 50 mm) standard attack fire hose
- Double male adapater
- 2½- by 1½-inch (65 mm by 38 mm) reducer adapter
- Nozzle with shutoff
- Pouch containing sprinkler tongs and wedges, pliers, lightweight spanner wrenches, and search and rescue tags

The pack uses an accordion load in a single stack of hose. If carried by two fire and emergency service responders, the total amount of hose can be increased (loaded in two stacks). **Figure 7.16** shows a crew carrying a standpipe pack along with other essential equipment.

CAUTION

Be aware that directing a fog stream into a highly heated room/compartment from a doorway often creates a situation where steam generates quickly and pressurizes the area. This pressure can push the fire out the doorway, surrounding emergency personnel in the stairwell, obstructing their visibility, and possibly burning them.

Figure 7.15 Typical components of a standpipe hose pack. *Courtesy of Rick Montemorra, Mesa (AZ) Fire Department.*

Wildland

The increasing fire risk presented by wildland fires spreading into urban areas that formerly were most often the concern of forest-service firefighters, has resulted in a greater role by community fire and emergency service organizations for these fire-suppression services. The concept of the standpipe pack and its deployment, though similar in nature, may not provide the most practical method of deployment during wildland fire-suppr ession operations. Fire and emergency service organizations faced with a wildland urban/interface fire situation must be prep ared to deploy fire and emergency service responders and equipment in a safe and efficient manner that allows fire-suppression activities in all types of terrain (often remote from the pumping apparatus). A wildland pack provides the deployment of a fire-suppression hoseline to these areas. A wildland pack can be configured as a tied bundle of hose that is extended by sequentially adding more hoseline. As one pack plays out, another is connected and extended (progressive hose lays).

Figure 7.16 Two firefighters carrying a standpipe hose pack, which includes equipment needed to start fire-suppression activities from a standpipe valve location. *Courtesy of Rick Montemorra, Mesa (AZ) Fire Department.*

Many fire and emergency service organizations have found that the traditional forest-service progressive hose pack can be replaced with the use of a hose pack bag. This pack bag is easily prepared by loosely placing the hose inside the pack bag **(Figure 7.17).** Storage is equally simple while allowing the same versatility as the forest-service progressive hose pack. All connections, fittings, and nozzles are contained in the pack.

This pack bag is specifically designed to not only transport hose across difficult terrain but also to progressively deploy the hose during a direct attack on a wildland fire. All equipment that may be required during the progressive lay are carried by the emergency responders. This equipment includes hose clamps, nozzles, inline tees, and adapters. The procedure for carrying and then progressively deploying hose from the wildland pack is given in **Skill Sheet 7-19**.

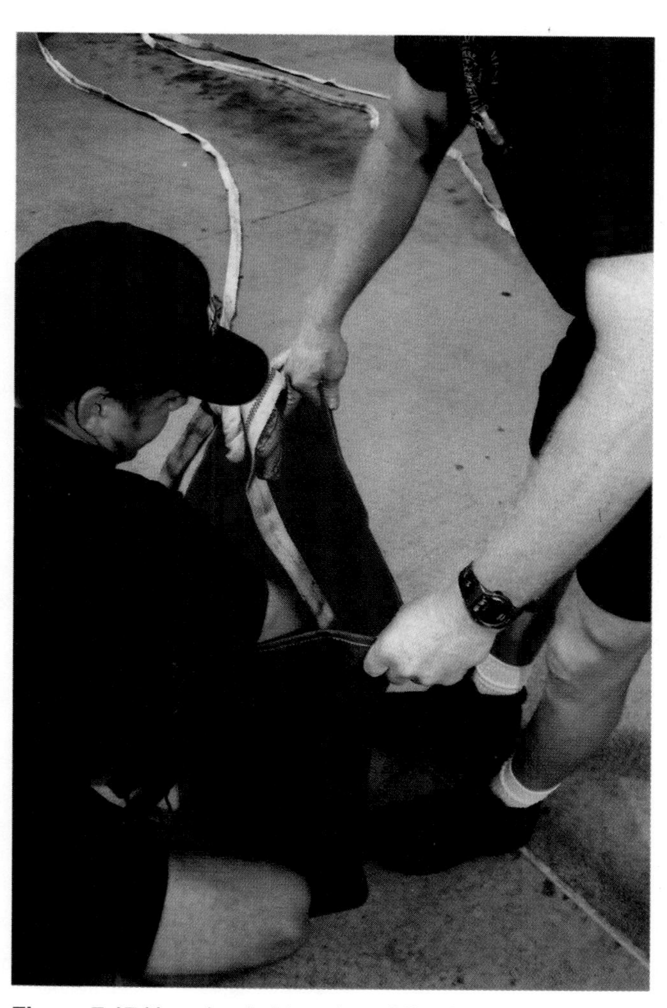

Figure 7.17 Hose loaded into the wildland hose pack bag allows for easy deployment.

Summary

This chapter explores the methods available to fire and emergency service organizations for loading and deploying fire attack hose. Organizations can select the methods that best suit their needs, depending upon local conditions such as the presence of high-rise or wildland fire challenges. For some organizations, an attack hoseline reverse finish is appropriate due to the need to locate a water source and pump the attack hoselines from one pumping apparatus. Other organizations that have an abundance of fire hydrants find that preconnected attack hoselines with a forward supply hose system meet their requirements. Still others find that a finish with a forward hose lay and a separate attack pumping apparatus provide the best method for fire-suppression or resolving other emergency situations. Ultimately, there is no one best attack hose load method. The attack hose load method (or methods) chosen by an organization must come as the result of an extensive evaluation of local conditions and fire-risk factors. Once in place, the chosen attack hose load configurations will assist with the development of pumping apparatus design and hose specifications as well as the establishment of fundamental fire and emergency service responder hose-evolution training programs.

Deploying Attack Fire Hose Forward Lay: Single Attack Hose

Skill Sheet 7-1

NOTE: Although ordinarily recommended for this type of evolution, a four-way valve is not shown in Steps 1 and 2 when a hydrant is available. Consider the use of this valve when using this procedure because it would allow additional flexibility when the first arriving pumping apparatus pressurizes the attack hoseline.

Step 1: ***Driver/Operator:*** Stop the apparatus at the hydrant.

Step 2: ***Hydrant Firefighter:*** Pull sufficient hose to reach and make the connection to the hydrant. Signal the Driver/Operator when hose is secure (see photo).

Step 3: ***Driver/Operator:*** Proceed to scene, deploying hose.

Step 4: ***Driver/Operator:*** Position at the scene at a location selected by the command officer. Come to a complete stop and chock wheels.

Step 5: ***Driver/Operator:*** Apply a hose clamp within 5 feet (1.5 m) of the last hose deployed from the hose bed on the water source side of the supply hose coupling (see photo).

Step 6: ***Driver/Operator:*** Signal the Hydrant Firefighter to charge the hose.

Step 7: ***Firefighter 1:*** Grasp two folds (one in each hand) of hose adjacent to the hose that extends from the hose bed to the ground. Pull the hose from the bed so that the folds clear the tailboard by 10 feet (3 m). Pull the hose parallel to the hose already deployed on the emergency-scene side of the ground.

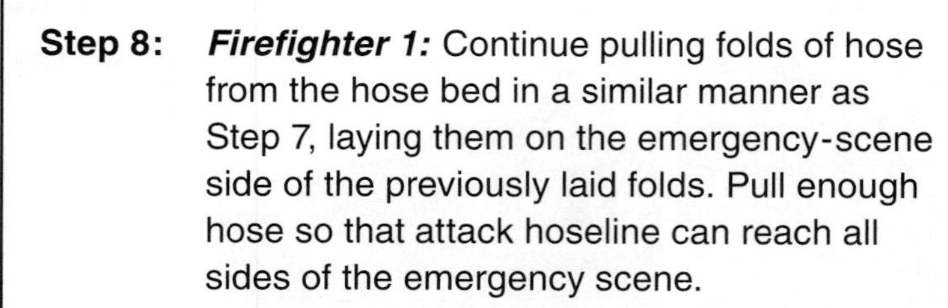

Step 8: ***Firefighter 1:*** Continue pulling folds of hose from the hose bed in a similar manner as Step 7, laying them on the emergency-scene side of the previously laid folds. Pull enough hose so that attack hoseline can reach all sides of the emergency scene.

Step 9: ***Firefighter 1:*** Disconnect the last coupling and return the female coupling to the hose bed.

Step 10: ***Firefighter 2:*** Remove all equipment, fittings and devices that may be needed at the emergency scene and place them in a location that is available and safe for use.

Step 11: ***Firefighter 1:*** Connect the nozzle or a hose appliance to the deployed hose. Attach a gated wye to split the hoseline into smaller attack hoselines if needed. Attach a master stream device to the extended hoselines if needed.

Step 12: ***Driver/Operator:*** Release the hose clamp to allow water to travel to the nozzle or hose device.

Deploying Attack Fire Hose — Reverse Lay: Single Attack Hose

Skill Sheet 7-2

Step 1: ***Driver/Operator:*** Stop the pumping apparatus in front of the emergency incident.

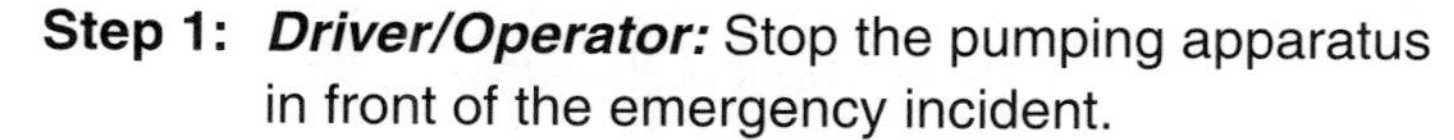

Step 2: ***Firefighter 1:*** Pull the first male coupling and the first fold of hose from the hose bed so that the folds clear the tailboard by 10 feet (3 m). Lay the folds on the ground in a position that allows them to be easily pulled toward the emergency scene.

Step 3: ***Firefighter 1:*** Pull the next two folds in a similar manner, laying them next to the previously laid folds on the side away from the emergency scene.

Step 4: ***Firefighter 1:*** Continue pulling attack hose in the same manner until enough hose is deployed to allow it to be advanced to all points around the emergency incident.

Step 5: ***Firefighter 2:*** Remove all equipment to support the emergency operation from the pumping apparatus and place it in a location where it is readily available.

7-2 Skill Sheet

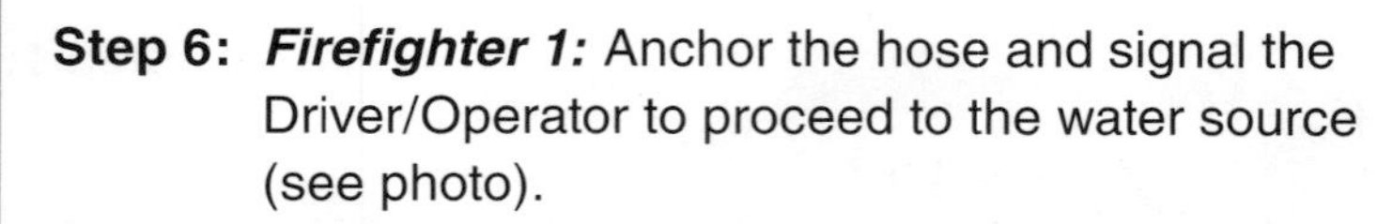

Step 6: ***Firefighter 1:*** Anchor the hose and signal the Driver/Operator to proceed to the water source (see photo).

Step 7: ***Other Firefighters:*** Stay at the scene to operate the attack hoseline.

Step 8: ***Driver/Operator:*** Proceed and make the final hose connections. Charge the line.

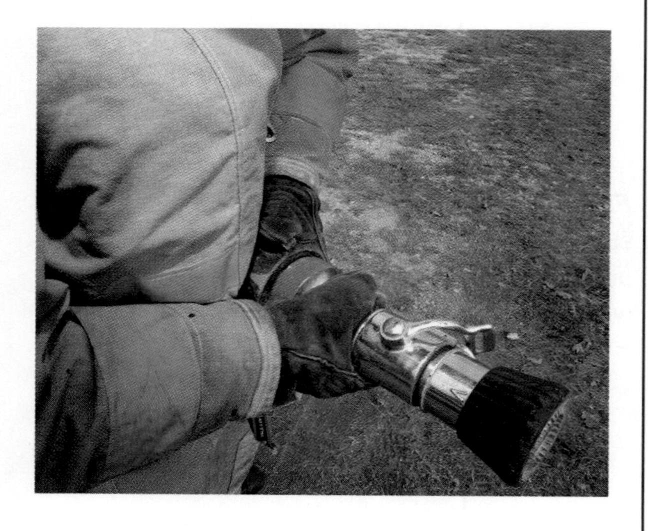

Step 9: ***Firefighter 2:*** Attach the nozzle and advance hoseline to the point of attack.

Finishing Reverse Lays Cisco Finish

Skill Sheet 7-3

Step 1: Connect a gated wye to the male coupling of the hose load at the rear of the hose bed.

Step 2: Position the gated wye in the center of the hose bed with the two male openings facing toward the rear of the bed.

Step 3: Connect one of the attack hoses to one side of the wye, and lay the first fold of hose on edge to the front of the hose bed.

Step 4: Make a fold at the front, and lay the hose on edge back to the rear of the bed.

Step 5: Make a fold at the rear near the wye connection, and then lay the hose back to the front next to the previous fold.

Step 6: Continue accordion folding the hose until the entire length is loaded.

7-3 Skill Sheet

Step 7: Connect and load the second length of hose in the same manner as the first, laying it progressively toward the opposite side of the bed.

Step 8: Attach a nozzle to each of the two hoses, and ensure that the nozzles and wye are in the closed positions.

Step 9: Lay the nozzles on top of the finish load, and tie each bundle at the rear with a rope or strap.

NOTE: The first fold on each side of the wye is *not* tied into the bundle.

Deploying Reverse Lays Cisco Finish

Skill Sheet **7-4**

Step 1: ***Firefighter 1:*** Grasp each side of one bundle of hose by the tie strap or rope, and pull it from the hose bed.

Step 2: ***Firefighter 1:*** Advance the bundle approximately 10 feet (3 m) from the tailboard of the pumping apparatus, and place it on the ground.

Step 3: ***Firefighter 2:*** Pull the opposite bundle from the hose bed in the same manner, and place it next to the first bundle, leaving room for the gated wye.

7-4 Skill Sheet

Step 4: ***Firefighter 1:*** Bring the gated wye and attached hose from the rear of the hose bed, and put the wye between the bundles near the rope or strap ties.

Step 5: ***Firefighter 1:*** Pick up the gated wye when the attack hose has been deployed to the ground along with enough hose to reach a location where the hose can be anchored.

Step 6: ***Firefighter 1:*** Anchor the hose, and signal the Driver/Operator (see photo).

Step 7: ***Driver/Operator:*** Proceed to the water source, deploying the hose from the bed.

Step 8: ***Firefighter 1:*** Continue to anchor the hose to ensure that it deploys from the bed as the pumping apparatus moves away.

Step 9: ***Firefighters 1 and 2:*** Untie the bundles after the pumping apparatus makes the lay, and begin the first hose advance.

NOTE: The second attack hose bundle remains for the second attack team.

Step 10: ***Firefighter 1:*** Open the gated wye when the attack teams are in position and ready for water.

7-5 Skill Sheet — Finishing Reverse Lays

Reverse Horseshoe Finish

Step 1: Connect the gated wye to the end (male) coupling of the hose load at the rear of the hose bed.

Step 2: Place the wye in the center of the hose load with the two male openings toward the rear of the hose bed.

Step 3: Connect one of the 1½-inch (38 mm) hoses to the gated wye. Place the hose on its edge to the front of the hose bed, and make a fold in the hose.

Step 4: Lay the hose back to the rear alongside the first length. Form a *U* at the edge of the bed, return the hose to the front, and make a fold.

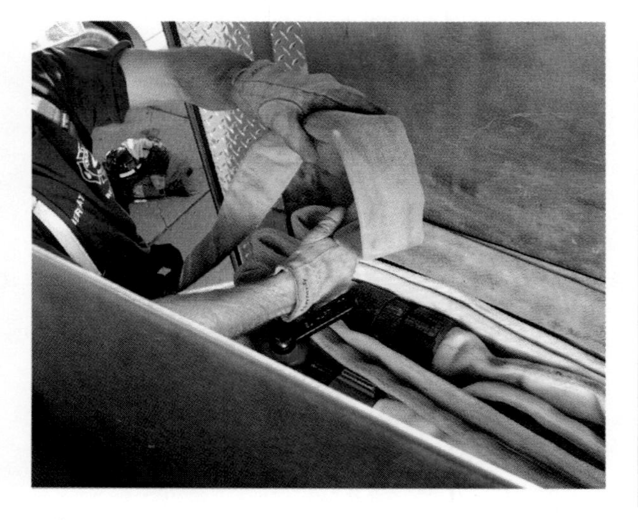

Step 5: Lay the hose back inside the previously laid length in the same manner, and continue until the entire hose length has been loaded.

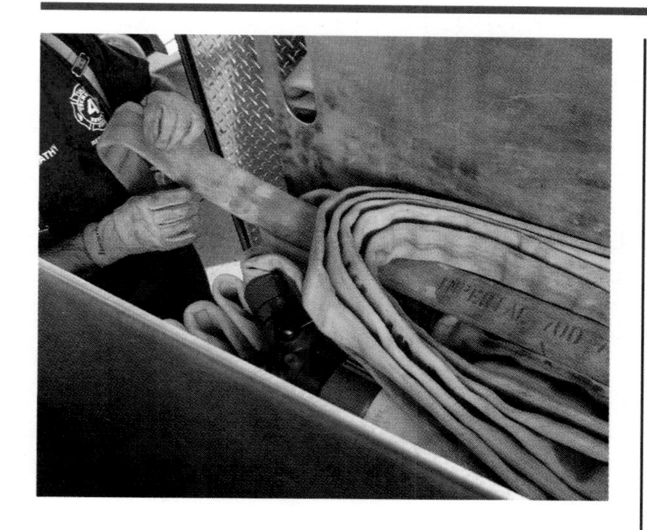

Step 6: Wrap the male end of the hose once around the horseshoe-shaped loops.

Step 7: Form a small loop by bringing the end back under the center of the loops, and then over the top.

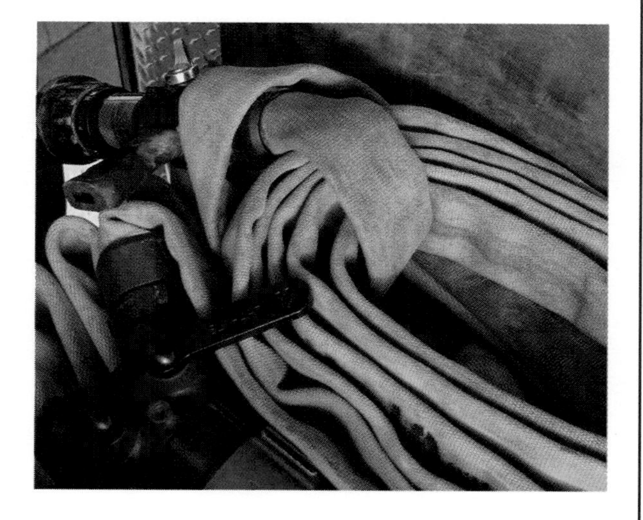

Step 8: Attach the nozzle, and place it inside the small loop.

Step 9: Pull the remaining slack hose back into the center of the horseshoe to tighten the loop against the nozzle.

Step 10 Load the second length of hose on the opposite side of the hose bed in the same manner.

Deploying Reverse Lays
Reverse Horseshoe Finish

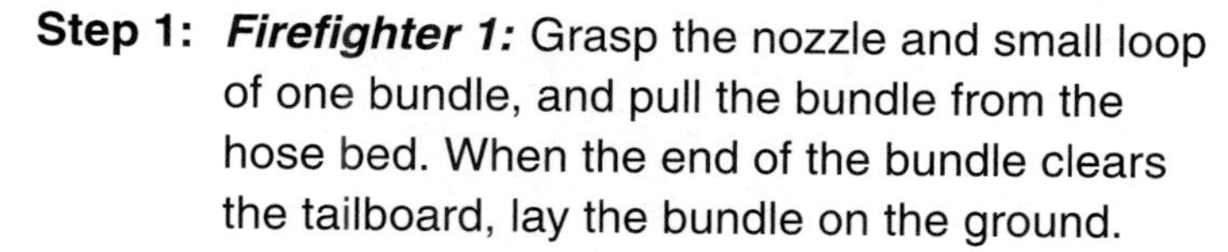

Step 1: ***Firefighter 1:*** Grasp the nozzle and small loop of one bundle, and pull the bundle from the hose bed. When the end of the bundle clears the tailboard, lay the bundle on the ground.

Step 2: ***Firefighter 1:*** Pull the opposite bundle in the same way.

Step 3: ***Firefighter 1:*** Pull the gated wye and attached hose from the bed, and place the wye between the bundles near the ties.

Step 4: ***Firefighter 1:*** Anchor the hose and wye, and signal the Driver/Operator to proceed to the hydrant.

Step 5: ***Driver/Operator:*** Proceed to the hydrant, deploying the hose.

Step 6: ***Firefighter 1:*** Continue to anchor the hose so that it deploys from the hose bed as the pumping apparatus proceeds toward the water source.

7-6 Skill Sheet

Step 7: ***Firefighter 1:*** Pick up one of the bundles, and advance toward the emergency scene, allowing the horseshoe loops to deploy one loop at a time until the entire bundle is in position.

Step 8: ***Firefighter 2:*** Deploy the second hose bundle in the same manner.

Finishing Reverse Lays — Skill Sheet 7-7

Skid-Load Finish

Step 1: Connect the first section of attack hose directly to the supply hose load.

Step 2: Place several folds, accordionlike, on edge across the front of the bed to provide some slack hose for pulling the attack hose finish from the hose bed.

Step 3: Turn the hose flat, and lay it to the rear of the hose bed approximately one-third of the way from one side.

Step 4: Form a loop that extends approximately 6 inches (152 mm) beyond the hose bed.

Step 5: Lay the hose back to the front of the hose bed, and then over to the other side of the bed approximately one-third of the way from the opposite side to form a supporting skid.

Step 6: Make the second skid in the same manner.

7-7 Skill Sheet

Step 7: Connect additional attack hose, and load it, accordionlike, back and forth across the two skids until all of the hose is loaded.

Step 8: Attach the nozzle, and place it on top of the load.

Deploying Reverse Lays Skid-Load Finish

Skill Sheet 7-8

Step 1: ***Firefighter 1:*** Grasp the small loops of each skid, and pull the finish from the bed. Maintain even tension on both skids to lower the finish from the hose bed to the ground.

Step 2 ***Firefighter 1:*** Pull the finish back from the tailboard approximately 15 feet (5 m).

Step 3: ***Firefighter 2:*** Place a hose clamp on the hose approximately 5 feet (1.5 m) from the first hose coupling on the supply hose coming from the hose bed.

NOTE: An inline quarter-turn valve or other water-controlling device can be placed at the beginning of the skid load and deployed as part of that load, eliminating the need for the hose clamp.

7-8 Skill Sheet

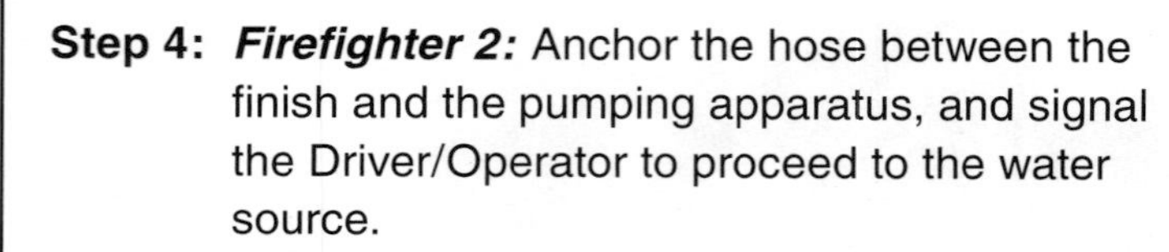

Step 4: ***Firefighter 2:*** Anchor the hose between the finish and the pumping apparatus, and signal the Driver/Operator to proceed to the water source.

Step 5: ***Driver/Operator:*** Proceed to the water source, deploying hose.

Step 6: ***Firefighter 1:*** Pick up the nozzle and advance the hose to the point where emergency operations are to begin.

Step 7: ***Firefighter 2:*** Open the clamp or inline gate when the emergency team is in position and ready for water.

Loading Preconnected Attack Hose — Skill Sheet 7-9

Reverse Hoseshoe Load: Large Attack Hose

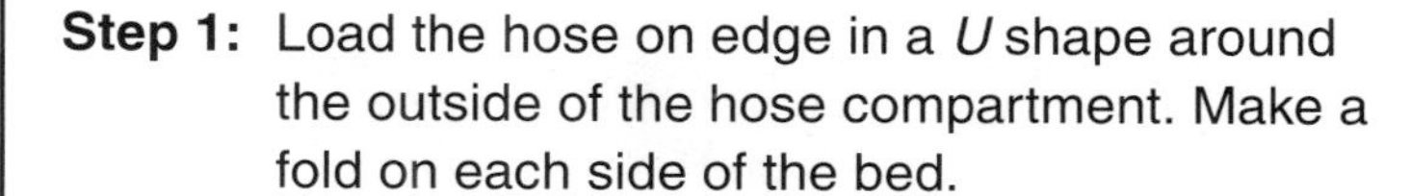

Step 1: Load the hose on edge in a *U* shape around the outside of the hose compartment. Make a fold on each side of the bed.

NOTE: These first folds provide slack when the load is pulled. The hose should go no farther than 5 feet (1.5 m) into the hose bed (adjust this length so that the hose will not touch the ground when carried over the shoulder).

Step 2: Form the first *U* of the horseshoe by placing the hose around the rear edge of the bed, and then return it to the front of the bed and make a fold.

Step 3: Place the next loop of hose inside the previously laid length in the same manner.

Step 4: Continue to build the horseshoe in this manner until the first tier is completely filled with hose.

Step 5: Load the second and succeeding tiers in the same manner as the first. Start each tier by making a fold of hose on either side of the hose bed.

NOTE: This step provides slack hose between each person when pulling the hose — one person per tier.

Step 6: Attach the nozzle when the desired amount of hose has been loaded, and place it in the center of the horseshoe loops.

Deploying Preconnected Attack Hose Reverse Horseshoe Load: Large Attack Hose

Skill Sheet 7-10

Step 1: ***Nozzle Firefighter:*** Grasp the nozzle with the outside hand while standing on the tailboard with one shoulder toward the load.

Step 2: ***Nozzle Firefighter:*** Bring the hose and nozzle behind the head and over the outside shoulder.

Step 3: ***Nozzle Firefighter:*** Pull the loops of the top tier far enough out of the bed to insert the nearest arm, and then place that arm through the loops (see photo).

Step 4: ***Nozzle Firefighter:*** Step down from the tailboard to pull the hose from the bed and walk away from the pumping apparatus until the slack hose between the tiers is fully extended.

7-10 Skill Sheet

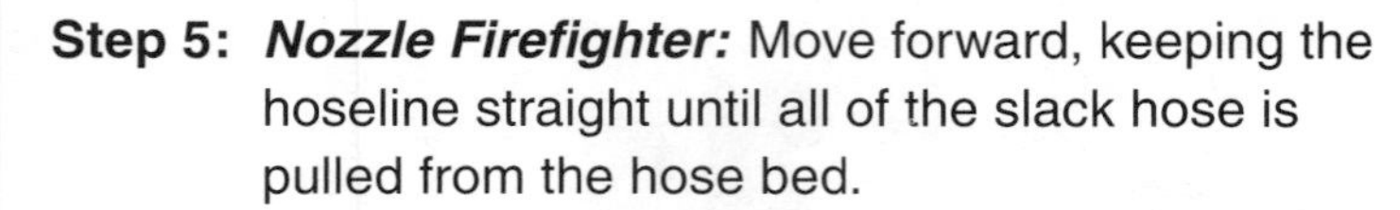

Step 5: ***Nozzle Firefighter:*** Move forward, keeping the hoseline straight until all of the slack hose is pulled from the hose bed.

NOTE: If deploying over 200 feet (60 m) of hose, each tier of hose should be deployed by all additional firefighters.

Step 6: ***All Firefighters:*** Advance toward the emergency scene when the load is completely pulled from the hose bed.

Step 7: ***Each Firefighter:*** Permit the hose to pay off the arm in succession, starting with the rear-most person. Lift each layer from the arm to aid in deploying the hose.

Loading Preconnected Attack Hose

Reverse Horseshoe Load: Small Attack Hose

Skill Sheet 

Step 1: Place the hose so that it lies flat down the middle of the hose bed, and then bring it up on edge along the sidewall of the hose compartment.

Step 2: Form the hose into a *U* shape just beyond the edge of the hose bed.

Step 3: Bring the hose to the front against the opposite side of the hose bed. Make a fold and bring the hose back to the rear of the bed.

7-11 Skill Sheet

Step 4: Place the hose around the inside of the previously laid length of hose in the same manner.

Step 5: Continue to build the hose horseshoe until the first tier is completely filled with hose.

Step 6: Start each new tier by bringing the hose up at the front of the bed (from the center of the horseshoe) and across to either sidewall (see photo). Load the second and succeeding tiers in the same manner as the first.

Step 7: Attach the nozzle, and place it in the center of the horseshoe loops.

7-12 Skill Sheet Deploying Preconnected Attack Hose Reverse Horseshoe Load: Small Attack Hose

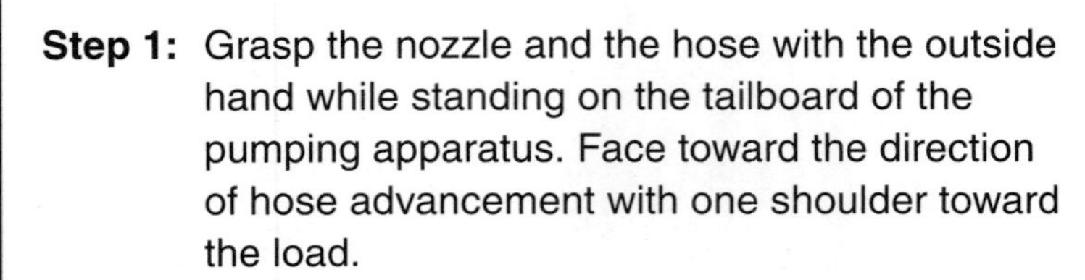

Step 1: Grasp the nozzle and the hose with the outside hand while standing on the tailboard of the pumping apparatus. Face toward the direction of hose advancement with one shoulder toward the load.

Step 2: Bring the hose and nozzle behind the head, and then over the outside shoulder so that the nozzle lies across the chest.

Step 3: Put the inside arm over the just-pulled length of hose, and pull the loops of the top tier far enough out of the bed to insert a forearm.

Step 4: Place a forearm through the loops and turn them so that they are perpendicular to the ground.

Step 5: Step down from the pumping apparatus tail-board to pull the hose from the hose bed, and begin walking away from the apparatus.

Step 6: Lift the outside loop of the horseshoe load away from the rest of the loops of hose.

Step 7: Drop the loop and reach for the next loop when the trailing hose tightens.

Step 8: Continue to pay off the hose in the same manner until the load is completely deployed.

Step 9: Throw each remaining loop off onto the ground in a separate place to eliminate kinking if a number of loops are still on the arm when arriving at the destination.

7-13 Skill Sheet

Loading Preconnected Attack Hose Flat Load

Step 1: Attach the female coupling of the attack hose to the discharge in the hose bed compartment.

Step 2: Lay the first length of hose flat in the hose bed.

Step 3: Angle the hose to lay the next fold adjacent to the first fold. Continue building additional tiers in this manner.

Step 4: Fold the hose at a point approximately one-third of the total length of the load. Form a loop in the hose that extends approximately 8 inches (203 mm) beyond the edge of the hose bed compartment. See loop closeup.

NOTE: This loop will later serve as a pull handle.

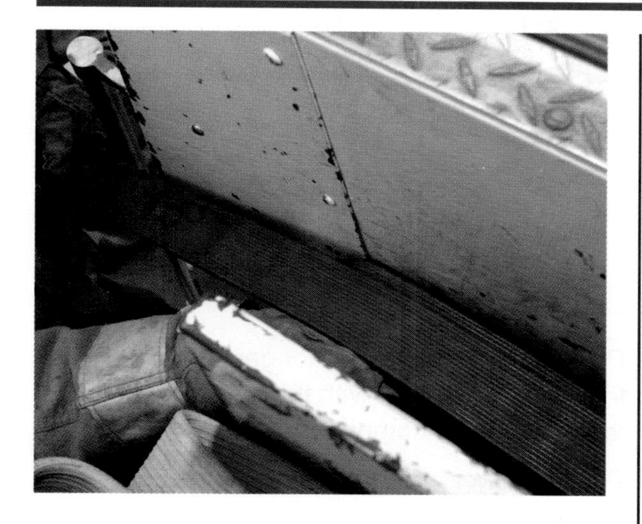

Step 5: Continue loading the hose in the bed as in Step 3, building each tier with folds evenly spaced across the bed.

Step 6: Make a second fold at a point approximately two-thirds of the total length of the load. Allow a 14-inch (356 mm) loop to hang over the rear edge of the hose bed. See loop closeup.

NOTE: This loop will also serve as a pull handle.

Step 7: Complete the load. Attach the nozzle, and place it in a location on the load that is easily reached.

7-14 Skill Sheet — Deploying Preconnected Attack Hose Flat Load

Step 1: Put one arm through the longer loop and grasp the shorter pull loop with the same hand while facing the hose compartment.

Step 2: Grasp the nozzle from the top of the hose bed with the opposite hand, ensuring that the nozzle is closed.

Step 3: Use the pull loops (using equal strength on each), and bring the attack hose out of the compartment to the ground.

Step 4: Advance the hose toward the emergency scene. Release the hand loop as the hose pulls taut in the hand.

Step 5: Continue to advance. Drop the second loop to the ground as the shoulder loop becomes taut.

Step 6: Advance until the hose is fully deployed.

7-15 Skill Sheet Loading Preconnected Attack Hose Triple-Layer Load

NOTE: Begin with the attack hose sections connected and stretched straight out and away from the hose bed with the nozzle attached.

Step 1: ***Firefighter 1:*** Connect the female coupling of the attack hose to the discharge in the hose compartment.

Step 2: ***Firefighter 1:*** Pick up the hose at a point two-thirds of the distance from the tailboard to the nozzle, carry it to the tailboard, and lay it back upon itself.

NOTE: This step forms three layers of hose when it is stacked one layer on top of the other with a fold at each end.

Step 3: ***Several Firefighters:*** Begin laying the hose into the hose bed by picking up the entire length of the three hose layers. Fold over the three hose layers as a unit and lay them into the hose bed.

Step 4: ***Several Firefighters:*** Fold the layers over at the front of the hose bed and lay them back to the rear on top of the previously laid hose. If the hose compartment is wider than one hose width, alternate folds on each side of the bed. Make all folds at the rear even with the edge of the hose bed.

Step 5: ***Several Firefighters:*** Continue to lay the hose into the hose bed in an *S*-shaped configuration until the entire length is loaded.

Step 6: ***Firefighter 1:*** Use a rope or strap to secure the nozzle to the first set of loops.

7-16 Skill Sheet — Deploying Preconnected Attack Hose Triple-Layer Load

Step 1: Face the hose bed, grasp the nozzle and folds of the first tier, and pull them from the compartment.

Step 2: Pivot as the hose slides free so that the nozzle and layer of hose are placed over the shoulder.

Step 3: Step down from the tailboard, and walk straight away from the rear of the pumping apparatus until the entire load is pulled from the hose bed compartment.

Step 4: Drop the folded end from the shoulder, and advance the nozzle to the emergency scene.

7-17 Skill Sheet

Loading Preconnected Attack Hose Minuteman Load

Step 1: Connect the first section of hose to the discharge in the hose compartment. Do not connect it to the other lengths of hose.

Step 2: Lay the hose flat in the hose bed to the front, and then lay the remaining hose out of the front of the bed to be loaded later.

NOTE: If the discharge is at the front of the hose bed, lay the hose to the rear of the bed, and then back to the front before it is set aside. This procedure provides slack hose for pulling the load clear of the hose bed.

Step 3: Couple the two remaining hose sections together, and attach a nozzle to the male end.

Step 4: Place the nozzle on top of the first length at the rear, angle the hose to the opposite side of the hose bed, and make a fold. Lay the hose back to the rear.

Step 5: Make a fold at the rear of the bed, angle the hose back to the other side, and make a fold at the front of the bed.

Step 6: Continue loading the hose in the same manner to alternating sides of the bed until the complete length is loaded.

Step 7: Connect the male coupling of the first section to the female coupling of the longer section.

7-17 Skill Sheet

Step 8: Lay the remainder of the first section in the hose bed in the same manner.

Deploying Preconnected Attack Hose Minuteman Load

Skill Sheet 7-18

Step 1: Grasp the nozzle, and pull the load halfway out of the hose bed.

Step 2: Face away from the apparatus, and place the load on the shoulder with the nozzle length on the bottom. Hold the stack firmly on the shoulder.

7-18 Skill Sheet

Step 3: Walk away from the pumping apparatus, and advance toward the emergency scene, pulling the remaining hose from the hose compartment.

Step 4: Allow the hose to deploy from the top of the stack, one fold at a time while advancing.

NOTE: This procedure prevents the hose from twisting or kinking.

Deploying Hose Packs Progessive Wildland

Step 1: ***Firefighters 1 and 2:*** Remove the wildland hose bag/pack from the pumping apparatus, and place on shoulders, allowing the hose weight to evenly distribute across the back.

Step 2: ***Firefighter 1:*** Connect or secure the female coupling of the hose to the pumping apparatus discharge outlet.

7-19 Skill Sheet

Step 3: ***Firefighter 1:*** Advance in the direction of the desired final deployment location.

Step 4: ***Firefighter 2:*** Assist with the deployment of hose from the pack of Firefighter 1.

Step 5: ***Firefighters 1 and 2:*** Connect a gated wye to the hoseline when the desired location is reached.

Step 6: ***Firefighter 2:*** Connect the second hose pack/ bag and continue extending the hoseline.

Alternate Step 6

Step 6: ***Firefighters 1 and 2:*** Deploy two attack lines from the gated wye and advance to the fire.

Step 7: ***Either Firefighter:*** Signal for the line to be charged once in position.

Appendices

Appendix A
Fire Hose Standards by Various Organizations

- **American National Standards Institute (ANSI)**
  1819 L Street, NW
  Suite 600
  Washington, DC 20036

ANSI/UL 92, *Fire Extinguisher and Booster Hose,* 1993.

- **American Society of Mechanical Engineers (ASME)**
  ASME International
  22 Law Drive
  Fairfield, NJ 07007-2900

B1.2-1983 (R2001) *Gages and Gaging for Unified Inch Screw Threads,* 1983, Revised 2001.

- **American Society for Testing and Materials (ASTM)**
  100 Barr Harbor Drive
  West Conshohocken, PA 19428-2959

ASTM B30-00e2, *Standard Specification for Copper Alloys in Ingot Form,* 1996.

ASTM B117-03, *Standard Practice for Operating Salt Spray (Fog) Apparatus,* 1995.

ASTM B584-00, *Standard Specification for Copper Alloy Sand Castings for General Applications,* 1996.

ASTM D380-94 (2000), *Standard Test Methods for Rubber Hose,* 1994, 2000.

ASTM D395-03, *Standard Test Methods for Rubber Property — Compression Set,* 1989.

ASTM D412-98a (2002) e1, *Standard Test Methods for Vulcanized Rubber and Thermoplastic Elastomers — Tension,* 1998, 2002.

ASTM D518-99, *Standard Test Method for Rubber Deterioration — Surface Cracking,* 1999.

ASTM D573-99, *Standard Test Method for Rubber — Deterioration in an Air Oven,* 1999.

ASTM D3183-02, *Standard Practice for Rubber — Preparation of Pieces for Test Purposes from Products,* 1984.

ASTM D380-94 (2000), *Standard Test Methods for Rubber Hose,* 2000.

ASTM D1149-99, *Standard Test Method for Rubber Deterioration-Surface Ozone Cracking in a Chamber.*

ASTM D1349-99, *Practice for Rubber-Standard Temperatures for Testing.*

ASTM D1415-88 (1999), *Standard Test Method for Rubber Property-International Hardness.*

ASTM D2240-03, *Standard Test Method for Rubber Property-Durometer Hardness.*

ASTM D3183-02, *Standard Practice for Rubber-Preparation of Pieces for Test Purposes from Products.*

ASTM D3767-01, *Standard Practice for Rubber-Measurement of Dimensions.*

ASTM D572-99, *Standard Test Method for Rubber-Deterioration by Heat and Oxygen.*

ASTM D573-99, *Standard Test Method for Rubber-Deterioration in an Air Oven.*

ASTM D865-99, *Standard Test Method for Rubber-Deterioration by Heating in Air (Test Tube Enclosure).*

ASTM E4-03, *Practices for Force Verification of Testing Machines.*

ASTM F1546/F1546M-96 (2001), *Standard Specification for Fire Hose Nozzles,* 2001.

ASTM F1121-87 (1998), *Standard Specification for International Shore Connections for Marine Fire Applications,* 1998.

ASTM F1546/F1546M-96 (2001) *Standard Specification for Fire Hose Nozzles,* 2001.

- **National Fire Protection Association (NFPA)**

  1 Batterymarch Park, P.O. Box 9101
  Quincy, MA 02269-9101

NFPA 1961, *Standard for Fire Hose* (2002).

NFPA 1962, *Standard for the Inspection, Care, and Use of Fire Hose, Couplings, and Nozzles and the Service Testing of Fire Hose* (2003).

NFPA 1963, *Standard for Fire Hose Connections* (2003).

NFPA 1964, *Standard for Spray Nozzles* (2003).

NFPA 1965, *Standard for Fire Hose Appliances* (2003).

NFPA 10, *Standard for Portable Fire Extinguishers* (2002).

NFPA 14, *Standard for the Installation of Standpipe, Private Hydrants and Hose Systems* (2003).

NFPA 24, *Standard for the Installation of Private Fire Service Mains and Their Appurtenances* (2002).

- **Underwriters Laboratories Inc. (UL)**

  333 Pfingsten Road
  Northbrook, IL 60062

UL 19, *Standard for Lined Fire Hose and Hose Assemblies,* 2001.

UL 219, *Standard for Lined Fire Hose for Interior Standpipes,* 1993.

UL 2167, *Standard for Water Mist Nozzles for Fire-Protection Service,* 2002.

UL 401.3, *Standard for Safety for Portable Spray Hose Nozzles for Fire-Protection Service,* 1993.

- **Underwriters Laboratories' of Canada (UL Canada)**

  7 Underwriters Road
  Toronto, ON, Canada
  M1R3B4

CAN/ULC-S515-1988, *Automobile Fire Fighting Apparatus,* 1988, Amended 2000.

ULC-S558-1995, *External-Lug, Quick-Connect Couplings and Adapters for Forestry Fire Hose,* 1995.

ULC-S551-1995, *Forged External-Lug, Quick-Connect Couplings and Adapters for Forestry Fire Hose,* 1995.

CAN/ULC-S503-1990, *Carbon Dioxide Hand and Wheeled Fire Extinguishers,* 1990, Amended 2002.

CAN/ULC-S504-2002, *Dry Chemical and Dry Powder Hand and Wheeled Fire Extinguishers,* 2002.

CAN/ULC-S514-1988, *Dry Chemical for Use in Hand and Wheeled Fire Extinguishers,* 1988.

CAN/ULC-S522-1986 (R 1999), *Fire Extinguisher and Booster Hose,* 1986, Revised 1999.

ULC-S554-1998, *Foam Fire Extinguishers,* 1998.

CAN/ULC-S507-1992 9 *Litre, Stored Pressure, Water-Type Fire Extinguishers,* 1992, Amended 2002.

CAN/ULC-S532-1990, *Regulation of the Servicing of Portable Fire Extinguishers,* 1990, Amended 2002.

CAN4-S543-1984, *Internal-Lug, Quick Connect Couplings for Fire Hose,* 1984.

CAN/ULC-S523-1991, *Light Attack Fire Fighting Apparatus (Mini Pumper),* 1991.

ORD-C822, *Listing and Labelling of Fire Department Pumpers,* 1982.

CAN/ULC-S518-1990, *Non-Percolating Forestry Hose,* 1990.

ULC-S518.1-1995, *Synthetic Non-Percolating Forestry Hose,* 1995.

CAN/ULC-S519-1990, *Percolating Forestry Hose,* 1990.

ULC-S519.1-1995, *Synthetic Percolating Forestry Hose,* 1995.

ULC-S511-1999, *Lined Fire Hose for Interior Standpipes and Municipal and Industrial Fire Protection Service,* 1999.

ULC-S513-1978, *Threaded Couplings for 1½- and 2½-Inch Fire Hose,* 1978.

ORD-C262, *Valves, Gate, for Fire Protection Service,* 1992.

ORD-C668, *Valves, Hose for Fire Protection Service,* 1975.

- **Factory Mutual (FM)**

  FM Approvals, an affiliate of FM Global Insurance Company
  1151 Boston-Providence Turnpike
  P. O. Box 9102
  Norwood, MA 02060

FM 1140, *Quick Opening Valves ¼-Inch Through 2-Inch Nominal Size,* 1998.

FM 1411, *Playpipes,* 1970.

FM 1515, *Hose Manifold,* 1970.

FM 1521, *Angle Hose Valves,* 1970.

FM 1522, *Straightway Hose Valve,* 1970.

FM 1530, *Fire Department Connections,* 1970.

FM 1531, *Wall Hydrants,* 1977.

FM 2111, *Fire Hose,* 1999.

FM 2121, *Unlined Linen Fire Hose,* 1970.

FM 2131, *Couplings for 1.5-Inch and 2.5-Inch Single- and Double-Woven Jacket Lined Fire Hose and Unlined Linen Fire Hose,* 1970.

FM 2141, *Hose Racks and Reels for Lined Lightweight and Unlined Fire Hose,* 1970.

FM 2151, *Hose Houses and Outdoor Hose Cabinets,* 1970.

- **U.S. Department of Agriculture (USDA)/U.S Forest Service (USFS)**

  United States Forest Service (USFS)
  San Dimas Technology and Development Center
  San Dimas, CA 91773
  (Specifications available from U.S. General Services Administration)

GSA-FS 5100-186c, *Fire Hose, Lined, Woven Jacket, 1-Inch and 1½-Inch*, 2003.

GSA-FS 5100-187b, *Lined, Woven Jacket, 1 Inch and 1½ Inch*, 2003.

GSA-FS 5100-190b, *Threads, Gaskets, Rocker Lugs, Connections and Fittings, etc.*, 1996.

GSA-FS 5100-185e, *Hose, Rubber, High Pressure, ¾ Inch Waterway*, 1997.

GSA-FS 5100-238c, *Shutoff, Valve, Ball*, 2000.

GSA-FS 5100-184c, *Class 4210, Suction Hose*. 1997.

GSA-FS 5100-245c, *Clamp, Shutoff, Fire Hose*. 1996.

GSA-FS 5100-108e, *Couplings, Lightweight, Fire and Suction Hose*, 1996.

GSA-FS 5100-243c, *Nozzle, Fire Hose, Mop-Up*, 1996.

GSA-FS 5100-105d, *Strainer, Suction Hose*, 1997.

GSA-FS 5100-102d, *Couplings, Fire and Suction Hose*, 1998.

GSA-FS 5100-240d, *Nozzle, Twin Tip, Shutoff, 1 Inch Base*, 2000.

GSA-FS 5100-239c, *Nozzle with Shutoff, Combination Barrel*, 2000.

GSA-FS 5100-340b, *Reel, Hose, Booster*, 1997.

GSA-FS 5100-241d, *Nozzle, Shutoff*, 2000.

GSA-FS 5100-382c, *Valve, Check and Bleeder*, 1997.

GSA-FS 5100-244d, *Nozzle Tips, Straight Stream and Spray*, 2000.

GSA-FS 5100-101c, *Wrenches, Spanner, Fire Hose*, 1997.

GSA-FS 5100-242d, *Nozzle, Screw Tip, 1½ Inch Inlet*, 1998.

GSA-FS 5100-107c, *Fire Hose Connections and Fittings*, 2000.

GSA-FS 5100-380d, *Valve, Wye*, 1996.

- **Rubber Manufacturers Association**

  1400 K Street, NW, Suite 900
  Washington, DC 20005

SP-710, *Product Description and Quality Control Report for Acrylic Elastomers*, Third Edition, 1999.

SP-410, *Product Description and Quality Control Report for Adhesives*, 1995.

SP-610, *Product Description and Quality Control Report for Carbon Black*, Second Edition, 1992.

SP-130, *Product Description and Quality Control Report for Emulsion Styrene Butadiene Copolymers (SBR)*, Second Edition, 1999.

SP-810, *Product Description and Quality Control Report for Ethylene Propylene Elastomers (EPDM)*, Second Edition, 1999.

SP-510, *Product Description and Quality Control Report for Fluorocarbon Elastomers*, Fourth Edition, 1999.

SP-110, *Product Description and Quality Control Report for Hydrogenated Acrylonitrile Butadiene (HNBR) Elastomers*, Second Edition, 1999.

SP-150, *Product Description and Quality Control Report for Liquid Plasticizers for Rubber Compounds*, Second Edition, 1999.

SP-241, *Product Description and Quality Control Report for NBR Elastomers*, Fourth Edition, 2001.

SP-310, *Product Description and Quality Control Report for Polychloroprenes*, 1992.

SP-210, *Product Description and Quality Control Report for Rubber Compounding Chemicals*, Second Edition, 1999.

SP-910, *Product Description and Quality Control Report for Silicone and Fluorosilicone Bases and Compounds*, 1991.

- **U.S. Federal Specifications, available from:**

  Naval Publications and Forms Center
  5801 Tabor Avenue, Attn: NPODS
  Philadelphia, PA 19120-5094

A-A 59226 (ZZ-H-452), *Hose Assembly, Nonmetallic, Fire Fighting, with Couplings*, May 18, 1998.

ZZ-H-421, *Chemical Engine Booster Hose.*

A-A-59566 (ZZ-H-561), *Water Suction and Discharge Hose.*

MIL-H-24606, *Hose, Fire, Synthetic Fiber, Double Jacketed, Treated for Abrasion Resistance, with Couplings, Fire Fighting and Other Water Service*, Revision B, August 12, 1994.

MIL-PRF53207, *Hose Assembly, Rubber: Lightweight Collapsible, 6-inch; for Drinking (Potable) Water*, Revision B, March 14, 1996.

NSF 61, *Drinking Water System Components — Health Effects, or 21 CFR, Food and Drugs*, Sec. 177.2600, "Rubber articles intended for repeated use."

Appendix B
Fire Hose Record Forms

Annual Hose Test Record

______________________________ Fire Department

Test Date: _____ / _____ / _____ Test Supervisor: ____________________

ID #	Location	Size	Condition	Last Tested	Manufacturer	Remarks

Hose Record Card

ID No. ______ City ______ Engine Co. ______

Size (dia.) ______ Length ______ Type hose ______ Const. ______ Cost ______

Mfg. ______ Part no. ______ Date rcd. ______ Date in service ______

Vendor ______ Cplg. mfg. ______ Type cplgs. ______

Repairs

Date	Kind	Cost	New length	New ID. no.

Remarks: ______

Test Record

Service test to ______ psi

Date	Service test pres. psi	Test OK	Reason failed	Date	Service test pres. psi	Test OK	Reason failed

Date	Exposed to possible damage		Date	Reason
		Removed From service		
		Condemmed		
		Sold		
		Wrnty, failure		

Example showing front and back of a hose record card. *Courtesy of Scandia Industries, Inc.*

Hose Repair Tag

Hose to be repaired must be tagged

ID number:	Company number
Picked up by:	Date picked up
Delivered by:	Date delivered

Repairs Needed:

______

______

______

Repairs Made:

______

______

______

Repaired by:	Date repaired
Service tested ______ psi	Date tested

☐ Hose is not repaired.
ID no. of replacement hose

➡ ______

Note: This tag must be filled out and returned with hose.
➡ Enter repairs on hose record card.

****

Example of a hose repair tag. *Courtesy of Memphis (TN) Fire Department.*

Appendix C
Foam Properties

Table C.1 Foam Properties					
Type	**Characteristics**	**Storage Range**	**Application Rate**	**Application Techniques**	**Primary Uses**
Protein Foam (3% and 6%)	• Protein based • Low expansion • Good reignition (burnback) resistance • Excellent water retention • High heat resistance and stability • Performance can be affected by freezing and thawing • Can freeze protect with antifreeze • Not as mobile or fluid on fuel surface as other low-expansion foams	35–120°F (2°C to 49°C)	0.16 gpm/ft^2 (6.5 L/min/m^2)	• Indirect foam stream; do not mix fuel with foam • Avoid agitating fuel during application; static spark ignition of volatile hydrocarbons can result from plunging and turbulence • Use alcohol-resistant type within seconds of proportioning • Not compatible with dry chemical extinguishing agents	• Class B fires involving hydrocarbons • Protecting flammable and combustable liquids where they are stored, transported, and processed
Fluoroprotein Foam (3% and 6%)	• Protein and synthetic based; derived from protein foam • Fuel shedding • Long-term vapor suppression • Good water retention • Excellent, long-lasting heat resistance • Performance not affected by freezing and thawing • Maintains low viscosity at low temperatures • Can freeze protect with antifreeze • Use either freshwater or saltwater • Nontoxic and biodegradable after dilution • Good mobility and fluidity on fuel surface • Premixable for short periods of time	35–120°F (2°C to 49°C)	0.16 gpm/ft^2 (6.5 L/min/m^2)	• Direct plunge technique • Subsurface injection • Compatible with simultaneous application of dry chemical extinguishing agents • Deliver through air-aspirating equipment	• Hydrocarbon vapor suppression • Subsurface application to hydrocarbon fuel storage tanks • Extinguishing in-depth crude petroleum or other hydrocarbon fuel fires

Continued on next page.

Table C.1 (continued)

Type	Characteristics	Storage Range	Application Rate	Application Techniques	Primary Uses
Film Forming Fluoroprotein Foam (FFFP) (3% and 6%)	• Protein based; fortified with additional surfactants that reduce the burnback characteristics of other protein-based foams • Fuel shedding • Develops a fast-healing, continuous-floating film on hydrocarbon fuel surfaces • Excellent, long-lasting heat resistance • Good low-temperature viscosity • Fast fire knockdown • Affected by freezing and thawing • Use either freshwater or saltwater • Can store premixed • Can freeze protect with antifreeze • Use alcohol-resistant type on polar solvents at 6% solution and on hydrocarbon fuels at 3% solution • Nontoxic and biodegradable after dilution	35–120°F (2°C to 49°C)	**Ignited Hydrocarbon Fuel:** 0.10 gpm/ft² (4.1 L/min/m²) **Polar Solvent Fuel:** 0.24 gpm/ft² (9.8 L/min/m²)	• Cover entire fuel surface • May apply with dry chemical agents • May apply with spray nozzles • Subsurface injection • Can plunge into fuel during application	• Suppressing vapors in unignited spills of hazardous liquids • Extinguishing fires in hydrocarbon fuels
Aqueous Film Forming Foam (AFFF) (1%, 3%, and 6%)	• Synthtic based • Good penetrating capabilities • Spreads vapor-sealing film over and floats on hydrocarbon fuels • Can use nonaerating nozzles • Performance may be adversely affected by freezing and storing • Has good low-temperature viscosity • Can freeze protect with antifreeze • Use either freshwater or saltwater • Can premix	35–120°F (-4°C to 49°C)	0.10 gpm/ft² (4.1 L/min/m²)	• May apply directly onto fuel surface • May apply indirectly by bouncing it off a wall and allowing it to float onto fuel surface • Subsurface injection • May apply with dry chemical agents	• Controlling and extinguishing Class B fires • Handling land or sea crash rescues involving spills • Extinguishing most transportation-related fires • Wetting and penetrating Class A fuels • Securing unignited hydrocarbon spills

Continued on next page.

Table C.1 (continued)

Type	Characteristics	Storage Range	Application Rate	Application Techniques	Primary Uses
Alcohol-Resistant AFFF (3% and 6%)	• Polymer has been added to AFFF concentrate • Multipurpose: Use on both polar solvents and hydrocarbon fuels (use on polar solvents at 6% solution and on hydrocar-bon fuels at 3% solution) • Forms a membrane on polar solvent fuels that prevents destruction of the foam blanket • Forms same aqueous film on hydrocarbon fuels as AFFF • Fast flame knockdown • Good burnback resistance on both fuels • Not easily premixed	25–120°F (-4°C to 49°C) (May become viscous at temperatures under 50°F [10°C])	**Ignited Hydrocarbon Fuel:** 0.10 gpm/ft^2 (4.1 L/min/m^2) **Polar Solvent Fuel:** 0.24 gpm/ft^2 (9.8 L/min/m^2)	• Apply directly but gently onto fuel surface • May apply indirectly by bouncing it off a wall and allowing it to float onto fuel surface • Subsurface injection	Fires or spills of both hydrocarbon and polar solvent fuels
High Expansion Foam	• Synthetic detergent based • Special-purpose, low water content • High air-to-solution ratios: 200:1 to 1,000:1 • Performance not affected by freezing and thawing • Poor heat resistance • Prolonged contact wiith galvanized or raw steel may attack these surfaces	27–110°F (-3°C to 43°C)	Sufficient to quickly cover the fuel or fill the space	• Gentle application; do not mix foam with fuel • Cover entire fuel surface • Usually fills entire space in confined space incidents	• Extinguishing Class A and some Class B fires • Flooding confined spaces • Volumetrically displacing vapor, heat, and smoke • Reducing vaporization from liquefied natural gas spills • Extinguishing pesticide fires • Supressing fum-ing acid vapors • Supressing vapors in coal mines and other subterranean spaces and con-cealed spaces in basements • Extinguishing agent in fixed extinguishing systems • Not recom-mended for outdoor use

Continued on next page.

Table C.1 (continued)

Type	Characteristics	Storage Range	Application Rate	Application Techniques	Primary Uses
Class A Foam	• Synthetic • Wetting agent that reduces surface tension of water and allows it to soak into combustible materials • Rapid extinguishment with less water use than other foams • Use regular water stream equipment • Can premix with water • Mildly corrosive • Requires lower percentage of concentration (0.2 to 1.0) than other foams • Outstanding insulating qualities • Good penetrating capabilities	25–120°F (-4°C to 49°C) (Concentrate is subject to freezing but can be thawed and used if freezing occurs)	Same as the minimum critical flow rate for plain water on similar Class A Fuels; flow rates are not reduced when using Class A foam	• Can propel with compressed-air systems • Can apply with conventional nozzles	Extinguishing Class A combustibles only

Appendix D
IFSTA Friction Loss Calculations (English System)

$$FL = CQ^2L \quad (A)$$

where FL = Friction loss in psi

C = Friction loss coefficient (from Table D.1)

Q = Flow rate in hundreds of gallons/minute (flow/100)

L = Hose length in hundreds of feet (length/100)

Table D.1 contains the commonly used friction loss coefficients for the various sizes of hose in use today. These coefficients are used by IFSTA and NPFA. They are only approximations of the friction loss for each size hose. The actual coefficients for any particular piece of hose vary with the condition of the hose and the manufacturer. When using the coefficients provided in this appendix, the results reflect a worst-case situation: The results are probably slightly higher than the actual friction loss. To require more exact friction loss calculations, obtain different coefficients from the manufacturer of the fire hose or make actual calculations.

Table D.1
Friction Loss Coefficient — Single Hoselines

Hose Diameter and Type (inches)	Coefficient C
Booster ¾	1,100
Booster: 1	150
Booster 1¼	80
1½	24
1¾ (with 1½-inch couplings)	15.5
2	8
2½	2
3 (with 2½-inch couplings)	0.8
3 (with 3-inch couplings)	0.677
3½	0.34
4	0.2
4½	0.1
5	0.08
6	0.05
Standpipes: 4	0.374
Standpipes: 5	0.126
Standpipes: 6	0.052

The steps for determining friction loss using Equation A are as follows:

Step 1: Obtain from Table D.1 the friction loss coefficient for the hose being used.

Step 2: Determine the number of hundreds of gallons of water per minute flowing (Q) through the hose by using Q = gpm/l00.

Step 3: Determine the number of hundreds of feet of hose (L) by using L = feet/100.

Step 4: Insert the numbers from Steps 1, 2, and 3 into Equation A to determine the total friction loss.

Table D.2
¾-Inch Rubber Hose
C-Factor: 1,100

Lay Length (feet)	Flow in gpm 20	30	40	50	60
	Friction Loss in psi				
50	22	50	88	138	198
100	44	99	176		
150	66	149			
200	88	198			
250	110				
300	132				
350	154				
400	176				

Table D.3
1-Inch Rubber Hose
C-Factor: 150

Lay Length (feet)	Flow in gpm 20	30	40	50	60
	Friction Loss in psi				
50	3	7	12	19	27
100	6	14	24	38	54
150	9	20	36	56	81
200	12	27	48	75	108
250	15	34	60	94	135
300	18	41	72	113	162
350	21	47	84	131	189
400	24	54	96	150	

Table D.4
1½-Inch Hose
C-Factor: 24

Lay Length (feet)	Flow in gpm 40	60	80	95	125	150
	Friction Loss in psi					
50	2	4	8	11	19	27
100	4	9	15	22	38	54
150	6	13	23	32	56	81
200	8	17	31	43	75	108
250	10	22	38	54	94	135
300	12	26	46	65	113	162
350	13	30	54	76	131	189
400	15	35	61	87	150	216
450	17	39	69	97	169	
500	19	43	77	108	188	

Table D.5
1¾-Inch Rubber Hose
C-Factor: 15.5

Lay Length (feet)	Flow in gpm				
	95	125	150	175	200
	Friction Loss in psi				
50	7	12	17	24	31
100	14	24	35	47	62
150	21	36	52	71	93
200	28	48	70	95	124
250	35	61	87	119	155
300	42	73	105	142	186
350	49	85	122	166	
400	56	97	140	190	
450	63	109	157		
500	70	121	174		

Table D.6
2-Inch Rubber Hose
C-Factor: 8

Lay Length (feet)	Flow in gpm				
	100	125	150	175	200
	Friction Loss in psi				
50	4	6	9	12	16
100	8	13	18	25	32
150	12	19	27	37	48
200	16	25	36	49	64
250	20	31	45	61	80
300	24	38	54	74	96
350	28	44	63	86	112
400	32	50	72	98	128
450	36	56	81	110	144
500	40	63	90	123	160

Table D.7
2½-Inch Hose
C-Factor: 2

	Flow in gpm						
	175	**200**	**225**	**250**	**275**	**300**	**350**
Lay Length (feet)	**Friction Loss in psi**						
50	3	4	5	6	8	9	12
100	6	8	10	13	15	18	25
150	9	12	15	19	23	27	37
200	12	16	20	25	30	36	49
250	15	20	25	31	38	45	61
300	18	24	30	38	45	54	74
350	21	28	35	44	53	63	86
400	25	32	41	50	61	72	98
450	28	36	46	56	68	81	110
500	31	40	51	63	76	90	123
550	34	44	56	69	83	99	135
600	37	48	61	75	91	108	147
650	40	52	66	81	98	117	159
700	43	56	71	88	106	126	172
750	46	60	76	94	113	135	184
800	49	64	81	100	121	144	196
850	52	68	86	106	129	153	208
900	55	72	91	113	136	162	221
950	58	76	96	119	144	171	233
1,000	61	80	101	125	151	180	245
1,050	64	84	106	131	159	189	
1,100	67	88	111	138	166	198	
1,150	70	92	116	144	174	207	
1,200	74	96	122	150	182	216	
1,250	77	100	127	156	189	225	
1,300	80	104	132	163	197	234	
1,350	83	108	137	169	204	243	
1,400	86	112	142	175	212		
1,450	89	116	147	181	219		
1,500	92	120	152	188	227		
1,550	95	124	157	194	234		
1,600	98	128	162	200	242		
1,650	101	132	167	206	250		
1,700	104	136	172	213			
1,750	107	140	177	219			
1,800	110	144	182	225			
1,850	113	148	187	231			
1,900	116	152	192	238			
1,950	119	156	197	244			
2,000	123	160	203	250			

Table D.8
3-Inch Hose
C-Factor: 0.8

	Flow in gpm					
	250	**325**	**500**	**750**	**1,000**	**1,250**
Lay Length (feet)	**Friction Loss in psi**					
50	3	4	10	23	40	63
100	5	8	20	45	80	125
150	8	13	30	68	120	188
200	10	17	40	90	160	250
250	13	21	50	113	200	
300	15	25	60	135	240	
350	18	30	70	158		
400	20	34	80	180		
450	23	38	90	203		
500	25	42	100	225		
550	28	46	110	248		
600	30	51	120			
650	33	55	130			
700	35	59	140			
750	38	63	150			
800	40	68	160			
850	43	72	170			
900	45	76	180			
950	48	80	190			
1,000	50	85	200			
1,050	53	89	210			
1,100	55	93	220			
1,150	58	97	230			
1,200	60	101	240			
1,250	63	106	250			
1,300	65	110				
1,350	68	114				
1,400	70	118				
1,450	73	123				
1,500	75	127				
1,550	78	131				
1,600	80	135				
1,650	83	139				
1,700	85	144				
1,750	88	148				
1,800	90	152				
1,850	93	156				
1,900	95	161				
1,950	98	165				
2,000	100	169				

Table D.9
3-Inch Hose with 3-Inch Couplings
C-Factor: 0.677

	Flow in gpm				
	250	350	500	750	1,000
Lay Length (feet)	Friction Loss in psi				
50	2	4	8	19	34
100	4	8	17	38	68
150	6	12	25	57	102
200	8	17	34	76	135
250	11	21	42	95	169
300	13	25	51	114	
350	15	29	59	133	
400	17	33	68	152	
450	19	37	76	171	
500	21	41	85		
550	23	46	93		
600	25	50	102		
650	28	54	110		
700	30	58	118		
750	32	62	127		
800	34	66	135		
850	36	70	144		
900	38	75	152		
950	40	79	161		
1,000	42	83	169		
1,050	44	87	178		
1,100	47	91	186		
1,150	49	95			
1,200	51	100			
1,250	53	104			
1,300	55	108			
1,350	57	112			
1,400	59	116			
1,450	61	120			
1,500	63	124			
1,550	66	129			
1,600	68	133			
1,650	70	137			
1,700	72	141			
1,750	74				
1,800	76				
1,850	78				
1,900	80				
1,950	83				
2,000	85				

Table D.10
4-Inch Hose
C-Factor: 0.2

	Flow in gpm					
	500	**750**	**1,000**	**1,250**	**1,500**	**1,750**
Lay Length (feet)	**Friction Loss in psi**					
50	3	6	10	16	23	31
100	5	11	20	31	45	61
150	8	17	30	47	68	92
200	10	23	40	63	90	123
250	13	28	50	78	113	153
300	15	34	60	94	135	184
350	18	39	70	109	158	
400	20	45	80	125	180	
450	23	51	90	141		
500	25	56	100	156		
550	28	62	110	172		
600	30	68	120			
650	33	73	130			
700	35	79	140			
750	38	84	150			
800	40	90	160			
850	43	96	170			
900	45	101	180			
950	48	107				
1,000	50	113				
1,050	53	118				
1,100	55	124				
1,150	58	129				
1,200	60	135				
1,250	63	141				
1,300	65	146				
1,350	68	152				
1,400	70	158				
1,450	73	163				
1,500	75	169				
1,550	78	174				
1,600	80	180				
1,650	83					
1,700	85					
1,750	88					
1,800	90					
1,850	93					
1,900	95					
1,950	98					
2,000	100					

Table D.11
4½-Inch Hose
C-Factor: 0.1

	Flow in gpm					
	500	**750**	**1,000**	**1,250**	**1,500**	**1,750**
Lay Length (feet)	**Friction Loss in psi**					
50	1	3	5	8	11	15
100	3	6	10	16	23	31
150	4	8	15	23	34	46
200	5	11	20	31	45	61
250	6	14	25	39	56	77
300	8	17	30	47	68	92
350	9	20	35	55	79	107
400	10	23	40	63	90	123
450	11	25	45	70	101	138
500	13	28	50	78	113	153
550	14	31	55	86	124	168
600	15	34	60	94	135	184
650	16	37	65	102	146	
700	18	39	70	109	158	
750	19	42	75	117	169	
800	20	45	80	125	180	
850	21	48	85	133		
900	23	51	90	141		
950	24	53	95	148		
1,000	25	56	100	156		
1,050	26	59	105	164		
1,100	28	62	110	172		
1,150	29	65	115	180		
1,200	30	68	120			
1,250	31	70	125			
1,300	33	73	130			
1,350	34	76	135			
1,400	35	79	140			
1,450	36	82	145			
1,500	38	84	150			
1,550	39	87	155			
1,600	40	90	160			
1,650	41	93	165			
1,700	43	96	170			
1,750	44	98	175			
1,800	45	101	180			
1,850	46	104	185			
1,900	48	107				
1,950	49	110				
2,000	50	113				

Table D.12
5-Inch Hose
C-Factor: 0.08

Lay Length (feet)	Flow in gpm: 750	1,000	1,250	1,500	1,750	2,000
	Friction Loss in psi					
50	2	4	6	9	12	16
100	5	8	13	18	25	32
150	7	12	19	27	37	48
200	9	16	25	36	49	64
250	11	20	31	45	61	80
300	14	24	38	54	74	96
350	16	28	44	63	86	112
400	18	32	50	72	98	128
450	20	36	56	81	110	144
500	23	40	63	90	123	160
550	25	44	69	99	135	176
600	27	48	75	108	147	192
650	29	52	81	117	159	208
700	32	56	88	126	172	224
750	34	60	94	135	184	240
800	36	64	100	144	196	
850	38	68	106	153	208	
900	41	72	113	162	221	
950	43	76	119	171	233	
1,000	45	80	125	180	245	
1,050	47	84	131	189		
1,100	50	88	138	198		
1,150	52	92	144	207		
1,200	54	96	150	216		
1,250	56	100	156	225		
1,300	59	104	163	234		
1,350	61	108	169	243		
1,400	63	112	175	252		
1,450	65	116	181	261		
1,500	68	120	188			
1,550	70	124	194			
1,600	72	128	200			
1,650	74	132	206			
1,700	77	136	213			
1,750	79	140	219			
1,800	81	144	225			
1,850	83	148	231			
1,900	86	152	238			
1,950	88	156	244			
2,000	90	160	250			

Table D.13
Two 2½-Inch Hose
C-Factor: 0.5

Lay Length (feet)	500	750	1,000	1,250	1,500	1,750
	Flow in gpm					
	Friction Loss in psi					
50	6	14	25	39	56	77
100	13	28	50	78	113	153
150	19	42	75	117	169	230
200	25	56	100	156	225	
250	31	70	125	195		
300	38	84	150	234		
350	44	98	175			
400	50	113	200			
450	56	127	225			
500	63	141	250			
550	69	155				
600	75	169				
650	81	183				
700	88	197				
750	94	211				
800	100	225				
850	106	239				
900	113					
950	119					
1,000	125					
1,050	131					
1,100	138					
1,150	144					
1,200	150					
1,250	156					
1,300	163					
1,350	169					
1,400	175					
1,450	181					
1,500	188					
1,550	194					
1,600	200					
1,650	206					
1,700	213					
1,750	219					
1,800	225					
1,850	231					
1,900	238					
1,950	244					
2,000	250					

Table D. 14
Three 2½-Inch Hose
C-Factor: 0.22

Lay Length (feet)	Flow in gpm: 250	500	750	1,000	1,250	1,500
	Friction Loss in psi					
50	1	3	6	11	17	25
100	1	6	12	22	34	50
150	2	8	19	33	52	74
200	3	11	25	44	69	99
250	3	14	31	55	86	124
300	4	17	37	66	103	149
350	5	19	43	77	120	173
400	6	22	50	88	138	198
450	6	25	56	99	155	
500	7	28	62	110	172	
550	8	30	68	121	189	
600	8	33	74	132		
650	9	36	80	143		
700	10	39	87	154		
750	10	41	93	165		
800	11	44	99	176		
850	12	47	105	187		
900	12	50	111	198		
950	13	52	118			
1,000	14	55	124			
1,050	14	58	130			
1,100	15	61	136			
1,150	16	63	142			
1,200	17	66	149			
1,250	17	69	155			
1,300	18	72	161			
1,350	19	74	167			
1,400	19	77	173			
1,450	20	80	179			
1,500	21	83	186			
1,550	21	85	192			
1,600	22	88	198			
1,650	23	91				
1,700	23	94				
1,750	24	96				
1,800	25	99				
1,850	25	102				
1,900	26	105				
1,950	27	107				
2,000	28	110				

Table D.15
One 3-inch and One 2½-Inch Hose
C-Factor: 0.3

	Flow in gpm					
	500	750	1,000	1,250	1,500	1,750
Lay Length (feet)	Friction Loss in psi					
50	4	8	15	23	34	46
100	8	17	30	47	68	92
150	11	25	45	70	101	138
200	15	34	60	94	135	184
250	19	42	75	117	169	
300	23	51	90	141	203	
350	26	59	105	164		
400	30	68	120	188		
450	34	76	135			
500	38	84	150			
550	41	93	165			
600	45	101	180			
650	49	110	195			
700	53	118				
750	56	127				
800	60	135				
850	64	143				
900	68	152				
950	71	160				
1,000	75	169				
1,050	79	177				
1,100	83	186				
1,150	86	194				
1,200	90	203				
1,250	94					
1,300	98					
1,350	101					
1,400	105					
1,450	109					
1,500	113					
1,550	116					
1,600	120					
1,650	124					
1,700	128					
1,750	131					
1,800	135					
1,850	139					
1,900	143					
1,950	146					
2,000	150					

Table D.16
Two 2½-Inch and One 3-Inch Hose
C-Factor: 0.16

	Flow in gpm					
	500	**750**	**1,000**	**1,250**	**1,500**	**1,750**
Lay Length (feet)	**Friction Loss in psi**					
50	2	5	8	13	18	25
100	4	9	16	25	36	49
150	6	14	24	38	54	74
200	8	18	32	50	72	98
250	10	23	40	63	90	123
300	12	27	48	75	108	147
350	14	32	56	88	126	172
400	16	36	64	100	144	196
450	18	41	72	113	162	
500	20	45	80	125	180	
550	22	50	88	138	198	
600	24	54	96	150		
650	26	59	104	163		
700	28	63	112	175		
750	30	68	120	188		
800	32	72	128	200		
850	34	77	136			
900	36	81	144			
950	38	86	152			
1,000	40	90	160			
1,050	42	95	168			
1,100	44	99	176			
1,150	46	104	184			
1,200	48	108	192			
1,250	50	113	200			
1,300	52	117				
1,350	54	122				
1,400	56	126				
1,450	58	131				
1,500	60	135				
1,550	62	140				
1,600	64	144				
1,650	66	149				
1,700	68	153				
1,750	70	158				
1,800	72	162				
1,850	74	167				
1,900	76	171				
1,950	78	176				
2,000	80	180				

Table D.17
Two 3-Inch Hose
C-Factor: 0.2

	Flow in gpm					
	500	**750**	**1,000**	**1,250**	**1,500**	**1,750**
Lay Length (feet)	**Friction Loss in psi**					
50	3	6	10	16	23	31
100	5	11	20	31	45	61
150	8	17	30	47	68	92
200	10	23	40	63	90	123
250	13	28	50	78	113	153
300	15	34	60	94	135	184
350	18	39	70	109	158	
400	20	45	80	125	180	
450	23	51	90	141		
500	25	56	100	156		
550	28	62	110	172		
600	30	68	120	188		
650	33	73	130	203		
700	35	79	140			
750	38	84	150			
800	40	90	160			
850	43	96	170			
900	45	101	180			
950	48	107	190			
1,000	50	113	200			
1,050	53	118				
1,100	55	124				
1,150	58	129				
1,200	60	135				
1,250	63	141				
1,300	65	146				
1,350	68	152				
1,400	70	158				
1,450	73	163				
1,500	75	169				
1,550	78	174				
1,600	80	180				
1,650	83	186				
1,700	85	191				
1,750	88	197				
1,800	90	203				
1,850	93					
1,900	95					
1,950	98					
2,000	100					

Table D.18
Two 3-Inch and One 2½-Inch Hose
C-Factor: 0.12

	Flow in gpm					
	500	**750**	**1,000**	**1,250**	**1,500**	**1,750**
Lay Length (feet)	**Friction Loss in psi**					
50	2	3	6	9	14	18
100	3	7	12	19	27	37
150	5	10	18	28	41	55
200	6	14	24	38	54	74
250	8	17	30	47	68	92
300	9	20	36	56	81	110
350	11	24	42	66	95	129
400	12	27	48	75	108	147
450	14	30	54	84	122	
500	15	34	60	94	135	
550	17	37	66	103	149	
600	18	41	72	113		
650	20	44	78	122		
700	21	47	84	131		
750	23	51	90	141		
800	24	54	96	150		
850	26	57	102			
900	27	61	108			
950	29	64	114			
1,000	30	68	120			
1,050	32	71	126			
1,100	33	74	132			
1,150	35	78	138			
1,200	36	81	144			
1,250	38	84	150			
1,300	39	88				
1,350	41	91				
1,400	42	95				
1,450	44	98				
1,500	45	101				
1,550	47	105				
1,600	48	108				
1,650	50	111				
1,700	51	115				
1,750	53	118				
1,800	54	122				
1,850	56	125				
1,900	57	128				
1,950	59	132				
2,000	60	135				

Appendix E
IFSTA Friction Loss Calculations (International System)

$$FL = CQ^2L \qquad (A)$$

where FL = Friction loss in kPa

C = Friction loss coefficient (from Table E.1)

Q = Flow rate in hundreds of liters/minute (flow/100)

L = Hose length in hundreds of meters (length/100)

Table E.1 contains the commonly used friction loss coefficients for the various sizes of hose in use today. These coefficients are used by IFSTA and NPFA. They are only approximations of the friction loss for each size hose. The actual coefficients for any particular piece of hose vary with the condition of the hose and the manufacturer. When using the coefficients provided in this appendix, the results reflect a worst-case situation: The results are probably slightly higher than the actual friction loss. To require more exact friction loss calculations, obtain different coefficients from the manufacturer of the fire hose or make actual calculations.

Table E.1
Friction Loss Coefficient — Single Hoselines

Hose Diameter and Type (mm)	Coefficient C
Booster: 20	1 741
Booster: 25	238
Booster: 32	127
38	38
45	24.6
50	12.7
65	3.17
70 (with 77 mm couplings)	2.36
77 (with 65 mm couplings)	1.27
77 (with 77 mm couplings)	1.06
90	0.53
100	0.305
115	0.167
125	0.138
150	0.083
Standpipes: 100	0.600
Standpipes: 125	0.202
Standpipes: 150	0.083

The steps for determining friction loss using Equation A are as follows:

Step 1: Obtain from Table E.1 the friction loss coefficient for the hose being used.

Step 2: Determine the number of hundreds of liters of water per minute flowing (Q) through the hose by using Q = liters/min/l00.

Step 3: Determine the number of hundreds of meters of hose (L) by using L = meters/100.

Step 4: Insert the numbers from Steps 1, 2, and 3 into Equation A to determine the total friction loss.

Table E.2
20 mm Rubber Hose
C-Factor: 1 741

Lay Length (meters)	Flow in L/min: 80	120	160	200	240
	Friction Loss in kPa				
15	167	376	669	1 045	1 504
30	334	752	1 337		
45	501	1 128			
60	669	1 504			
75	836				
90	1 003				
105	1 170				
120	1 337				

Table E.3
25 mm Rubber Hose
C-Factor: 238

Lay Length (meters)	Flow in L/min: 80	120	160	200	240
	Friction Loss in kPa				
15	23	51	91	143	206
30	46	103	183	286	411
45	69	154	274	428	617
60	91	206	366	571	823
75	114	257	457	714	1 028
90	137	308	548	857	1 234
105	160	360	640	1 000	1 439
120	183	411	731	1 142	1 645

Table E.4
38 mm Hose
C-Factor: 38

Lay Length (meters)	Flow in L/min: 240	320	400	500	600
	Friction Loss in kPa				
15	33	58	91	143	205
30	66	117	182	285	410
45	98	175	274	428	616
60	131	233	365	570	821
75	164	292	456	713	1 026
90	197	350	547	855	1 231
105	230	409	638	998	
120	263	467	730	1 140	
135	295	525	821		
150	328	584	912		

Table E.5
45 mm Hose
C-Factor: 24.6

Lay Length (meters)	Flow in L/min				
	380	**500**	**600**	**700**	**800**
	Friction Loss in kPa				
15	53	92	133	181	236
30	107	185	266	362	472
45	160	277	399	542	708
60	213	369	531	723	945
75	266	461	664	904	1 181
90	320	554	797	1 085	1 417
105	373	646	930	1 266	1 653
120	426	738	1 063	1 446	
135	480	830	1 196	1 627	
150	533	923	1 328		

Table E.6
50 mm Hose
C-Factor: 12.7

Lay Length (meters)	Flow in L/min				
	400	**500**	**600**	**700**	**800**
	Friction Loss in kPa				
15	30	48	69	93	122
30	61	95	137	187	244
45	91	143	206	280	366
60	122	191	274	373	488
75	152	238	343	467	610
90	183	286	411	560	732
105	213	333	480	653	853
120	244	381	549	747	975
135	274	429	617	840	1 097
150	305	476	686	933	1 219

Table E.7
65 mm Hose
C-Factor: 3.17

Lay Length (meters)	Flow in L/min: 700	800	900	1 000	1 100	1 200	1 400
	Friction Loss in kPa						
15	23	30	39	48	58	68	93
30	47	61	77	95	115	137	186
45	70	91	116	143	173	205	280
60	93	122	154	190	230	274	373
75	116	152	193	238	288	342	466
90	140	183	231	285	345	411	559
105	163	213	270	333	403	479	652
120	186	243	308	380	460	548	746
135	210	274	347	428	518	616	839
150	233	304	385	476	575	685	932
165	256	335	424	523	633	753	1 025
180	280	365	462	571	690	822	1 118
195	303	396	501	618	748	890	1 212
210	326	426	539	666	805	959	1 305
225	349	456	578	713	863	1 027	1 398
240	373	487	616	761	921	1 096	1 491
255	396	517	655	808	978	1 164	1 584
270	419	548	693	856	1 036	1 232	1 678
285	443	578	732	903	1 093	1 301	
300	466	609	770	951	1 151	1 369	
315	489	639	809	999	1 208	1 438	
330	513	670	847	1 046	1 266	1 506	
345	536	700	886	1 094	1 323	1 575	
360	559	730	924	1 141	1 381	1 643	
375	582	761	963	1 189	1 438	1 712	
390	606	791	1 001	1 236	1 496		
405	629	822	1 040	1 284	1 553		
420	652	852	1 078	1 331	1 611		
435	676	883	1 117	1 379	1 669		
450	699	913	1 155	1 427			
465	722	943	1 194	1 474			
480	746	974	1 232	1 522			
495	769	1 004	1 271	1 569			
510	792	1 035	1 310	1 617			
525	815	1 065	1 348	1 664			
540	839	1 096	1 387	1 712			
555	862	1 126	1 425				
570	885	1 156	1 464				
585	909	1 187	1 502				
600	932	1 217	1 541				

Table E.8
77 mm Hose with 65 mm Couplings
C-Factor: 1.27

	Flow in L/min					
	1 000	**1 200**	**1 400**	**2 000**	**3 000**	**4 000**
Lay Length (meters)	**Friction Loss in kPa**					
15	19	27	37	76	171	305
30	38	55	75	152	343	610
45	57	82	112	229	514	914
60	76	110	149	305	686	1 219
75	95	137	187	381	857	1 524
90	114	165	224	457	1 029	
105	133	192	261	533	1 200	
120	152	219	299	610	1 372	
135	171	247	336	686	1 543	
150	191	274	373	762	1 715	
165	210	302	411	838		
180	229	329	448	914		
195	248	357	485	991		
210	267	384	523	1 067		
225	286	411	560	1 143		
240	305	439	597	1 219		
255	324	466	635	1 295		
270	343	494	672	1 372		
285	362	521	709	1 448		
300	381	549	747	1 524		
315	400	576	784	1 600		
330	419	604	821	1 676		
345	438	631	859			
360	457	658	896			
375	476	686	933			
390	495	713	971			
405	514	741	1 008			
420	533	768	1 045			
435	552	796	1 083			
450	572	823	1 120			
465	591	850	1 157			
480	610	878	1 195			
495	629	905	1 232			
510	648	933	1 269			
525	667	960	1 307			
540	686	988	1 344			
555	705	1 015	1 382			
570	724	1 042	1 419			
585	743	1 070	1 456			
600	762	1 097	1 494			

Table E.9
77 mm Hose with 77 mm Couplings
C-Factor: 1.06

	Flow in L/min					
	1 000	1 200	1 400	2 000	3 000	4 000
Lay Length (meters)	Friction Loss in kPa					
15	16	23	31	64	143	254
30	32	46	62	127	286	509
45	48	69	93	191	429	763
60	64	92	125	254	572	1 018
75	80	114	156	318	716	1 272
90	95	137	187	382	859	1 526
105	111	160	218	445	1 002	
120	127	183	249	509	1 145	
135	143	206	280	572	1 288	
150	159	229	312	636	1 431	
165	175	252	343	700	1 574	
180	191	275	374	763	1 717	
195	207	298	405	827		
210	223	321	436	890		
225	239	343	467	954		
240	254	366	499	1 018		
255	270	389	530	1 081		
270	286	412	561	1 145		
285	302	435	592	1 208		
300	318	458	623	1 272		
315	334	481	654	1 336		
330	350	504	686	1 399		
345	366	527	717	1 463		
360	382	550	748	1 526		
375	398	572	779	1 590		
390	413	595	810	1 654		
405	429	618	841	1 717		
420	445	641	873			
435	461	664	904			
450	477	687	935			
465	493	710	966			
480	509	733	997			
495	525	756	1 028			
510	541	778	1 060			
525	557	801	1 091			
540	572	824	1 122			
555	588	847	1 153			
570	604	870	1 184			
585	620	893	1 215			
600	636	916	1 247			

Table E.10
100 mm Hose
C-Factor: 0.305

	Flow in L/min					
	2 000	**3 000**	**4 000**	**5 000**	**6 000**	**7 000**
Lay Length (meters)	**Friction Loss in kPa**					
15	18	41	73	114	165	224
30	37	82	146	229	329	448
45	55	124	220	343	494	673
60	73	165	293	458	659	897
75	92	206	366	572	824	1 121
90	110	247	439	686	988	
105	128	288	512	801	1 153	
120	146	329	586	915		
135	165	371	659	1 029		
150	183	412	732	1 144		
165	201	453	805	1 258		
180	220	494	878			
195	238	535	952			
210	256	576	1 025			
225	275	618	1 098			
240	293	659	1 171			
255	311	700	1 244			
270	329	741				
285	348	782				
300	366	824				
315	384	865				
330	403	906				
345	421	947				
360	439	988				
375	458	1 029				
390	476	1 071				
405	494	1 112				
420	512	1 153				
435	531	1 194				
450	549	1 235				
465	567	1 276				
480	586					
495	604					
510	622					
525	641					
540	659					
555	677					
570	695					
585	714					
600	732					

Table E.11
115 mm Hose
C-Factor: 0.167

	Flow in L/min					
	2 000	**3 000**	**4 000**	**5 000**	**6 000**	**7 000**
Lay Length (meters)	**Friction Loss in kPa**					
15	10	23	40	63	90	123
30	20	45	80	125	180	245
45	30	68	120	188	271	368
60	40	90	160	251	361	491
75	50	113	200	313	451	614
90	60	135	240	376	541	736
105	70	158	281	438	631	859
120	80	180	321	501	721	982
135	90	203	361	564	812	1 105
150	100	225	401	626	902	1 227
165	110	248	441	689	992	
180	120	271	481	752	1 082	
195	130	293	521	814	1 172	
210	140	316	561	877	1 263	
225	150	338	601	939		
240	160	361	641	1 002		
255	170	383	681	1 065		
270	180	406	721	1 127		
285	190	428	762	1 190		
300	200	451	802	1 253		
315	210	473	842			
330	220	496	882			
345	230	519	922			
360	240	541	962			
375	251	564	1 002			
390	261	586	1 042			
405	271	609	1 082			
420	281	631	1 122			
435	291	654	1 162			
450	301	676	1 202			
465	311	699	1 242			
480	321	721				
495	331	744				
510	341	767				
525	351	789				
540	361	812				
555	371	834				
570	381	857				
585	391	879				
600	401	902				

Table E.12
125 mm Hose
C-Factor: 0.138

	Flow in L/min					
	3 000	4 000	5 000	6 000	7 000	8 000
Lay Length (meters)	Friction Loss in kPa					
15	19	33	52	75	101	132
30	37	66	104	149	203	265
45	56	99	155	224	304	397
60	75	132	207	298	406	530
75	93	166	259	373	507	662
90	112	199	311	447	609	795
105	130	232	362	522	710	927
120	149	265	414	596	811	1 060
135	168	298	466	671	913	1 192
150	186	331	518	745	1 014	
165	205	364	569	820	1 116	
180	224	397	621	894	1 217	
195	242	431	673	969		
210	261	464	725	1 043		
225	279	497	776	1 118		
240	298	530	828	1 192		
255	317	563	880	1 267		
270	335	596	932			
285	354	629	983			
300	373	662	1 035			
315	391	696	1 087			
330	410	729	1 139			
345	428	762	1 190			
360	447	795	1 242			
375	466	828				
390	484	861				
405	503	894				
420	522	927				
435	540	960				
450	559	994				
465	578	1 027				
480	596	1 060				
495	615	1 093				
510	633	1 126				
525	652	1 159				
540	671	1 192				
555	689	1 225				
570	708	1 259				
585	727					
600	745					

Table E.13
Two 65 mm Hose
C-Factor: 0.789

Lay Length (meters)	Flow in L/min: 1 000	2 000	3 000	4 000	5 000	6 000
	Friction Loss in kPa					
15	12	47	107	189	296	426
30	24	95	213	379	592	852
45	36	142	320	568	888	1 278
60	47	189	426	757	1 184	1 704
75	59	237	533	947	1 479	
90	71	284	639	1 136	1 775	
105	83	331	746	1 326		
120	95	379	852	1 515		
135	107	426	959	1 704		
150	118	473	1 065			
165	130	521	1 172			
180	142	568	1 278			
195	154	615	1 385			
210	166	663	1 491			
225	178	710	1 598			
240	189	757	1 704			
255	201	805				
270	213	852				
285	225	899				
300	237	947				
315	249	994				
330	260	1 041				
345	272	1 089				
360	284	1 136				
375	296	1 184				
390	308	1 231				
405	320	1 278				
420	331	1 326				
435	343	1 373				
450	355	1 420				
465	367	1 468				
480	379	1 515				
495	391	1 562				
510	402	1 610				
525	414	1 657				
540	426	1 704				
555	438	1 752				
570	450					
585	462					
600	473					

Table E.14
Three 65 mm Hose
C-Factor: 0.347

	Flow in L/min					
	1 000	2 000	3 000	4 000	5 000	6 000
Lay Length (meters)	Friction Loss in kPa					
15	5	21	47	83	130	187
30	10	42	94	167	260	375
45	16	62	141	250	390	562
60	21	83	187	333	520	750
75	26	104	234	416	651	937
90	31	125	281	500	781	1 124
105	36	146	328	583	911	1 312
120	42	167	375	666	1 041	1 499
135	47	187	422	750	1 171	1 686
150	52	208	468	833	1 301	
165	57	229	515	916	1 431	
180	62	250	562	999	1 561	
195	68	271	609	1 083	1 692	
210	73	291	656	1 166		
225	78	312	703	1 249		
240	83	333	750	1 332		
255	88	354	796	1 416		
270	94	375	843	1 499		
285	99	396	890	1 582		
300	104	416	937	1 666		
315	109	437	984	1 749		
330	115	458	1 031			
345	120	479	1 077			
360	125	500	1 124			
375	130	520	1 171			
390	135	541	1 218			
405	141	562	1 265			
420	146	583	1 312			
435	151	604	1 359			
450	156	625	1 405			
465	161	645	1 452			
480	167	666	1 499			
495	172	687	1 546			
510	177	708	1 593			
525	182	729	1 640			
540	187	750	1 686			
555	193	770				
570	198	791				
585	203	812				
600	208	833				

Table E.15
One 77 mm and One 65 mm Hose
C-Factor: 0.473

	Flow in L/min					
	2 000	**3 000**	**4 000**	**5 000**	**6 000**	**7 000**
Lay Length (meters)	**Friction Loss in kPa**					
15	28	54	102	192	364	688
30	57	107	203	384	727	
45	85	161	305	577	1 091	
60	114	215	406	769	1 455	
75	142	268	508	961		
90	170	322	610	1 153		
105	199	376	711	1 345		
120	227	430	813	1 538		
135	255	483	914			
150	284	537	1 016			
165	312	591	1 117			
180	341	644	1 219			
195	369	698	1 321			
210	397	752	1 422			
225	426	805	1 524			
240	454	859	1 625			
255	482	913				
270	511	967				
285	539	1 020				
300	568	1 074				
315	596	1 128				
330	624	1 181				
345	653	1 235				
360	681	1 289				
375	710	1 342				
390	738	1 396				
405	766	1 450				
420	795	1 503				
435	823	1 557				
450	851	1 611				
465	880	1 665				
480	908	1 718				
495	937					
510	965					
525	993					
540	1 022					
555	1 050					
570	1 078					
585	1 107					
600	1 135					

Table E.16
Two 65 mm and One 77 mm Hose
C-Factor: 0.253

	Flow in L/min					
	2 000	3 000	4 000	5 000	6 000	7 000
Lay Length (meters)	Friction Loss in kPa					
15	15	34	61	95	137	186
30	30	68	121	190	273	372
45	46	102	182	285	410	558
60	61	137	243	380	546	744
75	76	171	304	474	683	930
90	91	205	364	569	820	1 116
105	106	239	425	664	956	1 302
120	121	273	486	759	1 093	1 488
135	137	307	546	854	1 230	1 674
150	152	342	607	949	1 366	
165	167	376	668	1 044	1 503	
180	182	410	729	1 139	1 639	
195	197	444	789	1 233		
210	213	478	850	1 328		
225	228	512	911	1 423		
240	243	546	972	1 518		
255	258	581	1 032	1 613		
270	273	615	1 093	1 708		
285	288	649	1 154			
300	304	683	1 214			
315	319	717	1 275			
330	334	751	1 336			
345	349	786	1 397			
360	364	820	1 457			
375	380	854	1 518			
390	395	888	1 579			
405	410	922	1 639			
420	425	956	1 700			
435	440	990				
450	455	1 025				
465	471	1 059				
480	486	1 093				
495	501	1 127				
510	516	1 161				
525	531	1 195				
540	546	1 230				
555	562	1 264				
570	577	1 298				
585	592	1 332				
600	607	1 366				

Table E.17
Two 77 mm Hose
C-Factor: 0.316

	Flow in L/min					
	2 000	**3 000**	**4 000**	**5 000**	**6 000**	**7 000**
Lay Length (meters)	**Friction Loss in kPa**					
15	19	43	76	119	171	232
30	38	85	152	237	341	465
45	57	128	228	356	512	697
60	76	171	303	474	683	929
75	95	213	379	593	853	1 161
90	114	256	455	711	1 024	
105	133	299	531	830	1 194	
120	152	341	607	948		
135	171	384	683	1 067		
150	190	427	758	1 185		
165	209	469	834			
180	228	512	910			
195	246	555	986			
210	265	597	1 062			
225	284	640	1 138			
240	303	683	1 213			
255	322	725				
270	341	768				
285	360	811				
300	379	853				
315	398	896				
330	417	939				
345	436	981				
360	455	1 024				
375	474	1 067				
390	493	1 109				
405	512	1 152				
420	531	1 194				
435	550	1 237				
450	569					
465	588					
480	607					
495	626					
510	645					
525	664					
540	683					
555	702					
570	720					
585	739					
600	758					

Table E.18
Two 77 mm and One 65 mm Hose
C-Factor: 0.189

	Flow in L/min					
	2 000	3 000	4 000	5 000	6 000	7 000
Lay Length (meters)	Friction Loss in kPa					
15	11	26	45	71	102	139
30	23	51	91	142	204	278
45	34	77	136	213	306	417
60	45	102	181	284	408	556
75	57	128	227	354	510	695
90	68	153	272	425	612	833
105	79	179	318	496	714	972
120	91	204	363	567	816	1 111
135	102	230	408	638	919	1 250
150	113	255	454	709	1 021	1 389
165	125	281	499	780	1 123	1 528
180	136	306	544	851	1 225	1 667
195	147	332	590	921	1 327	
210	159	357	635	992	1 429	
225	170	383	680	1 063	1 531	
240	181	408	726	1 134	1 633	
255	193	434	771	1 205	1 735	
270	204	459	816	1 276		
285	215	485	862	1 347		
300	227	510	907	1 418		
315	238	536	953	1 488		
330	249	561	998	1 559		
345	261	587	1 043	1 630		
360	272	612	1 089	1 701		
375	284	638	1 134			
390	295	663	1 179			
405	306	689	1 225			
420	318	714	1 270			
435	329	740	1 315			
450	340	765	1 361			
465	352	791	1 406			
480	363	816	1 452			
495	374	842	1 497			
510	386	868	1 542			
525	397	893	1 588			
540	408	919	1 633			
555	420	944	1 678			
570	431	970	1 724			
585	442	995				
600	454	1 021				

Glossary/Index

Glossary

A

Acceptance Test — *See* Proof Test.

Accessibility — Ability of pumping apparatus to get close enough to a structure or an emergency scene to deploy hose and conduct emergency operations.

Accident — Unplanned, uncontrolled event that results from unsafe acts of people and/or unsafe occupational conditions, either of which can result in injury.

Accordion Load — Arrangement of fire hose in a hose bed or compartment where the hose lies on edge with the folds adjacent to each other (accordion shape).

Adapter — Fitting for connecting hose couplings with dissimilar threads but with the same inside diameter. Also see Fitting, Increaser, and Reducer.

Adjustable Flow Nozzle — Nozzle designed so that the amount of water flowing through the nozzle can be increased or decreased at the nozzle; usually accomplished by adjusting the pattern of the stream. *Also see* Nozzle.

Aerial Apparatus — Fire fighting vehicle equipped with a hydraulically operated ladder, elevating platform or other similar device for the purpose of placing personnel and/or water streams in elevated positions. *Also see* Aerial Ladder.

Aerial Ladder — Power-operated (usually hydraulically) ladder mounted on a special truck chassis. *Also see* Aerial Apparatus.

Apparatus Bay (Apparatus Room) — Area of a fire station where fire apparatus are parked.

Appliance — Device (other than a coupling) that is used with hose and through which water must pass. The term is applied to any valve, nozzle, fitting, wye, siamese, deluge monitor, or other piece of hardware used in conjunction with fire hose for the purpose of delivering water. *Also see* Valve, Valve Device, and Fitting.

Applicator Pipe — Curved pipe attached to a nozzle for applying water precisely over a burning object.

Aspirating Foam Nozzle — Nozzle designed for the application of all types of foam concentrates; may be air-aspirating (air is introduced into the foam stream at the nozzle and draws in foam solution by the Venturi process) or water-aspirating (draws air into the back of the nozzle and forces the foam solution through a screen to create finished foam). *Also see* Nonaspirating Foam Nozzle, Nozzle, and Venturi Principle.

Aspiration — Adding air to a foam solution as it is discharged from a nozzle; also called *aeration*.

Attack Hose — Fire hose used by trained fire and emergency service responders to combat fires primarily beyond the incipient stage; fire hose between the attack pumping apparatus and the nozzles; also, any fire hose used in a handline to control and extinguish fire. Minimum size: 1½ inches (38 mm).

Automatic Hydrant Valve — Valve device that, when connected to a hydrant, opens automatically to permit water to flow into the supply line; may be mechanically operated or radio controlled. *Also see* Valve Device.

Automatic Nozzle — Nozzle that automatically adjusts itself to provide a proper stream at the proper nozzle pressure. *Also see* Nozzle.

B

Back Flushing — Cleaning of a fire pump or piping by flowing water through it in the opposite direction of normal flow.

Ball Valve — Valve having a ball-shaped internal component with a hole through its center that permits water to flow through when aligned with the waterway. *Also see* Valve.

Banding Method — Attaching a coupling to a fire hose with tightly wound strands of narrow-gauge wire or steel bands.

Blunt Start — *See* Higbee Cut.

Booster (Fire) Hose — Noncollapsible, preconnected hose having an elastomeric or thermoplastic tube, a braided or spiraled reinforcement, and an outer protective cover; used to extinguish low-intensity or small incipient-stage exterior fires; often carried on pumping apparatus on reels. Sizes: ¾ to 1½ inches (20 mm to 38 mm). *Also see* Booster Reel.

Booster Reel — Motorized or hand-powered reel on which booster hose is carried; reels are mounted on the front, rear, or deck above the pump of apparatus that are equipped with pumps or on skid units of wildland apparatus. *Also see* Booster (Fire) Hose.

Broken Stream — Stream of water that has been disrupted into coarsely divided drops.

Burst Test — Destructive test performed by the manufacturer on a 3-foot (1 m) length of fire hose to determine its maximum strength.

Butterfly Valve — Type of control valve that uses a flat baffle or circular plate operated by a quarter-turn handle to control water flow. *Also see* Valve.

C

Calendering — Fire hose inner-tube manufacturing process in which rubber is pressed between opposing rollers to produce a flat sheet; lapping and bonding together the edges of the sized sheet forms a tube.

Calibrate — To standardize or adjust the increments on a measuring instrument.

Capacity — Maximum ability of a pump or water distribution system to deliver water.

Cast Coupling — Coupling manufactured by a process in which molten metal is poured into a mold, allowed to cool, and then the mold is removed from the hardened coupling. *Also see* Coupling.

Cellar Nozzle/Applicator — Special nozzle for attacking fires in basements, cellars, and other spaces belowground; applicator tube can be attached.

C-factor — Factor that indicates the roughness of the inner surface of a fire hose.

Chain Hose Tool — Tool used to carry, secure, and otherwise aid in handling fire hose. *Also see* Hose Tool.

Charge — To pressurize a fire hose or fire extinguisher.

Charged Hoseline — Hoseline loaded with extinguishing agent under pressure and ready for use.

Check Valve — Automatic valve that permits liquid flow in only one direction. *Also see* Valve.

Clapper Valve — Hinged valve that permits the flow of water in one direction only. *Also see* Valve.

Class A Foam Concentrate — Foam fire-suppression agent that is designed for use on Class A combustible fires; essentially a wetting agent that reduces the surface tension of water and allow it to soak into combustible materials more readily than plain water. *Also see* Foam Concentrate.

Class B Foam Concentrate — Foam fire-suppression agent that is designed for use on unignited or ignited Class B flammable or combustible liquids. *Also see* Foam Concentrate.

Collar Method — Attaching a coupling to a fire hose with a two- or three-piece collar that is bolted into place.

Combination Lay — Fire hose lay in which two or more different-sized hoselines are deployed by one pumping apparatus in either direction: water source to fire or fire to water source. *Also see* Hose Lay (1, 2).

Combination Nozzle — Nozzle designed to provide a straight stream or fog stream. *Also see* Nozzle.

Coupling — Fitting permanently attached to the end of a fire hose; used to connect two hoselines together or a hoseline to such devices as nozzles, appliances, discharge valves, valve devices, or hydrants.

Curing — Manufacturing step in making fire hose; the process of applying heat and pressure to "set" the shape of the tube and to increase its smoothness.

D

Dead-end Hydrant — Fire hydrant that receives water from only one direction. *Also see* Hydrant.

Discharge Velocity — Rate at which water travels from an orifice.

Distributor Nozzle — Blunt-ended nozzle with numerous orifices set at various angles; usually lowered through holes in floors into basements or other confined spaces; also referred to as a *Bresnan distributor. Also see* Nozzle.

Donut Roll — Length of fire hose rolled for storage or transport where both couplings are on the outside of the roll; variations include twin donut and self-locking twin donut rolls.

Draft — Process of acquiring water from a static source and transferring it into a pump that is above the source's level; atmospheric pressure on the water surface forces the water into the pump where a partial vacuum was created.

Driver/Operator — Person who operates a fire apparatus to, during, and from the scene of a fire operation or any other time the apparatus is in use; also person responsible for routine maintenance of the apparatus and any equipment it carries; also called *engineer.*

Drop-Forged Coupling — Coupling made by raising and dropping a drop hammer onto a block of metal as it rests on a forging die, thus forming the metal into the desired shape. *Also see* Coupling.

Dry-barrel Hydrant — Fire hydrant that has no water in the barrel of the hydrant when it is not in use; its opening valve is at the main rather than in the barrel of the hydrant; used in areas where freezing weather conditions could occur. *Also see* Hydrant.

Dual Lay — One of any number of ways to deploy multiple hoselines. *Also see* Hose Lay (1, 2).

Dual Pumping — Operation where a strong (high-pressure) hydrant is used to supply two pumping apparatus by connecting them intake-to-intake; the second apparatus receives the excess water not being pumped by the first apparatus, which is directly connected to the water supply source; sometimes incorrectly referred to as *tandem pumping.*

Dutchman — Extra fold placed along the length of a section of fire hose as it is loaded onto the pumping apparatus hose bed so that the nearest coupling will deploy without lodging between the hose lengths.

E

Eductor — Portable proportioning device that injects a liquid such as foam concentrate into the water flowing through a hoseline. *Also see* Proportioner and In-line Proportioner.

Elbow — Fitting designed with a 30- to 45-degree angle from the horizontal that provides support for intake or discharge hoselines when they are attached to the pumping apparatus. *Also see* Fitting.

Elevated Master Stream — Fire stream in excess of 350 gpm (1 325 L/min) that is deployed from the tip of an aerial device.

Elevation Pressure — Gain or loss of pressure in a hoseline due to a change in elevation; also called *elevation loss.*

Elongation — Amount of additional length created by pressurizing a fire hose.

Expander — Device that enlarges the expansion rings used for securing threaded couplings to fire hose; types: manual, hydraulic, and power.

Expansion Ring Method — Attaching a threaded coupling to a fire hose in which a metal expansion ring is placed inside the end of the hose and then expanded to compress the hose tightly against the inner surface of the coupling.

Extinguishing Agent — Generic term used for materials that are used to extinguish fires.

Extruded Coupling — Coupling manufactured by the process of extrusion. *Also see* Coupling.

Extrusion — Fire hose manufacturing process that forms a seamless tube by forcing a heated mass of rubber or thermoplastic through the orifice of an extrusion machine die; this die is the diameter of the final hose product.

F

FDC — Abbreviation for Fire Department Connection.

Female Coupling — Threaded swivel device on a fire hose or appliance with internal threads designed to receive a male coupling having external threads of the same thread design and diameter. *Also see* Coupling.

Filler Yarn — *See* Weft Yarn.

Finish — Arrangement of fire hose usually placed on top of a hose load and connected to the end of the load.

Finished Foam — Extinguishing agent formed by mixing foam concentrate with water and aerating the solution for expansion; also simply known as *foam*. *Also see* Foam Solution and Foam Concentrate.

Fire Department Connection (FDC) — Point at which a fire and emergency services organization can connect into a sprinkler or standpipe system to boost the water flow in the system.

Fire Extinguisher Hose — Hose attached to portable fire extinguisher units; designed to withstand high pressure while remaining maneuverable for use. *Also see* Fire Hose.

Fire Flow — (1) Quantity of water available for fire-fighting operations in a given area; calculated in addition to the normal water consumption in the area. (2) Amount of water required to extinguish a fire in a timely manner.

Fire Hose — Specially constructed hose that gives a quick, flexible means of bringing water from a source and delivering it to an emergency incident; designed to withstand the hazards of a fire scene; also called *hoseline*. *Also see* Handline Stream.

Fitting — Device that facilitates the connection of hoselines of different sizes to provide an uninterrupted flow of extinguishing agent. *Also see* Appliance.

Flat Load — Arrangement of fire hose in a hose bed or compartment in which the hose lies flat with successive layers one upon the other.

Floating Valve — Valve with a spring-loaded, dome-shaped disk within the waterway that is held in the closed position by both spring tension and internal water pressure; when incoming pressurized water flows against the disk from the outside, it opens to permit water to flow through the valve. *Also see* Valve.

Foam Concentrate — Raw foam liquid as it rests in its storage container before the introduction of water and air. *Also see* Class A Foam Concentrate, Class B Foam Concentrate, Foam Solution, and Finished Foam.

Foam Solution — Result of mixing the appropriate amount of foam concentrate with water; it exists between the proportioner and the nozzle or aerating device that adds air to create finished foam. *Also see* Foam Concentrate and Finished Foam.

Fog Nozzle — Nozzle that produces either a fixed or variable spray pattern. *Also see* Nozzle.

Fog Stream — Stream made of small droplets of water that leave the nozzle in a spray or "fog" pattern; patterns range from straight to narrow fog to wide fog.

Forestry (Fire) Hose — Small-sized (1 to 1½ inches [25 mm to 38 mm]), lightweight, single-jacket fire hose that meets the specialized requirements for combating fires in forests and other wildland settings. *Also see* Fire Hose.

Forged Coupling — Coupling formed by pounding a hot metal pellet into a forging die, which forms the metal into the desired shape. *Also see* Coupling.

Forward Lay — Method of deploying fire hose from the water source to the fire scene. *Also see* Hose Lay (1, 2).

Four-way Hydrant Valve — Valve device that permits a pumping apparatus to boost the pressure in a supply line connected to a hydrant without interrupting the water flow. *Also see* Valve Device.

Friction Loss — Loss of pressure created by water turbulence moving against the interior walls of fire hose along with the roughness of interior hose surfaces.

Front Bumper Well — Hose compartment built into the front bumper of a pumping apparatus; carries attack hose for small fires.

G

Gate Valve — Control valve with a solid plate operated by a handle and screw mechanism; rotating the handle moves the plate into or out of the waterway. *Also see* Valve.

Gated Wye — Hose appliance with one female inlet and two or more male outlets with a gate valve on each outlet. *Also see* Appliance and Wye.

H

Handline Stream — Stream supplied by a hoseline sized from 1½ to 3 inches (38 mm to 77 mm) and flowing 95 to 350 gpm (360 L/min to 1 325 L/min).

Hard Suction Hose — Rigid, noncollapsible, rubberized length of fire hose with a steel core that operates under vacuum conditions without collapsing, allowing a pumping apparatus or pump to draft water from a static or nonpressurized source that is below the pump (lakes, rivers, wells, etc.); also called *hard sleeve* or simply *suction. Also see* Suction Hose.

Higbee Cut — Flattened angle at the end of the coupling thread that identifies the first thread to prevent cross-threading when couplings are connected; also known as *blunt start. Also see* Higbee Indicator.

Higbee Indicator — Notch or grove cut into coupling lug to identify the exact location of the Higbee cut; also known as *Higbee mark. Also see* Higbee Cut.

Horseshoe Load — Arrangement of fire hose in a hose bed where the hose lies on edge around the perimeter of the hose bed in a U shape or horseshoe configuration.

Hose Bed — Main hose-carrying area of a pumping apparatus or other piece of fire apparatus; designed for carrying hose.

Hose Bin — Tray or compartment often located on the running board or over a hose bed for carrying extra hose.

Hose Bridge — Device placed astride hose to prevent damage to hose from traffic passing over it; also known as hose ramp. *Also see* Hose Tool.

Hose Cap — Threaded female fitting used to close a hoseline or a pump outlet. *Also see* Fitting.

Hose Clamp — Mechanical or hydraulic device that compresses a fire hose to stop the flow of water. *Also see* Hose Tool.

Hose Dryer — Enclosed cabinet containing racks on which fire hose can be dried.

Hose Jacket — (1) Outer covering of a fire hose. (2) Device clamped over a fire hose to contain water at a rupture point or to join damaged or dissimilar couplings. *Also see* Hose Tool.

Hose Lay — (1) Connected lengths of hose from water source to pumping apparatus. (2) Layouts of hose from a fire pump to the place where the water needs to be. (3) Arrangement of connected lengths of fire hose and accessories on the ground at a wildland fire beginning at the first pumping unit and ending at the point of water delivery.

Hose Pack — Compact bundle of hose, usually bound to facilitate carrying by one person.

Hose Plug — Threaded male fitting used to close a hoseline. *Also see* Fitting.

Hose Rack — Portable or fixed unit for storing or drying fire hose.

Hose Ramp — *See* Hose Bridge.

Hose Record — Individual history of a section of hose from the time it is purchased until it is taken out of service.

Hose Reel — Cylindrical device upon which fire hose is manually or mechanically rolled for later deployment.

Hose Roller — Metal device having a roller that can be placed over a windowsill or roof's edge to protect a fire hose and make it easier to hoist. *Also see* Hose Tool.

Hose Test Gate Valve — Special valve designed to withstand test pressures and reduce the quantity of water available to the hose during hose testing; permits pressurizing the hose but will not allow water to surge through the hose if it fails. *Also see* Valve.

Hose Tool — Accessory that makes the handling of fire hose more efficient, easier, and safer. Examples: hose strap, rope, or chain; spanner and hydrant wrenches; chafing block; hose jacket, roller, or clamp; etc.

Hose Tower — Structure from which fire hose can be hung to drain and dry.

Hose Wringer — Device used to remove water and air from large diameter hose. *Also see* Hose Tool.

Hydrant — Upright metal casting that is connected to a water supply system and equipped with one or more outlets to which a hoseline or pumping apparatus may be connected to supply water for fire-fighting operations; also called *fire hydrant. Also see* Dry-Barrel Hydrant, Wet-Barrel Hydrant, and Dead-End Hydrant.

Hydrant Wrench — Specially designed tool used to open and close a hydrant valve.

I

Increaser — Adapter fitting used to connect a hose or appliances of differing sizes; allows larger hoselines to extend from small outlets. *Also see* Fitting, Adapter, and Reducer.

Industrial (Fire) Hose — Fire hose, usually of lighter construction than fire service hose; used by industrial fire brigades. *Also see* Fire Hose.

In-line Proportioner — Type of foam delivery device that is located in the water supply line near the nozzle. The foam concentrate is drawn into the water line using the Venturi Principle. *Also see* Proportioner and Venturi Principle.

In-line Relay Valve — Valve device placed along the length of a supply hose that permits a pumping apparatus to connect to the valve to boost pressure in the hose. *Also see* Valve Device.

Intake — Inlet for water into the fire pump.

Intake Hose — Hose that connects a pumping apparatus or portable pump to a nearby water source (either pressurized or nonpressurized). See Suction Hose for types.

Intake Relief Valve — Valve device with a spring-loaded internal component that diverts any sudden pressure surge away from the fire pump; designed to prevent damage to a pump from water hammer or any sudden pressure surge. *Also see* Valve.

K

Kink — Sharp bend in a fire hose that restricts water flow.

Kink Test — Test conducted by manufacturer of fire hose to ensure performance by folding the hose over on itself, securing it to maintain the kink, and pressurizing; test pressures vary with the type of hose.

L

Ladder Pipe — Master stream device mounted on the fly of an aerial ladder. *Also see* Master Stream.

Large Diameter Hose (LDH) — Fire hose that is 3½ inches (90 mm) or larger in diameter (used for supply and suction hose); used to move large volumes of water quickly; those over 6 inches (260 mm) in diameter are often called *aboveground water mains. Also see* Fire Hose.

LDH — Abbreviation for Large Diameter Hose.

Lined Hose — Fire hose composed of one or two woven outside covers or jackets and an inside rubber lining. *Also see* Fire Hose and Unlined Hose.

Longitudinal Hose Bed — Hose bed located to the side of the main hose bed; designed to carry preconnected attack hose. *Also see* Hose Bed.

Low-volume Stream — Stream that discharges less than 95 gpm (360 L/min). *Also see* Handline Stream and Master Stream.

M

Make a Hydrant — Procedure for connecting to and laying hose forward from a fire hydrant. Also called *catch a hydrant, dress a hydrant, wrap a hydrant,* and numerous other variations.

Male Coupling — Hose end with protruding threads that fits into the thread of a female coupling of the same pitch and appropriate diameter and thread count. *Also see* Coupling.

Manifold — Valve device that receives a supply of water and distributes it through valves to a number of hoses; also known as *portable hydrant. Also see* Valve Device.

Master Stream — Large-volume stream that discharges more than 350 gpm (1 325 L/min) and is fed by multiple 2½- or 3-inch (65 mm or 77 mm) or larger supply hoselines connected to a master stream nozzle. *Also see* Monitor, Handline Stream, and Low-volume Stream.

Mechanical Damage — Hose: Occurs when an object contacts the hose and cuts, tears, or stresses the material. Coupling: Becomes misshapen or bent when dropped or run over by vehicles. *Also see* Thermal Damage.

Minuteman Load — Hose load carried on the shoulder of and deployed and advanced by one person.

Monitor — Portable master stream appliance consisting of a manifold, stream straightener, and nozzle; stream direction can be changed while water is discharging; can be fixed, portable, or a combination. *Also see* Master Stream.

Mop-up — In wildland fire fighting, the act of ensuring fire extinguishment such as extinguishing or removing burning material along or near the control line, felling dead trees (snags), and trenching logs to prevent rolling.

Multiagent Foam Nozzle — Dispenses a variety of foam concentrates and other extinguishing agents in sequence or simultaneously. *Also see* Nozzle.

Multiple-jacket Hose — Type of fire hose construction consisting of a combination of two separately woven jackets (double jackets) or two or more interwoven jackets and lined with an inner rubber tube. *Also see* Fire Hose.

Multipurpose Nozzle — Nozzle used for specific stream operations with limited applications; nozzle examples include exposure, marine, piercing, distributor, cellar, etc. *Also see* Nozzle.

Multiversal — Master stream appliance that may be removed from the pumping apparatus and anchored on the ground for use. *Also see* Master Stream and Monitor.

N

National Standard Thread — Screw thread of specific dimensions for fire service use as specified in NFPA 1963, *Standard for Fire Hose Connections.*

Nonaspirating Foam Nozzle — Nozzle that does not draw air into the foam solution stream. The foam solution is agitated by the nozzle design causing air to mix with the solution after it has exited the nozzle. *Also see* Aspirating Foam Nozzle and Nozzle.

Nonthreaded Coupling — Coupling with no distinct male or female components; also called *Storz coupling* or *sexless coupling. Also see* Coupling.

Nozzle — Appliance on the discharge end of a hoseline that forms a fire stream of definite shape, volume, and direction.

O

Occupant-use Fire Hose — Fire hose used by a building's (or ship's) occupants to fight incipient-stage fires before the arrival of trained fire and emergency service responders; also known as standpipe hose, house line fire hose, or shipboard/marine fire station hose). *Also see* Fire Hose.

P

Piercing (Applicator) Nozzle — Nozzle with an angled, case-hardened steel tip that can be driven through a wall, roof, or ceiling to extinguish hidden fire. *Also see* Nozzle.

Piston Valve — Valve with an internal piston that moves within a cylinder to control the flow of water through the valve. *Also see* Valve.

Portable Hydrant — *See* Manifold.

Preconnect — (1) Connecting attack hose to a discharge when the hose is loaded; done to shorten the time it takes to deploy the hose for fire fighting. (2) Soft-sleeve intake hose that is carried connected to the pump intake. (3) Hard suction hose or discharge hose carried connected to a pump, eliminating delay when hose and nozzles must be connected and attached at a fire.

Proof Test — Test of coupled fire hose conducted by the manufacturer at the request of the purchasing agency to ensure that it meets standards or bid specifications. The hose is subjected to extremely high pressures to ensure its ability to withstand the most extreme conditions in the field; formerly called *acceptance test*.

Proportioner — Fixed mixing device that injects an extinguishing agent such as foam into a water stream. *Also see* In-line Proportioner and Proportioning.

Proportioning — The mixing of water with an appropriate amount of foam concentrate to form a foam solution.

Pumping Apparatus — Fire service apparatus that has the primary responsibility to pump water; also called *engine* or *pumper.*

Q

Quarter-Turn Coupling — Nonthreaded coupling with two hooklike lugs that slip over a ring of the opposite coupling, and then rotate 90 degrees clockwise to lock. *Also see* Coupling and Nonthreaded Coupling.

R

Reducer — Adapter fitting used to connect hose or appliances of differing sizes; allows smaller hoselines to extend from large outlets. *Also see* Fitting, Adapter, and Increaser.

Reducing Wye — Wye that has two outlets smaller in diameter than the inlet valve. *Also see* Wye.

Reel Load — Arrangement of fire hose, especially large diameter hose, on a pumping-apparatus-mounted reel.

Reinforcement — Structural support within a rubber-covered fire hose in the foam of woven, braided, or spiraled yard (previously called *jacketing* or *jacket*).

Relay Operation — Using two or more pumping apparatus to move water over a long distance by operating them in series. Water discharged from one apparatus flows through hoses to the inlet of the next apparatus, and so on. Also called *relay pumping.*

Reverse Lay — Method of deploying fire hose from the fire scene to the water source. *Also see* Hose Lay (1, 2).

Rise — Distance that a fire hose elevates above the test table while under test pressure.

S

Screw-in Expander Method — Attaching threaded couplings to rubber-jacket booster hose with expanders that are screwed into place.

Section — Individual coupled piece of fire hose.

Self-educting Nozzle — Handline or master stream nozzle that has the foam eductor built into it. *Also see* Nozzle and Eductor.

Service Test — (1) Fire hose design service test conducted by the manufacturer following production to ensure compliance with NFPA 1961, *Standard on Fire Hose.* (2) Annual service test conducted by the user to ensure that fire hose still meets minimum operational standards. *Also see* Proof Test.

Siamese — Valve device with two or more female inlets and one male outlet used to combine two or more hoselines into one. *Also see* Valve Device.

Single-jacket Hose — Type of hose construction consisting of one woven jacket and usually lined with an inner rubber tube. *Also see* Fire Hose.

Small Diameter Hose — Fire hose that is between ½ and 3 inches (13 mm and 77 mm) in diameter and most often used as attack hose (but also used for booster hose). *Also see* Fire Hose.

Snap Coupling — Coupling set with nonthreaded male and female components; when a connection is made, two spring-loaded hooks on the female coupling engage a raised ring around the shank of the male coupling.

Soft Sleeve Hose — Collapsible fire hose that connects a pumping apparatus to a pressurized water supply source (hydrant); generally a short length of supply hose with female couplings on both ends; often referred to as *soft suction. Also see* Suction Hose.

Solid Stream — Hose stream that stays together as a solid mass as opposed to a fog or spray stream; produced by a smoothbore nozzle; do not confuse with a straight stream produced by a fog nozzle.

Spanner Wrench — Small tool primarily used to tighten or loosen hose couplings; can also be used as a prying tool or to operate gas or water shutoff valves.

Split Lay — Fire hose deployed by two pumping apparatus: one making a forward lay and one making a reverse lay from the same point. *Also see* Hose Lay (1, 2).

Standpipe Hose — Single-jacket hose, lined or unlined, that is preconnected to a standpipe; used primarily by building occupants to mount a quick attack on an incipient fire. *Also see* Fire Hose.

Standpipe System — Wet or dry system of pipes in a large single-story or multistory building with fire hose outlets installed in different areas or on different levels to be used by firefighters and/or building occupants; used for quick deployment of hoselines during fire-fighting operations.

Storz Coupling — Nonthreaded or sexless coupling commonly found on large diameter hose. *See* Nonthreaded Coupling.

Straight Roll — Length of hose rolled for storage or transport where either the male or female coupling is exposed; also called *street roll.*

Suction Hose — Intake hose that connects pumping apparatus or portable pump to a water source. *Also see* Hard Suction and Soft Sleeve.

Suction Hose Strainer — Device attached to the drafting end of a hard suction hose to keep debris from entering the fire pump.

Supply Hose — (1) Fire hose that is designed to move water between a pressurized water source and a pump that is supplying attack hoselines. (2) Fire hose used to maintains a water system (either as a continuous conduit or by connecting supply sources). Minimum size: 3½ inches (90 mm). *Also see* Fire Hose.

T

Tension Ring Method — Attaching a coupling to hose using a tension ring and contractual sleeve.

Thermal Damage — Hose: Temperature extremes can weaken hose covers. Coupling: Temperature extremes can cause metal fatigue and distortion. *Also see* Mechanical Damage.

Threaded Coupling — Male or female coupling with a spiral thread. *Also see* Nonthreaded Coupling.

Tier — Layer of hose loaded in the hose bed of a pumping apparatus.

Trade Size — Generally recognized nominal size of fire hose; minimum internal diameter of the waterway as described by the manufacturer; or more accurately, the internal diameter of the hose when pressurized at operating pressures.

Transverse Hose Bed — Hose bed that lies across the pumping apparatus body, at a right angle to the main hose bed; designed to deploy preconnected attack hose to the sides of the pumping apparatus; also called *mattydale hose bed. Also see* Hose Bed.

Triple-layer Load — Hose load where hose folded in three layers is laid into the bed in an S-shaped fashion; one person can deploy the load.

Twist — Rotation of fire hose under pressure; specifically the number of revolutions a fixed length of hose rotates when it is pressurized to its test pressure.

U

Unlined Hose — Fire hose without a rubber lining most frequently used in interior standpipe systems. *Also see* Lined Hose and Fire Hose.

V

Valve — Water control device with an internal component that can be moved within the waterway to regulate the flow of a liquid or gas. *Also see* Valve Device.

Valve Device — Hose appliance device that allows the pressure and flow in hoses to be supplemented incrementally at intervals along the hose lay to compensate for pressure loss caused by friction. *Also see* Appliance and Valve.

Venturi Principle — Physical law stating that when a fluid (liquid or gas) is forced under pressure through a restricted orifice, there is an increase in the velocity of the fluid passing through the orifice and a corresponding decrease in pressure exerted against the sides of the constriction. Because the surrounding fluid is under greater pressure, it is forced into the area of lower pressure. When applied to a foam proportioner, water that is passed through the restricted orifice increases velocity and creates a vacuum, pulling foam concentrate up the foam tube into the water stream.

W

Warp — (1) Yarns or threads (called *ends*) of a hose reinforcement that run lengthwise down the hose. (2) Maximum distance any portion of a fixed length of hose deviates from the center of a fitting at one end to the center of a fitting at the other end with the hose pressured. *Also see* Weft Yarn.

Water Curtain — Fan-shaped stream of water applied between a fire and an exposed surface to prevent the surface from igniting from radiated heat.

Water Hammer — Force created by the rapid deceleration of water causing a violent increase in pressure that can be powerful enough to rupture piping or damage fixtures; generally results from closing a valve or nozzle too quickly.

Water Thief — Valve device with one female inlet and three gated outlets that allows the use of several attack lines; variation of a gated wye. *Also see* Wye and Valve Device.

Waterway — Path through which water flows within a fire hose.

Wear Jacket — Outer cover on a double-jacketed fire hose.

Weeping — Coupling leakage at the point of attachment.

Weft Yarn — Yarns (*filler yarn*) or threads that are helically wound throughout the length of a hose at approximately right angles to warp threads. *Also see* Warp (1).

Wet-barrel Hydrant — Fire hydrant that has water all the way up to the discharge outlets; may have separate valves for each discharge or one valve for all discharges; only used in areas where there is no danger of freezing weather conditions. *Also see* Hydrant.

Wild Line — Uncontrolled hoseline and nozzle or butt (coupling end of hoseline) that thrashes about from the reaction of highly pressurized, flowing water.

Woven-jacket Hose — Fire hose constructed with one or two outer jackets woven on looms from cotton or synthetic fibers.

Wrapped Hose — Nonwoven rubber fire hose manufactured by wrapping rubber-impregnated woven fabric around a rubber tube and encasing it in a rubber cover.

Wye — Valve device with one female inlet and two or more male outlets; usually gated. *Also see* Water Thief, Siamese, Gated Wye, Reducing Wye, and Valve Device.

Index

A

B

E

F

G

H

I

J

K

O

P

Q

R

S

T

U

V

NOTES

NOTES

NOTES

NOTES

FIRE HOSE PRACTICES
Eighth Edition

COMMENT SHEET

DATE ________________________ NAME ____

ADDRESS ____

ORGANIZATION REPRESENTED ____

CHAPTER TITLE ______________________________________ NUMBER _______________

SECTION/PARAGRAPH/FIGURE ___________________________ PAGE ____________________

1. Proposal (include proposed wording or identification of wording to be deleted), OR PROPOSED FIGURE:

2. Statement of Problem and Substantiation for Proposal:

RETURN TO: IFSTA Editor
Fire Protection Publications
Oklahoma State University
930 N. Willis
Stillwater, OK 74078-8045

SIGNATURE ___________________________________

Use this sheet to make any suggestions, recommendations, or comments. We need your input to make the manuals as up to date as possible. Your help is appreciated. Use additional pages if necessary.